P9-DMS-697

Evolutionary Biology

Stanley N. Salthe

Brooklyn College
of the City University
of New York

HOLT, RINEHART AND WINSTON, INC.

NEW YORK   CHICAGO
SAN FRANCISCO
ATLANTA   DALLAS
MONTREAL   TORONTO
LONDON   SYDNEY

To my parents,
to Walter Costa,
and
especially
to Barbara May, my wife.

Copyright © 1972 by Holt, Rinehart and Winston, Inc.
All rights reserved
Library of Congress Catalog Card Number: 76-183650
ISBN 0-03-082191-6
Printed in the United States of America
2 3 4   038   9 8 7 6 5 4 3 2 1

Preface

Evolutionary biology is not primarily an experimental science. It is an historical viewpoint about scientific data. It is also a construct of ideas, of which every biological datum is potentially a part.

The construct of evolutionary theory is organized chiefly around the question of how living systems have reached their present states. The construct is an attempt to provide a scientifically acceptable answer. There have been various mythic answers as well, and their existence will be pointed out in this context in order to show how science reflects and affects all aspects of our culture. Knowledge does not exist in a vacuum; a book about evolution should shed some light, even indirectly, on the major philosophical, political, social and economic problems facing people in our society today.

This book provides an overview. Ideas, viewed within the construct, are treated as functioning parts of a multifaceted whole, and not as independent beads on an evolutionary necklace. The purpose of the construct is to suggest how a temporary, seemingly improbable, order can have been produced out of statistically probable occurrences on a simpler level of organization without reference to forces outside the system. Specific examples of this general problem are the production of phenotype from the molecular genotype; the production of consciousness from neuronal interactions; the production of social behavior from collections of related individuals; and the production of directional evolutionary change from adaptation at the species level.

The concept of natural selection (together with various chance events) supplies a scientifically satisfying answer to the specific evolutionary aspect of this general problem. This answer, like many scientific answers is neither comforting nor immediately useful. Yet, understanding this answer is probably essential for the further spiritual and technological development of our culture. I hope the scope and grandeur of our concept of organic evolution will be made visible in this book. A course in evolution (or, indeed, any course) should aim at nothing less than an expansion of consciousness for the student.

Stanley N. Salthe

Contents

EVOLUTIONARY
BIOLOGY

CHAPTER 1

Introduction

In this chapter we will describe the nature of the field of evolutionary studies (or evolutionary biology). It will be a discussion, not of methodology, but of the philosophical and sociological import of the evolutionary viewpoint. It is hoped that this treatment will help to orient the student to the significance of that viewpoint.

Evolutionary biology is not a science as such, although it makes use of scientific data—often of data obtained specifically within the context of evolutionary studies. Evolution itself is a concept, or construct of ideas, centered around the problems of the origins of life and of man, and around the historical development of living systems. We intend to explore these ideas in this book, and to introduce the viewpoint around which they are organized in this chapter.

A convenient way to proceed is to note that evolutionary studies can be described as being of three different kinds: (1) comparative descriptive studies of different biological systems, (2) reconstructions of evolutionary history, and (3) a search for the forces (or principles) involved in evolutionary change. (Of course, any given study may involve more than just one of these.) These could also be described as being the three basic components of the discipline referred to as evolutionary biology.

COMPARATIVE STUDIES

Comparative studies of living or fossil biological systems provide the essential data without which the concept of evolutionary change could not have received credence. The fundamental point that emerges from these kinds of studies is that different biological systems display curious similarities of structure or function. For example, all vertebrate backbones have essentially similar construction; or all eucaryotic cytochromes c are of fundamentally the same basic molecular structure—ranging from molds to man. At the same time there are slight differences among different forms; structures in different biological systems are *similar,* but not identical. The question then arises as to how they became so similar, or how they became different, and which of these questions is the more interesting one to ask. In Chapters 3 and 4 arguments are given to the effect that these structures are similar because they were once identical in ancestral forms, and that they are somewhat different because they became so after different lineages became separate from each other—both because of the differential accumulation of random mutations and because the different lineages took up different ways of life.

Having taken up different ways of life, organisms in different lineages needed somewhat different backbones or cytochromes c. This points up another concept that came out of comparative studies—that biological systems are in a sense teleological, or show obvious adaptations to different ways of life, or, that they possess structures that are "useful" to them (Chapter 3). Thus, thick fur is clearly useful to mammals in cold climates; enzymes that are very stable to heat are clearly necessary adaptations for bacteria living in hot springs; finlike hindlimbs are useful for whales and porpoises during swimming. No other aspect of nature shows this property of having subcomponents or subprocesses that are of "use" to the larger structure or process. The velocity of a wind serves no purpose, nor does the potential energy of a stone on a hillside, nor the mass of the planet earth, nor the crystal structure of a rock. The aspect of purposiveness of much of biological structure is one important way in which biological systems differ from all other natural phenomena. Older teleological concepts of adaptation, no longer acceptable to modern thinkers, had it that functional considerations were not only necessary for understanding biological structure but that they were sufficient as well. As espoused, for example, by Lamarck (1744–1829), this viewpoint held that adaptation was directly imposed upon biological systems by the environment, and could subsequently be inherited as well. All biological phenomena were thought to be adaptive, and from such viewpoints arose notions of the "perfectness" of adaptations (see Chapter 11). Today we place as much importance on inherited limitations and predispositions as we do on functional considerations in explaining why a given structure or behavior has the particular form it has (Chapter 3).

The purposiveness or usefulness shown by the traits of living systems is today explained by the concept of adaptation. Cacti are seen to be "adapted to" living in deserts by various structural and chemical means (roundish stems to store water, spines to prevent animals from getting the water, bitter and psychotropic chemicals for the same purpose, deciduous roots so that no place of water loss can exist during the dry season, and so on); cats are adapted to being stealthy predators (solitary creeping behavior, sharp claws, cutting teeth, and so on). One way of explaining the necessity for adaptation is to see biological systems as energy exchange systems that need to trap a certain amount of energy per unit time from outside themselves in order to keep going; that is, only those biological systems that succeeded in trapping the requisite amounts of energy continued to exist, and these had to have had the means (or tools) required for that "purpose" in given environments. Another way of seeing this is to view biological systems as thermodynamically improbable, needing an energy input to maintain their improbable configurations. This energy is drawn from the surrounding environment, which is reduced to a lower energy state (characterized by an increase in entropy) as a result. For example, a plant is eaten by an herbivore. Some of the energy present in the plant is incorporated into the herbivore, maintaining its improbable high-energy configuration; the rest of the energy present in the plant is dispersed to a lower energy state (with loss of heat)—one of more probable configuration (CO_2, feces, and so on). It seems that, in order to transfer energy in this way—that is, contrary to entropic decay—two things are necessary: (1) some energy must be expended (catching prey, synthesizing digestive enzymes, and so on) and (2) specialized energy-trapping structures must be synthesized (also at energetic cost), such as claws, jaws, digestive enzymes, and so on. That is, organisms must be specialized in some way *in order to* obtain energy, and different energy sources necessitate different sorts of specializations (broad surfaces in plants, cutting tools in predatory carnivores, and so on). In short, organisms have the purpose of obtaining energy, and specialized adaptations to do this are expressions of that purpose.

The similarities of the adaptations of different living systems, then, can be viewed as a necessary outcome of their all having been derived by descent from a common ancestral form, and this is the view of modern evolutionists. Comparative biological studies uncover these similar patterns of structure in different systems by inductive reasoning, and the evolutionary biologist takes a specifically historical viewpoint concerning them.

HISTORICAL RECONSTRUCTION

The historical viewpoint toward biological similarities is not the only possible viewpoint that can be taken. For example, one could consider that there is no time at all, only a continuing present moment full of rich illusions, including a sense of the "past" built into the human mind. Such a viewpoint is basically unanalyzable, and no doubt that fact alone predisposed scientists toward the concept of time, in which context cause and effect have meaning. Causal explanations in the context of time are not restricted to analytical systems, however, but are shared by such constructionist systems as history and mythology. Indeed, the task of reconstructing the evolutionary history of living systems is comparable both in aim and in some ways in method to constructing a political history of some country or a mythology. This task is not a scientific one in that it does not (cannot) utilize the scientific method (observation–hypothesis–experimental test of hypothesis–new observation–new hypothesis, and so on) because experimental verification is not possible for any specific historical sequence. One can only compare the proposed history with the rules of history making, or with an ideology, or with derived contemporary facts, and judge whether it is plausible and internally consistent, or whether it adequately serves some ulterior purpose. This position, different from that of many evolutionary biologists, will be modified below.

Historical interpretations change as new information appears or new viewpoints or ideologies are used as bases from which to review old data. The sequence of fossils (continually added-to), the absolute dating information (periodically revised), as well as relative dating information devised from studies of primary gene products (amino acid sequences, immunology) form the hard data of biological history. The historical reconstruction based on these data has been gradually put together over the last century, but is still very incomplete concerning details in most lineages of biological systems. Even the rough overall picture is still changing very fast for the vast Precambrian period, during which the origin of life is conceived to have occurred and in which the earliest organic evolutionary changes occurred. New data have had less dramatic effects on the post-Cambrian picture, but even there rather drastic changes have to be made from time to time because of a new finding. There are many completely unsolved problems of some magnitude in this period as well; for example, whether the vertebrates were originally fresh water or marine organisms, or what the actual relationships are among the mollusks, arthropods, and annelid worms, or what the relationships are among the different kinds of molds and other eucaryotes.

An important difference between evolutionary history and mythology or some kinds of historical studies is that evolutionary history is always in principle incomplete, uncertain, and always being reworked. Mythologies, once formed in basic outline, may change slowly—for example, the meaning of one goddess may be usurped by another—but they do not change *in principle*. At any given moment they represent the absolute truth (or an absolute truth) for the individuals involved with them. Much the same can be said for some other historical enterprises. There is, for example, a particular Marxist viewpoint on the history of the social role of craftsmen. If one bases his historical viewpoint on a Marxist system, he must perforce take that viewpoint—or at least some variant of it. If, however, one believes in other principles, he is forced to espouse other viewpoints. Free of the constraints of other than a most general value system, evolutionists, like other scientists, have been able to explicitly see their interpretations as provisional; indeed, because of the nature of scientific inquiry (not actually the tool used in reconstructing a history, but forming the intellectual background of all evolutionary biologists), they are virtually forced to see them that way. Scientists, of course, are not free as individuals from value judgments, but the values they embrace—rationality, belief in causal relationships, and so on—are so general that they do not influence the choices made among different scientific theories or among different evolutionary reconstructions.

It should be pointed out that historical data are individually inaccessible to scientific inquiry. An historical event is nonrepeatable, and so no experiments can be done upon it as such. This is the same thing as describing it as unique. Unique objects or events are not as such the province of scientific research, which is aimed at generalizing and at verifying the generalizations with new samples of data. For example, there is a biological way of interpreting human fingerprint patterns, but it can never be possible to reconstruct exactly the genetic background and the epigenetic events that led to a given unique pattern; indeed, science is not concerned with any given pattern of that kind. Nor is it concerned with the actual sequence of events that led to the evolution of the earthworm, the flea, or the ostrich. Certain scientists (including the author) are interested in these evolutionary sequences, but they do not operate entirely as scientists when they try to reconstruct them.

Uniqueness is a property of every individual in a biological system. It is virtually impossible for two individuals to have the same genotype (even identical twins are now felt to differ in the quantities of different heritable cytoplasmic factors), nor will they undergo the same epigenetic histories. This is another way in which biological systems differ from other natural phenomena as we deal with them. For the purposes of physics and chemistry (until recently) the molecules of a gas can be treated as all identical; since it would make no difference if they were different, the simplifying assumption is made. Every carbon dioxide molecule might, however, be described as being in some way different from every other one, but then we would be leaving the province of science as it has been. It is undoubtedly true that each electron has an actual detailed history different from each other one that has ever existed, but this is irrelevant to physical science as we know it. Only the mass actions of electrons, or those properties common to all of them, are of scientific interest. The same may be said of lineages of organic beings or of individual human beings. Only those aspects that can be generalized, and verified, are of scientific interest.

Many biologists have become interested in the individual histories of certain lineages—say, that of man—and have striven to find scientific methods to study these histories. One way in which traditional scientific methods have been used to deal with unique evolutionary events is to ignore their uniqueness. Thus, there are immunological methods of comparing homologous proteins from different lineages. Enough work has been done in the area of primary gene products so that it is now possible to raise as an hypothesis the proposition that as two lineages have been separated for longer and longer periods of time since they last formed a single system, their proteins become increasingly different. We can then calculate average differences that would be expected to accumulate with given periods of time; or one can calibrate an immunological system by using proteins from living members of lineages for which there are generally accepted times of divergences known, and then test against this a given lineage that happens to be of interest, say, that of man. In this way we can derive data, for example, about the time of divergence of the lineage that led to man and that which led to the apes. To be sure, this point of divergence was a unique historical event involving only one lineage, but in this example, that event is treated as an average member of a class of events about which a generalization has been made. It is not the uniqueness of the event that is treated but what it supposedly has in common with other similar events.

Again, a given historical event will usually have many different results or repercussions. It is possible to study some of these separately with a view toward learning about their common cause. In this way it often happens that several lines of evidence lead back to a single interpretation about some historical event. Each line of evidence can be viewed as a separate verification of a postulated event, thereby fulfilling the requirement that hypotheses in science must be verified by independent or new observations. For example, studies of the changes in position of the earth's magnetic poles through time can be explained by the idea of continental drift. This idea can also explain similarities of sediments and crustal formations in far-flung continents. It can also explain how climate can have radically altered through time on a continent such as Antarctica. Each of these observations can be viewed as a separate verification of a postulated migration of continents inasmuch as each of them can be satisfactorily accounted for by this idea. As the number of phenomena that can be satisfactorily explained by a given postulated historical event increases, so does the probability that the event in question actually happened.

SEARCH FOR PRINCIPLES

The area of evolutionary biology that can be most truly described as scientific is that work devoted to searching out the forces involved in evolutionary change. This is undertaken partly by studying the patterns that emerge from comparisons of different organisms or their parts, or from comparisons of the kinds or classes of changes that appear to have occurred within different lineages during their histories (based on reconstructions combined with absolute dating information), and partly by making observations and controlled experiments on short-term changes in living biological systems. This search for the forces of evolutionary change is also a search for its principles. The nature of the forces imposes certain patterns on the changes that occur in many lineages and these each reflect upon or express those forces, thus revealing something about them. By somewhat different meanings of the word, both the underlying forces and their habitual or frequent expression (that is, the patterns they tend to cause) are referred to as principles. These principles are some of the criteria used to determine whether a given suggested evolutionary history is plausible or not—and, indeed, are used in the construction of the histories as well.

One might be concerned about the sense of logical circularity in the last statement. If a given pattern emerges from three different independent evolutionary reconstructions of three different lineages, it might come to be accepted as a provisional principle, and so may be used in reconstructing further evolutionary histories. In this way one could soon have seven different lineages whose reconstructed histories exemplify the principle. Such a procedure, as stated, would of course, be illogical. In practice, however, considerably more than a single principle would be used to order the historical data about any given lineage, including absolute dating, other patterns, and knowledge of the forces of change as obtained from experimental studies. Perhaps the best we can hope for in historical reconstruction is a coherent, internally consistent picture, regardless of whether or not some logically circular processes have been used in their construction.

Experimental studies on modern organisms are the province of population genetics and ecological genetics, with much assist from other fields of genetics, especially developmental genetics. Experiments uncover the types of organic changes that can be elicited, the rates at which they can be effected, and the conditions that favor structural or behavioral changes in populations of organic beings. Such experiments have been carried out by man for millenia in a semiconscious way in connection with the selective breeding of farm crops and animals. Darwin was aware of the structure of such experiments and with their results, as well as of a wide range of information about wild organisms and the places in which they live. From all of these he induced the principle of natural selection.

FORCES OF EVOLUTION

As currently viewed, the forces responsible for evolutionary change form a dichotomy of chance (random or stochastic events) *versus* determinism. (Actually, this dichotomy is here emphasized for convenience only. Deterministic events are those whose probability is either zero or 100%, and so are more properly seen as extremes of a range of probabilities. For heuristic purposes, however, we will treat these extremes as being qualitatively different from the rest of the range.) Natural selection is the only deterministic force recognized and is, indeed, the only mechanism seen to be involved in evolutionary change. Chance events are not properly mechanisms because they have no given structure and are not predictable in principle. They are, however, crucial to long-term evolutionary change. Throughout this book various chance events are discussed and their importance assessed. An important category of chance events that may be mentioned here is, for example, mutation, the source of all biological variability, without which evolutionary change would come to a halt (Chapter 9). It is pure chance what mutation—miscopying or copy error during gene replication—will occur at what time, and whether or not a given mutation will be advantageous in a given environment. Again, what is perhaps one of the most important evolutionary forces that we know of is a chance event—preadaptation (Chapter 4). In order to see this, it is first necessary to discuss natural selection.

Natural selection is fundamentally a process of differential reproduction (Chapter 8). It is a mechanism that distinguishes between individuals of a population or clone in terms of which will leave the most reproductive offspring in the next generation. It occurs for basically two reasons. (1) Some individuals will suffer from mutations that destroy some basic physiological process (they will have a genetic disease; of course, not *all* mutations will be destructive), and these will always be selected out of the population before they get a chance to breed because they are physiologically or reproductively unfit. (2) The environment is always limited; hence not all the individuals that are physiologically fit and reproductively capable will be able to acquire sufficient requisites (food, mates, predator escapes, nesting sites, and so forth) to breed successfully. That is, there will be competition among physiologically fit individuals. Some individuals will be more successful at reproduction than will others. Since the traits responsible for their success are in some degree heritable (that is, have some genetic components), these successful traits will be transmitted preferentially to the next generation. With time such traits will increase in frequency in the population, and that change will be an evolutionary change. In the temporary absence of environmental change (as over relatively few generations), traits that are well-adapted to the current environmental situation will be maintained by the selective process; that is, there will be no evolutionary change. Thus, in a changing environment, natural selection leads to evolutionary change, whereas in an unchanging environment it leads to maintenance of the phenotypic status quo. In both cases it operates by weeding out the relatively unfit individuals; in short, the mechanism is always the same, while the results of its operation differ according to the external environmental situation.

Now, given a broad lineage, say, the reptiles or the sunfishes, the process of natural selection can and will continue to produce well-adapted reptiles or sunfishes as the environment continues to change, or these lineages will become extinct. However, natural selection cannot produce, by itself, something entirely new, or what we will call a new adaptive zone (Chapter 4). It will not, by itself, produce birds from reptiles or penguins from birds. As near as we can tell at this point, such a transition always involves chance preadaptation for the radically new way of life. This is because natural selection cannot foresee the future; it can only produce adaptations to present environments, and it produces organisms whose ancestors were adapted in certain ways to past environments. These past adaptations are retained and/or elaborated upon during the continuing process of adaptation. In our concept of this process a wing can continue to be a good wing in new environments, or can become a better wing, but it cannot become a flipper. For natural selection to begin converting a wing into a flipper it is necessary to have had an important change in environment. For the organisms to survive such an important environmental change, they must have been adequately adapted to it—something that natural selection could not have produced as such, but which can occur by chance. (The environmental change, it should be borne in mind, could be brought about by behavioral changes in the organisms in question, and not necessarily by obvious changes in climate.) It may be pointed out that in general any structure has properties beyond those for which it was constructed. This applies as well to molecules, norms of reaction, equations, and so on. In new situations new properties appear in them, but what is being stressed here is that these new properties cannot be foreseen by natural selection and so are better handled conceptually as chance events.

Thus, if chance preadaptations (or prospective adaptations) were never produced (in fact, they frequently are), the origin of completely new ways of life could not take place, and evolution would be restricted at most to speciation, that is, to an ever greater specialization within a given adaptive zone—or to the elaboration of new ecological niches (see Glossary) within that adaptive zone. For example, some ancestral horned ungulate gave rise to many different antelopes, each with a unique ecological niche, but all within a single antelope adaptive zone. Note again that natural selection is always the mechanism by which any evolutionary change is effected, but that the arena within which it functions appears operationally to be largely determined by chance occurrences and that that arena determines what *directions* an evolutionary change will take. Thus, chance events will result in mutations producing new alleles at a given genetic locus, and selection will then "examine" them; chance events will alter the environment in such a way that individuals are forced into marginal situations that might uncover preadaptations and make them crucial to survival—or accessible to natural selection as adaptations.

Evolutionary change, then, is always the result of an interplay between determinism and chance. Natural selection is the major deterministic force and the only mechanism involved. The past products of selection, however (or the current adaptations of the lineage in question), enter into the adaptive process in a largely deterministic way as well. The present adaptations of a group of organisms will set narrow limits on what kinds of changes can be accepted by them because of the need for a harmonious whole organism. Thus, for an absurd example, elephants will not soon successfully incorporate adaptations for flight. One could say that their current adaptations are such that they are not preadapted for that activity. Such limitations must be present at every level of the phenotype, from that of primary gene products up; hence the influence of the past on present possibilities must be enormous. Consequently, the range of possible phenotypic changes that can occur within a short period of time associated with each possible environmental change must be quite limited. Faced with drastic environmental changes, this could be a primary factor in extinction.

NATURAL SELECTION

Given a big enough universe and a long enough period of time, random forces would produce by chance every conceivable configuration of energy and matter permissible within given ranges of temperature and pressure. The result would be a high degree of entropy. At any given time and place in this universe there will be some probability distribution, determined by the conditions prevailing, such as temperature, of various randomly generated events that may take place. Some will inevitably occur; some will not occur at all; others may, or may not, occur with assigned probabilities. Natural selection seems to make more probable the presence of certain configurations than would be predicted from knowledge of only the various environmental parameters. For instance, occurrence of the exact sequence of amino acids in wild-type human hemoglobin beta chains on a pure chance basis has a low probability (it is only one of a vast number of possible arrangements of 20 amino acids in a linear row of 146, 20^{146} or 10^{179}, none of which is thermodynamically more probable than any other), and its preservation as such in large numbers on earth at this time cannot be accounted for by any known physical or chemical forces affecting peptide bond formation. In the context of the human physiological environment, however, we may guess that natural selection has preferentially preserved this configuration because individuals in which it occurred out-reproduced those in which it did not occur. This is a fair inference because we find that each vertebrate species has its own characteristic beta chain sequence, with each presumably adapted to function in a different specific internal environment. In addition, the exact sequence cannot be fully explained only from the point of view of present function. It is the result of many millions of generations of independent adaptations made by many different kinds of ancestral animals in a wide variety of different environments, thus embodying historical accident as well. We use the term "accident" because the past causes of certain ancient selective pressures resulting in sequence alterations in the past have disappeared with the past environments, and because the latter had no causal relationship with respect to present environments, even though the structural alterations they then caused remain to be confronted by these present environments.

Leaving the realm of biology as such, we continue in a more general vein to examine the nature of selective forces, one of which would be natural selection in biological systems. We have again to deal with the concept of unique events. Randomly generated structures and events lack uniqueness because they do not stand out from the background of the probability distributions defining such events. Each such event is conceptually the equivalent of every other, regardless of whether it is structurally distinct or not. In the absence of some selective force, no event will stand out as being present in significantly more (or less) frequency than was predicted by the probability distribution, and so no event will be unique. Thus, as a specific example, the probability of the appearance of the human beta chain sequence in a system that would generate such structures at random is $\frac{1}{20}^{146}$, whereas the probability that two such will be formed (assuming, for simplicity, independence of the events) and exist side by side is $\frac{1}{20}^{146} \times \frac{1}{20}^{146}$ (or 10^{-358}), which is so infinitesimally small as to be virtually impossible. If the system operates sufficiently many times, however, improbable events become increasingly probable. Thus, the chances that a head will appear once in a hundred coin tosses are much greater than those that a head will appear in one toss. What selection can do is to vastly increase the probability that a given event or configuration will not be degraded as readily as it would normally be under prevailing conditions. If, as in biological systems, there is feedback from the product (configuration, event) to the system that produces it, selection can ultimately increase the probability that a given event or configuration will be produced as well, in which case the system is no longer one of random generation (this would be one way to summarize the conception of the origin of life; see Chapter 2).

Natural selection, the simplest definition of which is perhaps the perpetuation or preservation of a given unlikely configuration (either by increasing its duration or its production), may be the only force in the universe capable of promulgating uniqueness. It is, however, by no means clear that mechanisms like it are restricted to biological systems. Our current thinking about the origin of life involves visualizing selective forces operating upon prebiological systems in the sense that some were pre-adapted to endure longer than others, and that these formed the basis for more complex systems (the idea being that complex systems cannot be built upon ephemeral configurations).

Continuing in this vein, we must note that one's frame of reference is extremely important in determining where selective (or deterministic) forces may act. In an infinite universe all events will be repeated an infinite number of times, and none will be unique. Differential survival (duration) or differential production would no longer exist since all configurations would be produced equally frequently and survive equally well. They would all become "functionally" equivalent. Thus, if there is an infinity of electrons, then we are justified in treating them, as we now do for convenience, en masse, ignoring their possible individual histories. By the same token, if the universe of galaxies and stars is infinite, then the earth would not be unique. It seems absurd, however, to conclude that we would not be justified in treating the earth historically if the universe is infinite. The choice here clearly rises out of one's frame of reference. Our earth is to us a finite system—we perceive only one unique earth. If a given electron were the seat of some system of conscious beings, they would be justified in studying their own history even though it is of no consequence to us in our frame of reference. Indeed, we can declare that that history does not exist, or that natural selection does not operate at the level of electrons, or that electrons are not unique events, but this comes only from our point of view.

There is, then, a sense in which uniqueness, history, and natural selection are interrelated in any finite aspect of the universe. Uniqueness and history are not independent—all unique structures or events had a history, and all historical processes lead to unique configurations. One thing that makes an entity unique is its history. The question remains whether, as in biological systems, natural selection is the only mechanism that ever generates this condition in any finite portion of the universe.

One must ask whether chance alone would not generate a unique history. Theoretically one could store information, for example, about the vectors and coordinates of a small particle undergoing Brownian motion on a microscope slide, and declare it to be the unique history of that particle. But such a declaration would be entirely arbitrary and non-functional in our frame of reference because as far as we can see (or, for our purposes) there is no meaningful difference between the path of one particle and that of any other. We can conceive of an infinite number of such functionally equivalent pathways. For this reason we choose to consider that such motion is subject largely or only to chance variations in the particles' environment. This example, then, is not a genuine example of a history within our frame of reference, the problem being that it is treated and perceived in this frame of reference as one of a class of events rather than as a unique path. Thus, it appears that chance events can never lead to unique histories because we *define* them as events which could have occurred, as far as we know, in any of an infinite number of ways but that—for entirely unaccountable reasons—did occur as we have measured them.

We might also ask whether histories, such as that of furniture-making in eighteenth century North America, can, in principle at least, be explained fully by something like natural selection interacting with various chance events. The complexities of this matter, in any case outside the scope of this book, may be such as to render this question insoluble, leaving the answer in the realm of preference. It can be suggested that natural selection had a role here (in that only certain types of wood could be used for certain purposes because of their tensile properties), but it is not clear what other forces (besides chance, that is, deterministic forces) it would be useful to invoke in this context.

SOCIOLOGICAL IMPLICATIONS AND ROLES OF EVOLUTIONARY BIOLOGY

European society in the late nineteenth and early twentieth centuries underwent an expansion of historical consciousness that was expressed in different ways, for example, by Darwin, Marx, Einstein, and Freud. The profound implications of this concern with the etiology or ontogeny of situations can be appreciated, perhaps, by asking oneself what a history of art course would have been like in 1670. Would there, indeed, even have been one? Historical consciousness has been at the heart of Western thinking for about the past 200 years. One could probably make a good case for the idea that during this period much of the meaning or significance of life has been translated into change with time (the idea of progress, manifest destiny, the idea of working through time to set up an ideal society, and so on). This is not to imply that time-oriented viewpoints were expressed only in this place and during this time. Clearly, there have always been time-oriented ideologies (for example, various notions of personal spiritual redemption involving stages in a passage toward enlightenment). But this has been a time when virtually every important aspect or field of thought has become time oriented.

During this period there has also been an increase in the prestige and influence of science, which can be broadly described as cause-and-effect thinking grounded in, or based upon, those aspects of the world that can be measured by man or his instruments. The rise in prestige of this viewpoint inevitably led to a weakening of the prestige of other, basically mystical, viewpoints, at least in some sectors of Western society. This created a need, in those sectors, for alternative explanations of phenomena long explained only by mystical systems. Evolutionary reconstructions (based upon measurements of solid, hard fossils and upon scientifically acceptable principles) represent scientifically grounded expressions of time consciousness. (We note again that science is not itself intrinsically involved with the concept of time.) Sociologically, these reconstructions are designed to answer the preeminent questions concerning the origin of life and the origin of man, just as other historical reconstructions have attempted to do.

For the present purposes it is sufficient to give a short review of the nature of life and of man as seen through the viewpoint of organic evolution. Life is seen as arising, probably deterministically, out of nonliving matter and energy. The transition is one of long duration, and there is in principle no point of demarcation between life and nonlife. Thus, living systems are seen as differing from the nonliving partly quantitatively, that is, the difference is partly one of complexity. Living phenomena are seen to need a certain threshold of complexity in order to appear. The other crucial difference between living and nonliving systems is seen to be order. Indeed, the complexity itself is seen to be an aspect of orderliness, and vice-versa. Thus, a string of nucleic acids is simply a big molecule, but the string that codes for chicken cytochrome c is a gene. But it is a gene only inside the nucleus of a living chicken cell; that is, its special properties are manifest only in a special environment. By itself this molecule is nothing special; as part of a unique complexity of molecules it is a gene.

Living systems exhibit no special physical or chemical properties unique to themselves, and are seen to depend on these for their existence. The special properties shown by living systems are viewed as emerging from complexity itself. Thus, memory is a phenomenon that depends on various commonplace physical and chemical forces, but more importantly grows out of the interactions of such forces in an incredibly complex substratum of neurons in a special environment. In other words, memory is phenomenon that simply does not exist at the molecular level as such—it is a phenomenon that grows out of higher levels of organization. Evolution is seen, in part, as the gradual accumulation of complexity and new levels of organization, the end of which is not in sight.

Since life is viewed as an expression of a certain level of molecular complexity and order, it is entirely possible that it has originated elsewhere in the universe where the special conditions needed for generating complexity exist. Thus, life on earth is seen as probably only one manifestation of these properties of chemical complexity. It is not seen as being unique as a phenomenon, although the actual forms taken on earth may be unique, depending upon whether the universe is infinite or not. The general properties of life may be in the process of originating or evolving at many places in the universe. Thus, a general progression is set up as follows: atomic, molecular, organismic—with both ends left open in principle, the entire sequence applies anywhere in the universe. In any given expression of it, say, here on earth, the sequence is allowed to be represented as a unique historical sequence, with the present moment being a given expression of the living level of organization. But in a more universal sense this is not a fundamentally historical formulation. The sequence is conceptual only; that is, given the right environment in terms of energy flux, we would be at one point in the sequence regardless of whether the others had preceded it or not. Thus, on the sun we are basically at the atomic level of organization. The moon actuates a molecular situation. There is no need to imagine that either of these is actually undergoing a specific progression from one level of organization to another. Individual history can be incorporated into the sequence of levels of organization, or imposed upon it, but is not necessitated by it. Also, it would follow that there is no prescription concerning which direction the sequence might take in moving from one level to the next. There is apparently no theoretical reason for supposing that it must always move toward greater complexity; this would depend entirely on the change in entropy. In connection with the earth we think in terms of ever-increasing complexity and a forward flow of time, but, even if appropriate, this is seen as perhaps nothing more than historical accident.

Man is in one sense seen as a current expression of the continually altering (evolving) organization of organic being—simply one of many lineages continuing to evolve under pressure of changing environments. This view expresses the obvious connection man has, via the primates as a group, with all other living systems. Man's lineage, like others, is seen as continuing to evolve, with the current moment only accidentally in view and not representing in any fundamental sense a culmination. This view is strongly implicit in the grand scheme just outlined. On the other hand, man is often viewed from the point of view of his uniqueness, from the point of view of what makes him unique in comparison with other species. Such a discussion usually hinges upon various adaptations that are unique, and especially the adaptation of symbolic culture (Chapter 17) or those aspects of his brain that allowed that kind of adaptation. Symbolic culture is sometimes considered to represent another step in the progression—atomic, molecular, organismic, symbolic cultural, and so on, thereby placing man as the vehicle for the origin of not only an entirely new adaptive zone but a new mode of being, thus increasing his importance in the scheme.

Nevertheless, man's importance is one feature that really is sacrificed, as it were, by the organic evolutionary point of view. Thus, if the mind exists, it has to have been an adaptation, presumably useful for catching prey or building shelters. As such there is no guarantee—and, indeed, the proposition is highly unlikely—that that mind can be considered as a tool that can accomplish the feat of understanding the universe. However, it is just here that chance, in the form of preadaptation, enters the picture. The mind, an adaptation for flexible adaptability, might have been *pre-*

adapted for universal understanding by way of the instrument of symbolic culture.

The foregoing is essentially a paradigm for a mythology, and it is interesting to consider in what respects it satisfies both the criteria of mythology and the mythological needs of people living in our society. Without going into much detail, we note that in many myths man is seen, as here, to arise from interactions between nonhuman forces. Many myths also place man among other animals. Frequently, however, he is seen as very special, as in a way he can be here too. "Can be" is interesting. The organic evolution "myth" is unusual in that it leaves open many important interpretations. This is the influence of science, a cardinal principle of which is that new information may modify old interpretations. This is not characteristic of nonscientific myths, although they may be so vague on some points as to approach indeterminacy. Scientific .mythology would have always to be provisional in principle; there is always an unsettled air about it in comparison with supernatural ones. Instead of stating that X happened, it is suggested that X might have happened, or probably happened.

Since it is clear that there are many points of similarity between supernatural myths and the world history presented by evolutionary biology as one analyzes them, it is also of some interest to speculate about the possibility of similar functions within society. Thus, mythological beings and events often function as symbolic expressions of various culturally important attitudes and processes. For example, a dragon might symbolize or represent evil, or impediment, or challenge. What, indeed, might dinosaurs represent to the nonevolutionary biologist in our society in mythic terms? Could they represent transience, or obsolescence, or might they be a rise-and-fall paradigm, or can they represent failure? These questions are, of course, not the province of evolutionary biologists at all, and they are asked here only to suggest that the story of evolution does not exist in or arise out of a vacuum. The study of this question is a proper one, even now, for social scientists.

Nothing could be more certain, however, that, regardless of the scientific sophistication that went into constructing it, the organic evolutionary history of the world and its explanation of the origin of life and of man will function, if it does not already do so, as modern Western society's creation myth. (It must be realized that a myth does not have to be imaginary; it is any explanation of the origin of life and man.)

REFERENCES

Blum, H. F., *Time's Arrow and Evolution*. Princeton University Press, Princeton, N.J., 1968, 232 pp.

Morowitz, H. J., *Energy Flow in Biology*. Academic, New York, 1968, 179 pp.

Schrödinger, E., *What Is Life?* Doubleday, New York, 1956, 263 pp.

CHAPTER 2

Prebiotic and early evolution

This chapter is concerned with the origin of life and the earliest stages of organic evolution. It will become clear that it is not possible to discuss the evolution of living systems without considering also the evolution of the environment in which they operate. Organic evolution can be most fully understood as one aspect of the evolution of the earth itself.

PHILOSOPHICAL AND HISTORICAL BACKGROUND

To discuss the origin of life implies a belief that living systems had an origin. This is by no means the only possible assumption. Life, and the earth, can be thought of as having always existed, possibly going through periodic cycles of change without beginning or end. But it is natural, and perhaps even necessary, that the assumption of a beginning and an end should grow out of the historical point of view; if life has had a history, it could certainly have had a beginning and may eventually come to an end. Immersed in the analogous birth and death of individual organisms, historically minded biologists come to this notion quite naturally.

But living organisms always arise only from other living organisms. Louis Pasteur, during the latter part of the nineteenth century, was at great pains to demonstrate this. He succeeded so well that it was some 50 years before it again became scientifically respectable to consider the problem of the origin of life. Basically, his demonstration consisted of showing that no microorganisms would appear and grow in a sterile nutrient broth exposed to the air by way of the long neck of a flask containing it. This neck was so bent that individual microbes or reproductive spores would be stopped from reaching the broth from the outside. Actually, Pasteur was not here concerned with the origin of life at all but with showing that it was possible to exclude microbial life from a given environment, because they could appear in such environments only as the result of their reproductive processes and not as a result of spontaneous generation. The spontaneous generation of highly evolved modern microbes during a tiny fraction of geological time in a given experimental flask is not how we now picture the origin of life. In fact, Pasteur himself remarked that his experiments had no bearing on the possible origin of living forms in some quite different past environment. In a curious way the force of his experimental work and his determined publicizing of his results during a long public controversy had a damping effect on thinking directed toward the possibility of living systems arising from nonliving matter.

It is not surprising that the issue was raised again by men determinedly opposed to theological or mystical explanations of the phenomena of life—explanations that gain currency to the degree that scientific-historical explanations are missing. In 1924 in the Soviet Union, A. I. Oparin published the first edition of his book *The Origin of Life on the Earth,* and five years later the brilliant English biologist J. B. S. Haldane, who was then a Marxist, published an extremely influential article on the same subject. Together these treatments set the tone and limits for much of the work that was to follow. Today, with the governments of wealthy nations staking much of their prestige on space exploration, including a hope of the possible discovery of extraterrestrial life, the problem of the origin of life has undergone a renaissance. Persons working in this area now lack neither prestige nor research grants.

Assuming that life on earth had an origin on earth, the scientific point of view seeks to find a scientifically satisfying picture of how that origin came about. As Haldane pointed out, a supernatural explanation is adequate for many purposes but precludes scientific investigation by making it unnecessary. Scientists prefer a naturalistic explanation, and we take this point of view here. There are actually two different philosophical positions on this question that are referred to as naturalistic. One holds that the origin of living systems is an automatic result above a certain level of chemical complexity, with the corollary that wherever and whenever this level of complexity is reached, life spontaneously arises. The other point of view maintains that, although the origin of life is a spontaneous result at certain levels of chemical complexity, so many independent events must occur at just the right time and in the right place and in the right sequence and at appropriate energy levels that the origin of life is a very improbable occurrence, however natural. From this point of view, life may have originated only once in the entire universe. Given a long enough time, an improbable event becomes increasingly probable and even inevitable. Thus the probability of getting heads when flipping a coin is 1/2; the probability of getting heads once in two flips is 3/4; the probability of getting heads once in ten tosses is 999/1000—that is, it is virtually inevitable. Or, to use an example of Haldane's, the mere shuffling of the letters *A, C, E, H, I, M, N* produces the word MACHINE once in only 5040 tries, although it is very improbable in one try.

Most laboratory scientists working in this area prefer the first point of view for the obvious reason that we simply cannot do experiments on very rare events. Although it is doubtful that at this stage of our knowledge anyone is getting ready to synthesize life in the laboratory, it is encouraging to workers in the area to believe in the ultimate possibility of achieving this result. In the absence of knowledge, beliefs are chosen that are by and large functional for those who hold them. Thus, it is not surprising that those who have written from the point of view of the great rarity of life in the universe are humanists: If life is rare, so is man, and if man is a rarity, he is precious. It seems to the present author that if the origin of life is such a rare occurrence as to have happened only once in the universe, this makes it effectively a supernatural event and therefore not susceptible to scientific inquiry. The problem of extraterrestrial life is taken up again at the end of this chapter.

THE EARLIEST ENVIRONMENTS

Present estimates of the age of the earth range from 4.5 to 4.7 billion years (aeons). Whatever the manner of formation—and it is not felt to be relevant to the origin of life which of the current theories applies—the oldest rocks are dated 3.3 to 3.6 aeons, and because these are partly metamorphic, there must have been a solid crust (lithosphere) for some time before this. The first sedimentary rocks seem to be 2.7 to 3.2 aeons old, providing evidence of the weathering of rocks exposed to the atmosphere by that time. The very earliest rocks show evidence of having been solidified under water, suggesting that the hydrosphere antedates the lithosphere. By the time that the first sediments were being deposited, it is felt that the original atmosphere of the earth, including methane (CH_4), ammonia (NH_3), and hydrogen (H_2), had become broken down or escaped into space, being replaced by a secondary one produced by outgassing from the interior of the earth during volcanic activity. This secondary atmosphere was primarily water vapor (H_2O), carbon dioxide (CO_2), carbon monoxide (CO), nitrogen (N_2), sulfur dioxide (SO_2), and hydrochloric acid (HCl). The seas are considered to have become about as saline as they now are at least by 2 aeons ago but may always have been so. Some carbonates, however, were precipitated from them right from the beginning. Whatever ammonia remained after its loss from the atmosphere was dissolved in the seas. A silicate buffering system must have maintained the seas at an alkaline pH (the pH of water after a rock has been in it for a while is between 8 and 10).

Because the oldest known fossils—blue-green algal stromatolites (see Figure 2-1)—are found in sediments in South Africa dated at 3.2 aeons, the origin of life must have taken place between then and the formation of the hydrosphere just before 3.6 aeons ago, a period of time spanning 400 million years. The general conditions on the earth during these times can be summed up as follows: Moderately saline, shallow alkaline seas covered at first all the surface until orogenic activity lifted some of the crust above sea level about 3.2 aeons ago. The atmosphere was a reducing one (able to contribute H^+ or electrons) rather than, as now, an oxidizing one (able to capture hydronium ions and absorb electrons); there was no free oxygen, for, although this was being produced by the photolytic splitting of water ($2H_2O$ + light energy $\rightarrow 2H_2 + O_2$), it combined immediately with carbon, silicon, or sulfur to form CO, CO_2, SiO_2, and SO_2, that is, it became reduced, and so the atmosphere was an anaerobic one. Radiation from the sun was much more intense than it is now at the surface of the earth; the presence of ozone (O_3) in the ionosphere today screens out most of the ultraviolet, and ozone cannot be formed in the absence of free oxygen. There were probably a fair number of electric storms and rains. Volcanic activity was probably greater than it now is.

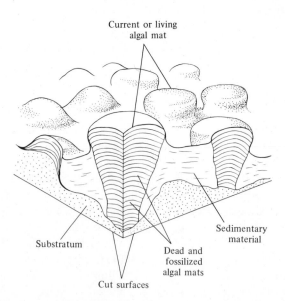

Figure 2-1 Two types of stromatolites that are formed by mats of blue-green algae in intertidal situations.

Concerning more specific historical events during this early period little is known. About 4 aeons ago meteorites of a kind known as carbonaceous chondrites may have collided with the earth, as others have done since. These seem to contain simple carbon compounds, and are discussed at the end of this chapter. By 3.2 aeons ago, lime-secreting ($CaCo_3$—chalk) blue-green algae were already present. These are of further interest in that the dome-shaped calcareous sedimentary structures— stromatolites (see Figure 2-1)—organized around them indicate that these blue-greens were intertidal organisms. This indicates that the moon was present at this time.

DEFINITIONS OF LIFE

Some time during this period of 400 million years, somewhere in the environment described in general terms above, the phenomenon we know as life came into being on earth. Although it is not possible to define life rigorously, we can note some of the important characteristics of living systems—characteristics that we must be able to derive conceptually from the inorganic world in imagining how life could originate. Living systems on earth are all composed of carbon compounds—that is, they are made of organic matter as opposed to inorganic, and have the following basic structure:

$$
\begin{array}{c}
\mid \\
-\text{C}- \\
\mid \quad \mid \quad \mid \quad \mid \quad \mid \\
-\text{C}-\text{C}-\text{C}-\text{C}-\text{C}-\text{C}- \\
\mid \quad \mid \quad \mid \quad \mid \quad \mid \quad \mid \\
-\text{C}- \\
\mid
\end{array}
$$

Many of these carbon compounds function as catalysts in chemical pathways that are arranged into metabolic systems, some of which derive energy from other organic material (catabolic systems) and some of which use the energy so derived to construct more organic matter (anabolic systems). The organic matter so constructed is not simply any kind of organic matter but specific compounds characteristic of the given system, and so these systems grow. This growth is not of the simple kind found when a crystal "grows" by the accretion of preexisting molecules; the specific organic compounds are constructed step by step according to directions encoded in one kind of these compounds (the hereditary material—nucleic acids). Thus, an integral part of growth in living systems is reproduction (or replication). At the other end of the scale is the nonliving world of crystals, liquids, and gases, and we must imagine means of transition from the one level to the other, given the setting described above.

During the eighteenth and early nineteenth centuries, life was defined as a principle residing in organic compounds, and, indeed, the two were considered to be synonymous. After Friedrich Wöhler synthesized urea in his laboratory in 1828, however, the notion that organic compounds could be made only within biological systems was no longer tenable. Since that time many organic syntheses have been worked out, culminating recently in the synthesis of human insulin, a polypeptide consisting of 52 amino acids. These laboratory syntheses, however, provide no evidence of the possible prebiotic synthesis of similar compounds; man is, after all, alive, and we are not here interested in whether he is capable of constructing organic chemicals outside his body. We must inquire whether organic compounds can be formed spontaneously in the absence of any living agent whatever, and the answer is yes.

ABIOTIC SYNTHESIS OF ORGANIC MATTER

In 1953, Stanley L. Miller, then a graduate student at the University of Chicago, performed what is now seen to be the classic experiment in this area. A flask quite like that used by Pasteur, but closed after sterilization, containing water and a mixture of the gases methane, ammonia, and hydrogen (a reducing atmosphere) had a high-frequency spark discharged in it for a week during which the water was boiling to produce water vapor. At the end of the week milligram quantities of the two amino acids alanine and glycine were found, plus smaller amounts of other amino acids, fatty acids, and other organic compounds, as well as CO_2, CO, and N_2. In the two decades since then, hundreds of such experiments have been performed using a variety of hydrospheres, atmospheres, and energy sources. The results can be roughly summarized as follows: In reducing atmospheres, moderate energy sources are sufficient to produce, depending on the mixtures used, most major classes of organic compounds important to living systems on earth; in oxidizing atmospheres, similar results can be achieved only with more intense energy input, and the products are mostly rapidly oxidized. Some controversy does exist as to which of the many conditions used represent fair models of primitive conditions on the early earth. In dilute, alkaline aqueous systems under reducing atmospheres (conditions most workers are willing to grant as primitive) there can be formed amino acids, sugars, fatty acids and other lipids, purines, peptides, and polypeptides. The spontaneous synthesis of pyrimidines, nucleotides, ATP, and porphyrin has been accomplished under conditions not everyone is willing to accept as possibly primitive. Nucleic acids and proteins have not yet been produced spontaneously. It is clear, however, that in time conditions will be found in which the latter syntheses will occur. After all, no enzymatic reaction found in living systems was created anew by these systems. Enzymes do not make reactions go; they simply speed up preexisting reactions that do not occur with any great rapidity in inorganic nature.

Among the energy sources that were available on the early earth, ultraviolet radiation from the sun is considered to have been the most important. Without an ozone screen this radiation must have effectively penetrated the seas to a depth of some 10 meters. With ammonia in solution, this radiation can produce one of the important intermediates in the spontaneous syntheses, cyanide (HCN). Other essential intermediates are formate (HCOOH) and formaldehyde (HCHO). One problem with this energy source is that it is continuous; there is no reason that it should not supply energy to break down some of the organic compounds as soon as they are built up. Formaldehyde, for example, is very unstable. We must keep in mind, however, the equilibrium constants of the various reactions; if they favor the side of synthesis, that is, this side has the lowest free energy, then at equilibrium (as much product is broken down as is built up) there will still be some synthetic product where there was none before, assuming that sufficient activation energy was supplied, say, by UV radiation, to make the reaction go. Also, less stable compounds could accumulate if they were formed by short wavelength UV radiation at the limit of its penetration into water and then diffused below this limit where longer wavelength (less energetic) UV radiation could perhaps more gently and more slowly stimulate their participation in further syntheses. There is also the possibility of ledges of some kind providing shaded areas at least during part of the day. There is, of course, also the night, during which compounds formed at sundown could further react under gentler conditions.

Other possible energy sources are of more intermittant nature and may for this reason have been important at certain stages. Lightning is an obvious example, and was mimicked in Miller's experiments. Ultrasonic energy has been considered a possible candidate and would be generated during the collision of a meteorite with the earth. In an experiment mimicking this situation, a bullet was fired through a mixture of CO_2, NH_3, and H_2 into water, and traces of organic compounds were subsequently recovered. Energy generated by radioactive decay has not been considered a suitable candidate, because it is estimated that during the first billion years of the earth's history most of the radioactive materials would have decayed to the point of being unimportant at the surface of the earth, where the important reactions began to occur at the beginning of the second billion years. In one of the earliest experiments in the modern period, however, it was shown in Melvin Calvin's laboratory, at the University of California at Berkeley, in 1951 that ionizing radiation was capable of supplying energy to reduce CO_2 to formic acid.

It has been demonstrated many times that thermal energy can be used in the abiotic synthesis of organic compounds, but its historical role is controversial. Natural sources of heat energy would be volcanic activity and infrared radiation from the sun. The sites of activity of this energy source and of some of the other intermittant sources would perforce be localized (IR radiation, being dissipated in the seas, could only act in small, shallow pools); the action of UV radiation in the seas is general and continuous. It has been asserted by some that it would not be scientifically sound to construct our ideas on this matter so as to include the necessity of visualizing a possibly rare topographic situation. Thus, if shallow pools are necessitated by the argument, any information demonstrating that there were no shallow pools in the environment would destroy the argument. Those holding this viewpoint, such as Philip H. Abelson, of the Carnegie Institution of Washington, suggest that we should consider only mechanisms capable of occurring in dilute, alkaline seas—the only environment that we can claim to have evidence of the existence of in the earliest rocks. The first sedimentary rocks already hold fossils, and so there is no evidence of land prior to life. Proponents of special sites argue that it has not been possible to demonstrate the synthesis of all necessary compounds in the prevalent aqueous environment, and so we are forced to imagine something else. They further argue that the fossil and geological records never show a complete picture of the earth at any given time and that negative evidence cannot disprove a theory. This controversy is related to the one concerning the rarity of life originations. The viewpoint insisting that we should visualize the origin in the most plentiful and homogeneous environment is apposite to that which holds life to be a common phenomenon in the universe; that which is willing to invoke specialized or even bizarre sites is concordant with the view that life is a rare occurrence in the universe.

After a long period of abiotic synthesis, the seas are visualized as gradually changing into what Haldane has dubbed a "hot, dilute soup" of organic chemicals; today, following Abelson, most workers think more in terms of a cool, dilute soup. Up to this point, it is fair to say that we have a more or less satisfactory overall view of what transpired and how it did so; there are only details to be filled in. The next step, however, that of visualizing the organization of separate chemicals into living systems is a different matter; it would be fair to say that here we can visualize a few details of the very earliest stages but have not even a dim concept of, let alone evidence pertaining to, how the later stages of the transition to life occurred.

MOLECULAR ASYMMETRY OF LIVING SYSTEMS

As a transition from the topic of the purely molecular basis of living systems to their organization, a few words should be said about a phenomenon whose significance is not entirely clear, namely, the optical activity shown by those molecules which are asymmetric in these systems. The inorganic synthesis of any kind of asymmetrical molecule produces roughly half left-handed (levorotatory) and half right-handed (dextrorotatory) enantiomorphs (see Figure 2-2). (A solution of pure levo (l) molecules bends the path of a beam of plane-polarized light in one direction, that is, is optically active, whereas a solution of pure dextro (d) molecules rotates it in the other direction. A mixture does not rotate the beam at all.) Not only do the asymmetric molecules of living systems show optical activity, but all living systems are characterized by the same enantiomorphs, important examples being the sugars (dextro) alpha-phosphatides (levo), and the amino acids (levo). (d-amino acids are rare and are used as chemical-warfare agents by many microorganisms; they are made by special enzymes—amino acid racemases—from levo molecules.) The importance of optical activity resides in the architecture of the polymers constructed with asymmetric molecules as the monomers. An orderly structure, such as the alpha-helix found in many proteins and polypeptides, is not possible if both enantiomorphs are involved. This seems to be the selective basis on which living systems became associated with only single optical isomers. But this does not explain how the first living systems could incorporate only pure enantiomorphs out of mixtures. Nor does it explain why the specific isomers that were chosen were chosen.

J. D. Bernal, of the University of London, has suggested that, in fact, the inorganic synthesis of compounds, such as amino acids, does not have to be considered always to produce an equal mixture of optical isomers. A synthesis taking place on a catalytic surface could produce mostly one enantiomorph. Indeed, l-alanine has now been synthesized in this way

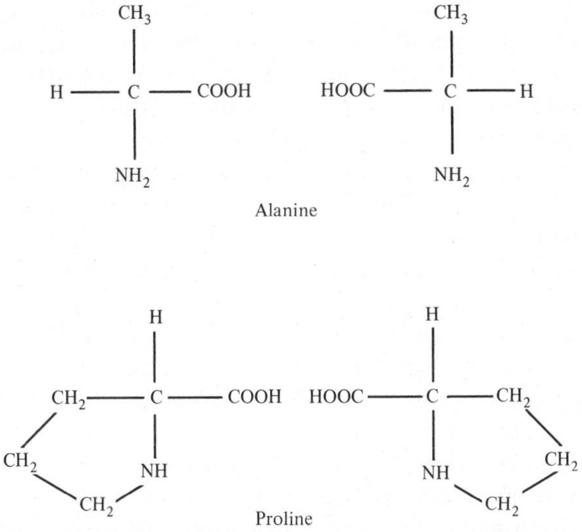

Figure 2-2   Optical isomers of two enantiomorphic amino acids.

about 95 percent pure. This explanation for the original incorporation of optically active compounds into living systems depends on the assumption that such systems arose in discrete, special sites, because otherwise, after synthesis, the optically resolved chemicals would diffuse into the seas, mixing there with the other enantiomorph. George Wald, of Harvard University, has proposed a more general solution. If equal amounts of *d*- and *l*-amino acids are produced, but, by chance, some *l*-acids form peptide bonds first, differential survival of the two enantiomorphs can be visualized. This is because amino acids are more stable after polymerization than they are in free solution and because a polymer of compounds like these "selects" from the available pool monomeric units that best fit the structure of the polymer, and these are units of the same optical activity. Units of different optical activity from the polymer have trouble fitting on sterically and do so only associated with a high free energy, and the bond is, therefore, unstable. This argument is an interesting example of the concept of levels of organization. The presence of only *l*-amino acids is meaningless other than in the context of protein structure. This structure demands optical resolution, but such resolution is also a phenomenon that grows out of it. In Wald's formulation, optical resolution occurred because of polypeptide formation, although resolution is a phenomenon that can exist by itself (but perhaps only experimentally, that is, in the minds of men).

As to why the particular isomers now found in living systems were selected, all analyses invoke chance. It seems highly improbable that a living system could not occur involving *d*-amino acids or levo sugars. Because a number of protoliving systems can be visualized as starting independently, each with a different spectrum of optical isomers, some explanation must be gotten as to why all present living systems on earth have the same spectrum. One idea is that for some unknown reason, perhaps irrelevant to optical isomerism, only one of these systems has actually survived. A corollary of this idea is that all living forms are descended from a single ancestral system. This idea is also consistent with the notion that life began only once on earth. If there were multiple origins and survivals—and John Keosian, of Rutgers University, has argued that biogenesis may continually occur—they could all wind up with the same spectrum of enantiomorphs by means of selection. Thus, the first successful living system begins initially to incorporate and then to synthesize optical isomers characteristic of it. In a rather short time, because the synthetic capabilities of living systems are greater than those of the inorganic world, the environment comes to be characterized by the same optical isomers as the living system. Any neobiogenesis that now occurs has to occur within this context, and, therefore, the new systems, by using what is available, will show the same spectrum of asymmetries. This idea well illustrates the proposition that living systems alter their environment, and we shall see this again in other examples.

THE PROBLEM OF CONCEIVING AN ORIGIN

Before launching into a discussion of biogenesis (or biopoesis) proper, we should give some more attention to the problem of the definition of life. No one disputes that a system having the properties given above—metabolism, growth, and reproduction—is alive. The problem arises in connection with what else, if anything, we can consider to have life. H. J. Muller, of Indiana University, and, more recently, Norman H. Horowitz, of the California Institute of Technology, have considered that any self-replicating system, or even molecule, should be considered to have the property of life. On the other hand, some, including Haldane and Bernal, have felt that any self-maintaining or continuing chemical process in a limited volume, that is, a protometabolic system, could be considered to be alive. Interestingly, by neither of these definitions is a virus alive; it cannot reproduce without the machinery of a cell, notably ribosomes, and it is certainly not a metabolic system. (The consensus on viruses today leans toward the idea that they are degraded or simplified forms—reduced to a genetic code, a protective boundary, and a mechanism for effecting entry into cells—descended from more complex parasites or from the genomes of the hosts that harbor them. They cannot be considered to represent any kind of primitive system.)

It is now clear that attempts to define life rigorously are no more than exercises in semantics. Unless we wish to believe that life arose full-blown at one moment, we must imagine a transition from the nonliving to the living. At some point along this transition there occur systems that are not fully alive nor yet simply inorganic. It does not matter whether we wish to call these alive or not. What are important in the controversy just described are the viewpoints on how life developed that are implicit in the definitions. One point of view has it that some kind of genetic code was primal and that it later incorporated more and more enzyme systems; the other maintains the primacy of energy-deriving and energy-using processes that later became more precise by adding a genetic code.

There are really no clues that allow us to choose between these alternatives in extant living systems. On the one hand, the genetic code cannot function, or even replicate, without enzymes; on the other, metabolic systems composed of enzymes cannot long continue without a code specifying the structure of the enzyme(s), because existing enzymes, inevitably destroyed sooner or later, cannot be replaced in the system by just any enzyme. They must be replaced by the same kind of enzyme, and this can only be guaranteed if there is a code specifying that structure. Thus, growth is not possible without replication, and growth is essential for the replacement of lost parts. Living systems represent Gordian knots in this sense, and the problem is similar to the old one in logic about the primacy of the chicken or the egg. As it now exists on earth, life is a complex system completely dependent on all its included mechanisms; remove any of them, and the system becomes nonliving even if it continues in some respects to function for a short period.

Under these circumstances it is tempting to suggest an origin deriving from a chance coming together of a potential genetic code with a proto-metabolic system. Such an occurrence can only have been rare, not predictable, and deprives us of the need to say anything further. It would be preferable from the scientific point of view to erect a number of hypotheses concerning the means by which the system could have evolved from simpler one, possibly still composed of the forerunners of both essential units—code and contained metabolism. In order to facilitate analysis, however, we first examine the possibilities for the evolution of the three main characteristics of living systems—discrete boundedness, controlled metabolism, and replication—from inorganic systems. Perhaps after that we can suggest ways in which these characteristics could have evolved simultaneously from much simpler systems.

ORIGIN OF MEMBRANES: THE PROBLEM OF SEPARATION FROM THE SURROUNDING ENVIRONMENT

Living systems today are without exception composed of one or more discrete units physically separate from the rest of nature but remaining open to it and dependent on it for raw materials, energy, and information. Because it is not always clear just what the fundamental autonomous units in any particular system are—are tapeworms individuals or colonies? are the free amoebae of cellular slime molds individual organisms or the free cells of a multicellular organism, the slug?—the statement above may simply reflect a limitation on our concepts due to perceptual bias. Perhaps living systems need not necessarily be composed of individual units. It may be noted, however, that present evolutionary theory is absolutely dependent on the existence of autonomous organisms that are genetically individual, that is, no two are exactly alike. For this reason we accept our perceptual bias and consider that living systems must be systems of discrete individuals. We must then imagine the origin of this discreteness from molecules and processes dispersed in the environment. Thus, we need initially to imagine how molecules could spontaneously—that is, not violating thermodynamic restrictions—aggregate from (or in) the cool, dilute soup.

There is at least one mechanism known by which dehydration reactions can occur in dilute aqueous solution to produce by condensation the biologically important C—O, C—N, and P—O bonds involved in the major polymerizations leading to proteins, nucleic acids, and other biological polymers. The removal of water from two molecules, such as two amino acids, thereby uniting them, can be achieved under these conditions in the presence of cyanide (HCN) and similar compounds, which are strong dehydrating agents. As indicated above, such compounds are thought to have been present in the prebiotic seas. Freely diffusing polymers, however, are only a step closer to living systems than are freely diffusing simple molecules, and it is difficult to see how, in the absence of further restrictions on their motility, they could produce any supramolecular form. What is needed is some means of concentrating the molecules and macromolecules formed in the dilute soup.

In order to visualize the aggregation of molecules counter to the natural process of diffusion, it seems to become necessary to visualize some kind of asymmetry in the environment, some microhabitat in contact with the seas or within them. One suggestion is that aggregates might have formed by adsorption onto claylike minerals with base-exchange ratios, resulting in a film of material at the bottom of the seas. These formations could be optically active as well. Another problem is the depth of the seas. If they are too shallow (as on shores, if any) wave action might continually scour the material off the clay, and no accumulation could occur. On the other hand, if too deep, there are problems with subsequent evolutionary events, unless we postulate a later rising of the sea bottom in the place in question. Surfaces in general are catalytically active, and because we must imagine some kind of "metabolism" ultimately arising, this model has another point in its favor. This and the following model also have in their favor the fact that they do not postulate a very unusual and therefore rare microhabitat. A similar idea suggests that the earliest aggregates accumulated in the pores of porous lava, by a kind of entrapment. Lava is felt to have been produced at or before the time in question during the extensive volcanic activities then going on.

An interesting possibility is that organic material could have accumulated in a layer at one level of the thermocline in the seas. Surface-active organic material, being hydrophilic at one end and hydrophobic at the other (see below), could be carried by diffusion or bubbling to the surface, where it would accumulate as a monomolecular layer being held by the hydrophobic ends sticking out of the water (see Figure 2-3). This material could subsequently become packed together by wave action (Langmuir circulation), so that the monomolecular layer becomes increasingly folded up. After reaching a certain weight, this material sinks, as it does today, either to the bottom if the water is shallow or, in deep water, to a level where the density of the water equals that of the sinking material, at some position in the thermocline. Any large body of water that is not circulating vigorously develops a temperature gradient from the warm surface to the cold bottom. The density of water increases with decreased temperature, and so the thermocline represents a density gradient as well. This model postulates, then, that the seas either were shallow or were not circulating vigorously everywhere—conditions that are not extremely unlikely at least.

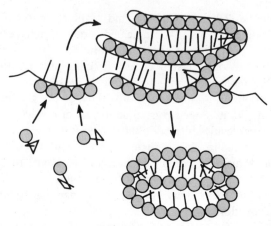

Figure 2-3   Spontaneous formation of supramolecular form by surface active (amphoteric) molecules at the air-water interface, and subsequent sinking of the folded-up surface film after it has reached a weight great enough to overcome the surface tension of the water.

There does seem to be some problem with the accumulation of surface-active material, however.

The prebiotic formation of surface-active materials of the kind found in living systems today—phosphatides or phospholipids—is not a serious problem. Fatty acids ranging in size from C_2 to C_{19} have been synthesized inorganically under possibly primitive conditions. The synthesis of glycerol from C_3 sugar, which forms spontaneously under primitive conditions, is possible, given some metallic catalyst, such as iron, which was plentiful (see below). In the presence of phosphoric acid, glycerol undergoes esterification to form α-glycerophosphate, which can then combine with fatty acids to form the l-α-phosphatides. Only the C_3 sugar and the fatty acids have actually been synthesized spontaneously as yet, but this does not mean that the rest cannot occur under as yet unknown conditions. The problem seems to be that phosphate is much too likely to be mostly precipitated as $CaPO_4$ and so not available for reaction with the glycerol. Phosphate, however, is so important in biological systems that it must have been made available somehow, possibly in the more soluble hypophosphates. In addition, the long, straight-chain fatty acids are also likely to become sequestered in salts and buried. The matter of surface-active molecules is taken up again below.

Another, and rather obvious, suggestion for increasing the concentration of organic materials locally is to imagine that the material was present in rather small ponds that became increasingly concentrated as they evaporated. The problem with this notion is that there is no evidence that any dry land was present during the time in question (of course, there is no evidence to the contrary, either, but in science one ought not to make ad hoc assumptions).

Given some method by which the prebiotically formed organic molecules and their polymers could become concentrated, we need to find some means by which they could become organized into a supramolecular structure of some kind. Oparin early addressed himself to this matter and came up with the coacervate model. A coacervate is a solution of high-molecular-weight chemicals—proteins, carbohydrates—that at certain conditions of temperature, pH, ionic composition and strength spontaneously separates into two phases, one of which is very dilute with respect to the macromolecules in question (the sol phase) and the other of which, usually dispersed in microdroplets (the gel phase), contains mostly macromolecule and very little water. Organic chemicals other than proteins and carbohydrates can be included in a coacervate, but the fundamental properties of this kind of system are determined by these two classes of compounds. In contact with an aqueous solution the gel phase of coacervates is osmotically active. Furthermore, because the constituent molecules usually have a net negative charge, positive ions can be pulled into the gel, followed by free negatively charged ions from the medium, by diffusion, followed by the entry of even more water to balance the increased concentration of osmotically active particles inside the gel. In this way the gel swells when contacting an aqueous solution, pulling into itself all manner of charged particles from the outside in the process. Oparin has shown that if an enzyme is incorporated into a coacervate, substrate free in the medium contacting the gel phase is converted into end product, the low-molecular-weight substances presumably diffusing in and out of the gel.

Sidney W. Fox, of the University of Miami, has extended this approach further. Addressing himself to the questions of prebiotic sources of high concentrations of amino acids and of their polymerization, he has shown that if heated dry to 150 to 200°C or, in the presence of phosphoric acid, only to 70°C, amino acids polymerize. The polymers are unlike true proteins in a number of ways. Most important, they are branching structures; they also do not show helical secondary structure; again, they are not antigenic. Fox refers to this substance as proteinoid rather than as protein. The conditions needed for this anhydrothermopolymerization are not, however, known to have been present at the appropriate time. Fox has suggested that the amino acids could have become concentrated dry in evaporated pools on land. These pools could have formed near or over sites of volcanic activity, providing a source of thermal energy. Subsequent flooding during rains could place the proteinoid back into an aquatic environment. This concept invokes specialized sites for which there is no evidence.

If the water flooding the proteinoid substance is, or can become, hot, then on cooling, the system forms a kind of coacervate, the proteinoid being contained in what Fox calls microspheres, ranging from 0.5 to 80 microns in diameter (see Figure 2-4). These proteinoid microspheres have a number of interesting properties. They can metabolize glucose (catalyze its breakdown) at a rate that is very low but significantly greater than can occur without a catalyst. Increasing the pH of the medium in which

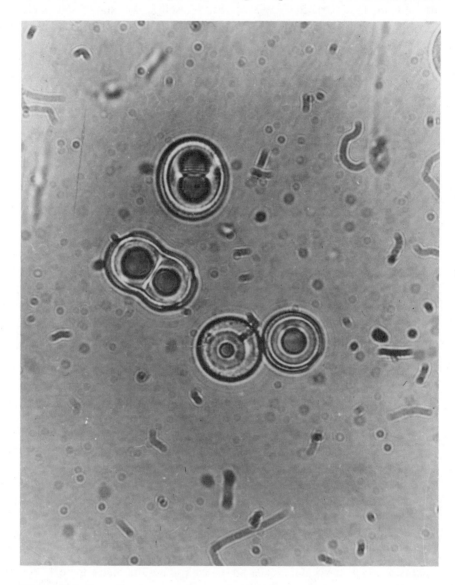

Figure 2-4   Proteinoid microspheres, showing their division upon changing the pH of the medium from 3 to 6.5. Courtesy of Dr. Sidney W. Fox, Institute of Molecular Evolution, University of Miami, and Academic Press.

they occur from 3 to 6.5 results in many of them undergoing a division that superficially resembles the cleavage of a cell such as the sea urchin ovum.

Both Oparin and Fox consider that the first structures in the ancestry of living systems could well have been protein coacervates. They feel that such structures could be imagined to grow by the incorporation of amino acids or peptides from the environment, and that after reaching a certain size, under certain conditions, they could divide, approximating a crude kind of reproduction. They could also have been the sites of the evolution of metabolism, because catalytic activity is, or can become, associated with them. Fox has pointed out that the microscopic structures of some of his microspheres are very similar to some of the few formed structures in carbonaceous meteorites that are not clearly terrestrial contaminants. Of course, they could have been formed from terrestrial organic compounds during impact with the earth, much as they are formed in Fox's laboratory (see below for further discussion of meteorites).

If some way can be imagined for the prebiotic concentration of surface-active amphoteric molecules (with a long, hydrophobic hydrocarbon chain at one end and a hydrophilic ionic group at the other), an interesting alternative for the spontaneous generation of form presents itself. Amphoteric molecules, such as soaps, detergents, phospholipids, or phosphatides, have the interesting property of being only half soluble in water, at the ionic end. In free solution the hydrocarbon chains fold up to reach the lowest possible free energy. (All thermodynamically possible structures and processes occur during the movement of a system to its lowest free energy or after this has been achieved.) On contacting the surface of the aqueous medium—in a bubble or at the atmosphere-water interface or a water-rock interface—the amphoteric molecule finds a configuration giving it an even lower free energy than it had folded up in solution, that is, with the hydrocarbon end protruding out of the water and the ionic end remaining in solution (see Figure 2-3). For this reason these compounds are referred to as surface-active. In this way they can form a monomolecular layer at a water surface, as referred to above.

If one measures any colligative property—that is, deriving only from the concentration of particles present—such as osmotic pressure, of a solution of amphoteric molecules, their initial high free energy in contact with water results in some curious effects as the concentration is increased. Adding more molecules results in an increased number of particles being present only up to a certain point, after which adding even more molecules of the amphoteric substance does not result in increases in the measurable effects caused by the concentration of molecular particles (see Figure 2-5). This is caused by the fact that, after a certain concentration has been reached, there are enough amphoteric particles present for them to achieve a free energy lower than that which they have in free solution by coming

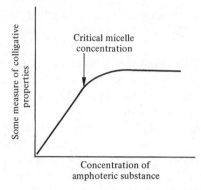

Figure 2-5 The behavior of amphoteric molecules with increasing concentration.

together to form multimolecular units called micelles. The micelles are characteristically composed of 60 to 100 molecules, with the hydrophobic chains tucked inside, away from contact with water, and the hydrophilic ends on the surface remaining in solution (see Figure 2-6). The point at which they begin to form is referred to as the critical micelle concentration (CMC). In a general sense, an amphoteric system that has gone to two or more phases above the CMC is a coacervate, although this term is usually restricted to protein and carbohydrate systems.

The formation of simple micelles is only the first event in a long series that occurs as one continues to increase the concentration of amphoteric substances. Figure 2-7 shows a phase diagram typical of substances of this kind. At a given temperature, as one increases the concentration, the structures representing the lowest free energy continue to change until, in the neat soap phase, the system can become an endlessly folded up bimolecular leaflet—the fundamental structure of the cell membranes of all living forms. Different amphoteric molecules show different phase diagrams and differ in the details of the structures produced, some of which have been studied by electron microscopy (see Figure 2-8). Mixing different amphoteric molecules gives even more variety to the forms that can be produced, and the addition of other kinds of surface-active substances still further modifies the forms produced. Thus, in this sort of system we have a process that spontaneously produces an almost infinite variety of different kinds of forms without any code.

These forms have been called liquid crystals, referring to the fact that, although the hydrophobic ends are long and flexible, the ionic ends are held rigidly together, as in a crystal lattice. This has important implications for early living or protoliving systems, continually under attack by various forms of energy, which can break as well as make bonds, in that the flexibility can act as an energy sink to disperse in motion the potentially de-

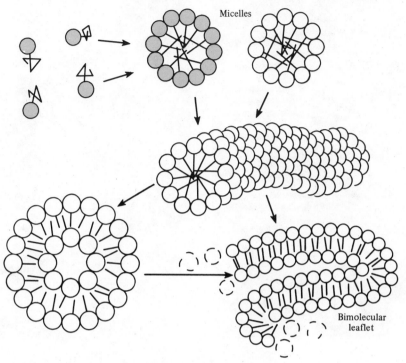

Figure 2-6 The sorts of liquid crystalline structures that can form from amphoteric molecules with increasing concentration.

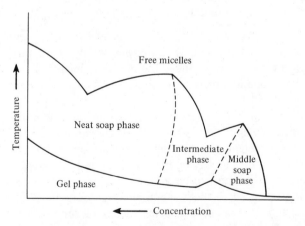

Figure 2-7 Simplified phase diagram of the soap potassium palmitate. Each phase has its own characteristic forms of liquid crystalline structures. The neat soap phase is characterized by lamellar structures essentially like the bimolecular leaflet.

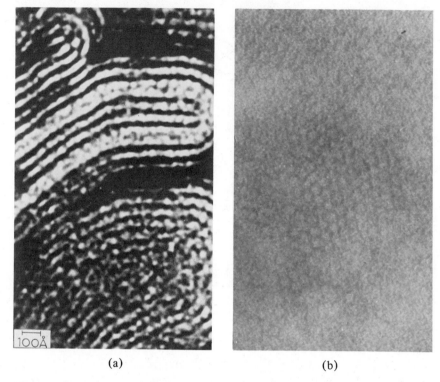

(a)   (b)

Figure 2-8 Liquid crystalline structures. (a) Lecithin liquid crystals in the neat soap phase ($\times$ 1,450,000) showing a slice through an endlessly folded bimolecular leaflet structure. From Green and Perdue, 1966. By permission of H. Fernández-Morán. (b) Liquid crystals of sodium linolate in the middle soap phase ($\times$ 800,000). From Stoeckinius, *J. Cellular Biology* **12**, 221 (1962).

structive energy without seriously disrupting—only temporarily deforming—the form of the systems. Indeed, since all known living systems are bounded by phosphatide membranes, this flexibility is characteristic of them.

The problems involved in conceiving how amphoteric molecules could accumulate in the prebiotic seas have been discussed above. Fundamentally they involve the problem of supplying phosphate and fatty acids, to form phosphatides, without their having been sequestered in insoluble salts. Because phosphoric acid seems to be a major component in a variety of biological pathways as well as in the spontaneous syntheses of organic compounds, we can only suppose that some means of supplying this material was present, and, therefore, could have been used to make phosphatides as well. We can assume that some of the fatty acids that were formed reacted with the phosphoric acid to produce phosphatides rather than becoming trapped in insoluble salts. Once formed, the surface-active material would become relatively stabilized if it reached a surface, and could possibly accumulate in the way envisioned above at a position in a thermocline or on the sea bottom.

All the coacervate structures mentioned so far—microspheres, micelles, bimolecular leaflets—have one thing in common that differentiates them sharply from living systems: The complexity of their structure is in direct equilibrium with their environment. What this means is that, for example, as an amphoteric system becomes more dilute, or its temperature increases, it could pass directly from the bimolecular leaflets of the neat soap phase into some other structures in the intermediate soap phase. If lifelike properties accrued to a system only in the neat soap phase, a rainstorm or a drop in temperature could eliminate these phenomena, only to have them reappear when the sun again evaporated the body of water. Indeed it has been suggested by the English biophysicist J. W. S. Pringle and by Cösta Ehrensvard, of the University of Lund, in Sweden, that the earliest living systems went through a stage that might be called a "living pond," in which some of the phenomena of life were present in an entire aqueous system but only at certain temperatures and concentrations, the system itself having no individual boundaries. This would be true even of a system composed of protocell membranes.

Although living systems are not immutable, they do not change their properties instantaneously with every small change in the environment. Put another way, they are not simply part of their environments. Large changes can, of course, lead to disorganization and death, but significant changes can be withstood with no fundamental alteration of phase. In terms of the liquid crystalline systems, what is needed is some mechanism that can stabilize them at the neat soap phase over wide ranges of temperature and concentration. In living systems some stabilization may be achieved by combination with polyamines or proteins; all cell membranes have these substances associated with the phospholipid membrane. When the phospholipids are extracted pure, they traverse their phase diagrams like any soap. Perhaps at some stage the protocell membranes in a living pond achieved some independence from the environment by becoming combined with peptides or proteins. Bernal has suggested that, once the bimolecular leaflet has become relatively stabilized, the evolution of intracellular membrane organelles—vacuoles, endoplasmic reticulum, Golgi apparatus, and the like—could have had its first impetus by means of selection for the greatest surface area in the smallest possible space, resulting in various infoldings, which is what these organelles fundamentally are.

The phosphatide systems are stressed here rather than the protein coacervates for three reasons. First, the former are capable of giving rise to far more structural variety in the absence of a genetic code than are the latter. This provides more flexibility in imaging the first living systems. Second, all living systems are fundamentally systems of phosphatide membranes. Fox has published photomicrographs showing that what seems to be a double membrane can be produced around the outside of his protein microspheres under certain conditions. This information is irrelevant, because all known biological membranes importantly involve phospholipids. Third, the liquid crystalline structures are especially resistant to destruction by environmental energy fluxes in that they have a built-in energy sink in their flexibility. This would have been of selective advantage in an environment much "hotter" than ours in terms of the density of energy transactions.

Micelles, as well as microspheres, can become associated with enzyme activity. Prepared in certain ways, micelles can be formed that include in their hydrophobic interiors many different kinds of compounds, including enzymes. These are called filled micelles, and they can catalyze the breakdown of substrates placed in the liquid phase. Thus, in one experiment, starch-filled micelles also incorporating enzymes were capable of taking glucose diffused from the liquid phase and converting it to starch and also of breaking down the starch into maltose, which then diffused into the liquid phase, where it could be detected. This little toy was clearly a self-contained metabolic system. We should now proceed to a discussion of the origin and evolution of metabolism.

ORIGIN OF METABOLISM AND OF CHEMICAL PATHWAYS

Biological metabolism is an orderly system of enzyme-catalyzed chemical reactions. Catalysts cannot initiate thermodynamically impossible reactions; they can only speed up reactions that are already proceeding in nature. Enzyme-catalyzed metabolic reactions, then, existed prior to the metabolic systems, although possibly proceeding at very slow rates. This has been strikingly confirmed by Degani and Halmann, of the Hebrew University in Israel, when they demonstrated the spontaneous breakdown of glucose-6-phosphate to pyruvate by way of the intermediates of the glycolytic pathway (Figure 4-2). Clearly, biological systems have exploited a preexisting pathway for use as their fundamental energy-deriving pathway. Seen from this point of view, biological metabolism seems to be a pattern of emphasis imposed on a chaotic world of probable chemical reactions. By emphasizing (catalyzing) only some of the possible reactions, biological systems initiated an asymmetry in nature that is characteristically associated with them. Any pattern picked out of the maze of thermodynamically permissible reactions would be improbable as such and stands out from the background of entropy. It is not that any of the emphasized reactions themselves are improbable or even that some pattern might become manifest temporarily in the mass of permissible reactions (although this would be only a momentary asymmetry); it is any *particular* pattern that is improbable. The persistence of a particular pattern for long periods of time, instead of its immediate obliteration followed by another temporary pattern, is what produces the asymmetry, or locally decreased entropy, associated with life.

We do not know on what basis the particular chemical pathways characteristic of metabolic systems came to be the ones emphasized. In view of the fact that the glycolytic pathway operates abiotically at a detectable rate, it may be that those reactions later incorporated into pathways were those which occurred at rates above a certain threshold value. A reaction working at an almost undetectable rate would, after all, have little chance of becoming part of a prebiological (nonenzymatic) system. From this point of view, the biochemical pathways found in living systems are simply the sum of many or all of the chemical reactions that could occur abiotically above a certain rate under the geochemical conditions present during the origin of life. If this is so, it presents a ready-made explanation of the "unity of biochemistry"—the fact that most fundamental chemical reactions are common to all forms of life on earth. An alternative explanation is that all present forms of life have descended from a single ancestral system. Both propositions are plausible enough and do not contradict each other.

We now examine various ideas concerning the origins of enzymes. An idea that follows quite naturally from what has already been explored here has been formulated by Melvin Calvin and by Sam Granick, of Rockefeller University, namely, that the functional groups at the active sites of enzymes are the same as those which abiotically catalyze the reaction in question—groups such as iron and sulfhydryl—and that enzyme evolution is the accumulation of polypeptide material around these groups adapting them more and more to the catalytic environment (ultimately the intracellular milieu). The idea is one of weak catalytic activity associated with some inorganic element being gradually transformed, by selection for increased efficiency, into an effective catalytic activity. It has been estimated that typical enzymes catalyze reactions at rates 10^{14} to 10^{18} times what might be expected without a catalyst. Granick has given the interesting example of iron, which is capable of catalyzing, by itself, the reactions performed by modern enzymes involving iron in their active sites. Thus, iron alone can activate oxygen to serve as a strong electron acceptor (cytochrome oxidase), can decompose peroxide by making it an electron acceptor (catalase), can activate water to serve as an electron acceptor (peroxidase), and can serve to transport electrons (cytochromes). The enzymes, of course, accomplish these reactions much more efficiently than does iron alone, but each one is specialized to perform only its specific function. It may be noted that the presence of ferrous iron in solution and of ferrous oxides has been established as far back as 3 billion years. From then until 1.8 billion years the so-called banded iron formations

(BIF) are found, and the earliest microfossils are associated with sedimentary iron formations. Therefore, systems based on iron catalysis probably did occur during these times. Indeed, Preston Cloud, of the University of California at Los Angeles, has noted that without invoking biological oxidations it is difficult to account for the presence of Fe_2O_3 and Fe_3O_4 in the BIF formed in an anoxic environment, because the reactions forming them are of the following kinds:

$$O_2 + Fe^{++} \rightarrow FeO + Fe_2O_3$$

He suggests that the oxygen was formed locally by photosynthetic organisms capable of splitting water into hydrogen and oxygen (see below).

Although the metabolic pathways can be visualized as being present in the abiotic world, we are not justified in assuming that whole pathways were incorporated entirely into the earliest protoliving systems, although in some cases this might have been true. Norman H. Horowitz has presented a plausible scheme—his "Garden of Eden" theory—for the gradual incorporation of more and more pathway steps into the earliest metabolic systems. Beginning with a system that is more or less fully alive or almost so (the eobiont stage), we can visualize them as deriving energy as heterotrophs from the abiotically formed complex organic chemicals in the cool, dilute soup. We may assume that the system that lasted longest in this situation was one that happened to have incorporated into itself the means of using one of the more common energy sources in the soup. If this system can grow, it sooner or later reaches the point of exhausting its energy supply. Assuming that other systems also arose using the common energy source, the point of exhaustion is reached even sooner. At this point intersystem selection comes into action. Those systems which are unable to incorporate some means of using another energy source simply stop functioning and become extinct. One way in which the problem can be avoided is to become able to use a new energy source. Although this is possible, it is unlikely, because the system has already committed itself, as it were, or has become specialized, to using the now scarce energy source. A more probable occurrence is that a few of the systems incorporate means of converting another molecule found in the soup into the previously used, and now scarce, energy source. This would be the first step in the building up of a biochemical pathway. The particular chemical that is chosen for conversion into the energy source is determined by (1) its relative abundance and (2) whether or not there is a thermodynamically permissible single reaction converting it into the needed energy source. A system based on a scarce chemical probably would not have enough time to incorporate a third, new, reaction into the developing pathway before becoming extinct. With these restrictions in mind it becomes clear that the selection of a new reaction in the developing pathway would have to be from the restricted number of possibilities present in those reactions and pathways which are capable of occurring with detectable rates inorganically, that is, that the energy-deriving metabolic pathways were fairly rigidly predetermined in the abiotic world. Given this, it does not matter too much conceptually whether these pathways were incorporated entirely or slowly built into the living systems in the Horowitz manner.

Seymour Cohen, of the University of Pennsylvania, has pointed out that, although the Horowitz scheme is adequate to account for the evolutionary development of energy-deriving metabolic pathways, it cannot explain the evolution of biosynthetic pathways. If the latter were evolved "backwards," as in the Horowitz scheme, we should have to assume that all end products of biosynthesis—including some very complex items indeed, such as chlorophyll, various vitamins, and anthocyanins—were present in the prebiotic soup and were used by the first eobionts. This is clearly absurd; it is probably safe to say that, for example, penicillin was not present in the earliest systems. It has a specific chemical-warfare function vis-à-vis bacteria, and selection has elicited it, and many other agents in molds, to serve this specific function.

Granick has discussed the evolution of biosynthetic pathways from the point of view of selection favoring the increased efficiency of enzyme activity. Those systems which had more efficient enzymes increased in mass or numbers above those which were not so efficient and ultimately replaced them. Granick has been able to show that a number of intermediates in the biosynthesis of chlorophyll are capable of reacting with light in ways similar to chlorophyll, and he has spoken of "biosynthesis recapitulating biogenesis." If it were true that uroporphyrin (see Figure 2-9) was once the homolog of the chlorophyll in modern systems, it would have to be capable of performing similar activity, and it does. Thus, we can visualize a "plant" in the past using uroporphyrin, and a later plant using protoporphyrin-IX, and finally a plant using chlorophyll, as members of a lineage gradually improving their ability to use light in photosynthesis. Each addition to the biosynthetic pathway represents the "invention" of a new enzyme capable of increasing the rate of a reaction that can produce a more efficient catalyst. Again, the reactions must be thermodynamically permissible and, therefore, implicit in the abiotic world. Calvin has made a further suggestion on the evolution of biochemical pathways. He has noted that several of the steps in the biosynthesis of chlorophyll involve oxidations and that the products themselves are, as one proceeds along the pathway, increasingly better oxidizing agents. These could theoretically make the preceding steps more probable, resulting in a positive feedback situation of autocatalysis—the product of step six functioning to make step one go faster. The idea here is that selection favors the invention of a new step in the pathway, because this invention makes the entire pathway more efficient, rather than simply adding a step to make the end product function more efficiently in another pathway.

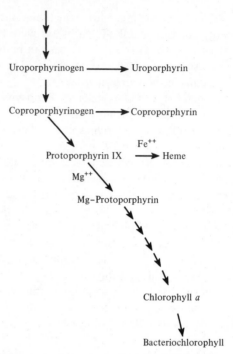

Figure 2-9   Simplified biosynthetic pathway of heme and chlorophyll.

GENE DUPLICATION

The propositions of Horowitz, Granick and Calvin involve the invention of new enzymes. As such they can only be conceived as occurring after the origin of eobionts having growth and replication and, therefore, a genetic code. Given present-day hereditary systems, there is evidence that genetic raw material can be made available to code for new enzymes. First, amino acid sequencing studies of proteins have shown us that gene duplication had to have been an important evolutionary phenomenon. Thus, trypsin and chymotrypsin from the same species show some 50 percent of their homologous amino acid residue to be identical (see Figure 2-10). Similar observations can be made on the vertebrate hemoglobin subunits—alpha, beta, gamma, and delta—which in addition show significant sequence similarities to myoglobin. Furthermore, many proteins show internal, serial homologies in their sequences suggesting that their codes are the result of the multiple duplication of a basic code unit (see Figure 2-11). From this kind of information it seems fair to conclude that the source of genetic material for new genes is some unknown (extramitotic) mechanism by which the genetic material duplicates itself, resulting in two or more copies of a gene, some of which are then free to mutate until they achieve, by chance, a new structure that would be favored by natural

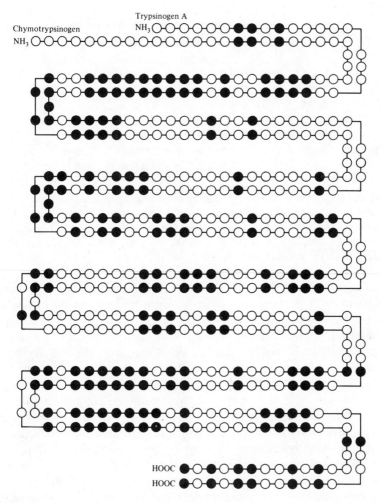

Figure 2-10   Alignment of bovine trypsinogen with bovine chymotrypsinogen A. The dark circles represent amino acids that are identical in the two sequences; the empty circles are residues that are different.

selection. If this is the source of genetic raw materials, the new enzymes would be expected to show some similarity to those already present; thus, a dehydrogenase locus, duplicated, could give rise most probably to another kind of dehydrogenase. In view of the fact that there are indeed series of similar enzymes, such as dehydrogenases, a number of different kinds of which have been shown to have very similar sequences of 10 amino acids around the active site, the proposed mechanism is plausible. (For further discussion of gene duplication see Chapter 7.)

Horowitz has made another kind of argument in favor of gene dupli-

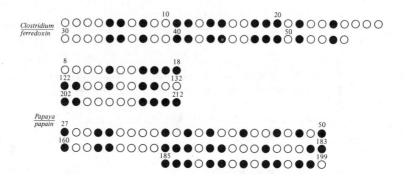

Figure 2-11   Internal homologies in the amino acid sequences of two primary gene products, suggestive of the origin of some enzymes by process of gene duplication. The dark circles represent identical amino acids; the empty circles are different ones. The numbers refer to the number of the amino acid residue counting from the amino end of the protein.

cation as the source of raw materials for evolution. In bacteria a number of biosynthetic pathways are coded for by genes all located in one part of the chromosome and under control of the same operator gene. Such a grouping is termed an operon. Horowitz suggests that the genes of an operon originated by tandem duplication of one original gene. To support this, he notes that the substrates for each of the separate enzymes are fairly similar, and, therefore, the active sites of the enzymes could have had a common descent. He also notes that these pathways frequently show end-product inhibition (see Chapter 8), and so the first (or other) enzymes in the pathway have some way of recognizing the final end product, which they do not actually catalyze. This recognition could, he contends, be the result of similar structure due to community of descent. Although these considerations may apply to some pathways, they cannot be considered generally valid. First, the allosteric sites (see Chapter 8) involved in end-product inhibition are special sites, not active sites. Second, not all pathways are composed of a series of similar reactions. Thus, an oxidation might be followed by a condensation which might, in turn, be followed by a reduction reaction. The kind of active site needed for each would be different, and it is unlikely that duplication of one kind of enzyme could give rise to a completely other kind.

There is also some information concerning the presence of extra DNA in the genomes of modern organisms. If gene duplication is an important evolutionary phenomenon, there is no reason to believe that it is not going on right now, in extant forms. If so, there must be some genetic material that is not being used, or, there must be more DNA than is needed to account for the estimated numbers of genes in living systems. Calculations of the numbers of genes present in the genomes based on the known structure of DNA do in fact give results at least one order of magnitude greater than calculations based on indirect genetic tests, such as crossing-over (see Chapter 7). In lower vertebrates and in insects and arachnids, the locus coding for ribosomal RNA is known to be reduplicated some 130 times. Some estimates suggest that in eucaryotes there is a part of the DNA that is repeated hundreds of thousands of times. There is, in short, at least in eucaryotes, much more DNA than is needed to account for the known number of enzymes, including isoenzymes, and there is present in eucaryotic genomes a great many multiply-repeated DNA sequences. Both of these data may be germane to the possible mechanism of gene duplication; thus, there appears to be an undescribed mechanism that could supply the genetic raw materials needed in the Horowitz and Granick formulations, given the presence of genomes like those found in modern living systems.

THE PROBLEM OF CATALYTIC SPECIFICITY

Returning to the possibilities of pathway origins and development in more primitive systems, we may take note of an interesting experiment reported by Sidney Fox and Gottfried Krampitz. They found that proteinoid material made abiogenically, as discussed above, was capable of catalyzing the breakdown of glucose, and, to a lesser extent, other sugars, to CO_2 at a rate significantly greater than when the microspheres were absent. It seems that amino acid polymers intrinsically have catalytic abilities. Clearly, in the absence of a genetic code, the earliest "enzymes" must have been of this kind—accidental enzymes arising out of the mass of statistical proteins (amino acid polymers with random sequences of amino acids) associated with the earliest protoliving systems. Some of them may have been combined in some way with inorganic catalysts, such as iron, as well.

Considering that a given accidental enzyme becomes degraded in time by the random flux of energy in the environment, then in the absence of a genetic code a system based on this enzyme will come to a halt. For this reason it is desirable to imagine the first systems as arising in association with some inorganic catalyst that was present in large quantity. The first enzymes could then be any of a number of different sequences of amino acids that had some single factor in common—perhaps closely spaced cystein residues capable of combining with the inorganic catalyst. Any catalytic activity that depended on the overall structure of one of the accidental enzymes would then be seen as irrelevant to the existence of the system, which would have been based on catalytic abilities that could arise out of a great number of different amino acid sequences. These would not have been very efficient enzymes, but the power of survival of the system as a whole would have been selectively superior to more efficient but more ephemeral systems. Also, if a certain kind of enzyme activity could be achieved by a great number of different statistical proteins, the system could grow simply by generating more statistical protein, there then being no need for a genetic code and the precise polymer synthesis that it allows. It may be pointed out that much of the specificity of modern enzymes is involved with adapting them to specific intracellular milieux. Thus, the functional sites of all eucaryotic cytochromes *c* are identical in sequence but other portions of the molecules are very different. In view of the fact that mutations altering the sequences in the non-active site regions can result in any specific system in molecular diseases, it is clear that these "nonfunctional" regions cannot in given specific extant living system contain just any sequence of amino acids. Typically, the sequence is almost as tightly controlled in these regions as in the active sites. But in order to qualify only as a "eucaryotic cytochrome *c*," there is no specific sequence needed in the nonfunctional sites. Instead we find restrictions of a more general kind, such as "residues three to five must contain at least two hydrophobic amino acids," allowing a great deal of latitude. In early eobionts we can by extrapolation further generalize the qualifications needed to perform electron transfer to simply a polypeptide of a size within certain limits that contains a hemelike active site. Proceeding further back to the earliest protoliving systems, we may generalize even further and reduce the strictures to simply "any polypeptide combined with iron" (see Figure 2-12). We can, therefore, imagine a polypeptide-inorganic catalyst protometabolic system capable of growth without a genetic code. And for good measure we can associate it with phosphatide-protein boundary membranes.

Although such systems would be restricted to the sites of inorganic catalysts, two such sites, separated in space or time would have to be considered to be the same, just as two identical solutions of table salt on different continents would be the same. Thus, at the level of generalized metabolic capabilities without a genetic code, there is no more individuality to a system—even if bounded by a membrane whose structure is stable over wide changes in environment—than there is to two sodium chloride solutions. The origins of such systems may be considered to be automatic and predetermined, and so very probable. The only difference here is that the protoliving system is composed to molecules no two of which are alike; the salt solution is composed of molecules all of which are identical, we think. From the point of view of functional individuality of structure, both extremes are alike in having none. Both a salt crystal and a statistical protein coacervate can grow by the nonbiological method of incorporation from the external environment.

Different systems could, of course, arise in connection with different inorganic catalysts, and on this basis alone there could be selection based on differential survival. The system that survived longest—possibly deter-

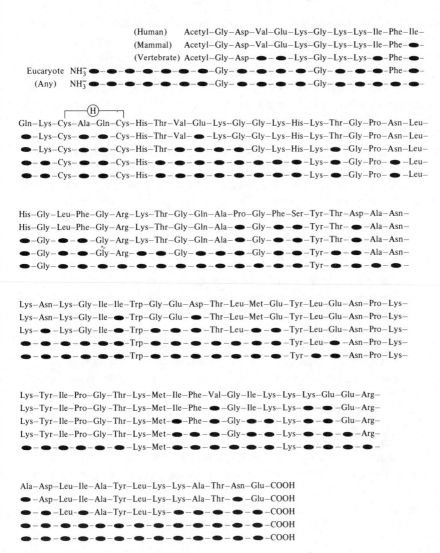

(Human)   Acetyl–Gly–Asp–Val–Glu–Lys–Gly–Lys–Lys–Ile–Phe–Ile–
(Mammal)   Acetyl–Gly–Asp–Val–Glu–Lys–Gly–Lys–Lys–Ile–Phe–●–
(Vertebrate) Acetyl–Gly–Asp–●–●–Lys–Gly–Lys–Lys–●–Phe–●–

Eucaryote NH₃⁻ ●–●–●–●–●–●–●–Gly–●–●–●–●–●–Gly–●–●–●–Phe–●–
(Any)   NH₃⁻ ●–●–●–●–●–●–●–Gly–●–●–●–●–●–Gly–●–●–●–●–●–

Gln–Lys–Cys–Ala–Gln–Cys–His–Thr–Val–Glu–Lys–Gly–Gly–Lys–His–Lys–Thr–Gly–Pro–Asn–Leu–
●–Lys–Cys–●–●–Cys–His–Thr–Val–●–Lys–Gly–Gly–Lys–His–Lys–Thr–Gly–Pro–Asn–Leu–
●–Lys–Cys–●–●–Cys–His–Thr–●–●–●–●–Gly–Lys–His–Lys–●–Gly–Pro–Asn–Leu–
●–●–Cys–●–●–Cys–His–●–●–●–●–●–●–●–Lys–●–Gly–Pro–●–Leu–
●–●–Cys–●–●–Cys–His–●–●–●–●–●–●–●–Lys–●–Gly–Pro–●–Leu–

His–Gly–Leu–Phe–Gly–Arg–Lys–Thr–Gly–Gln–Ala–Pro–Gly–Phe–Ser–Tyr–Thr–Asp–Ala–Asn–
His–Gly–Leu–Phe–Gly–Arg–Lys–Thr–Gly–Gln–Ala–●–Gly–●–●–Tyr–Thr–●–Ala–Asn–
●–Gly–●–●–Gly–Arg–Lys–Thr–Gly–Gln–Ala–●–Gly–●–●–Tyr–Thr–●–Ala–Asn–
●–Gly–●–●–Gly–Arg–●–●–Gly–●–●–●–Gly–●–●–Tyr–●–●–Ala–Asn–
●–Gly–●–●–●–●–●–●–●–●–●–●–●–●–●–Tyr–●–●–●–●–

Lys–Asn–Lys–Gly–Ile–Ile–Trp–Gly–Glu–Asp–Thr–Leu–Met–Glu–Tyr–Leu–Glu–Asn–Pro–Lys–
Lys–Asn–Lys–Gly–Ile–●–Trp–Gly–Glu–●–Thr–Leu–Met–Glu–Tyr–Leu–Glu–Asn–Pro–Lys–
Lys–●–Lys–Gly–Ile–●–Trp–●–●–●–Thr–Leu–●–●–Tyr–Leu–Glu–Asn–Pro–Lys–
●–●–●–●–●–●–Trp–●–●–●–●–●–●–●–Tyr–Leu–●–Asn–Pro–Lys–
●–●–●–●–●–●–Trp–●–●–●–●–●–●–●–Tyr–●–●–Asn–Pro–Lys–

Lys–Tyr–Ile–Pro–Gly–Thr–Lys–Met–Ile–Phe–Val–Gly–Ile–Lys–Lys–Lys–Glu–Glu–Arg–
Lys–Tyr–Ile–Pro–Gly–Thr–Lys–Met–Ile–Phe–●–Gly–Ile–Lys–Lys–●–●–Glu–Arg–
Lys–Tyr–Ile–Pro–Gly–Thr–Lys–Met–●–Phe–●–Gly–●–●–Lys–●–●–Glu–Arg–
Lys–Tyr–Ile–Pro–Gly–Thr–Lys–Met–●–●–●–Gly–●–●–Lys–●–●–●–Arg–
●–●–●–●–●–●–Lys–Met–●–●–●–●–●–●–Lys–●–●–●–●–

Ala–Asp–Leu–Ile–Ala–Tyr–Leu–Lys–Lys–Ala–Thr–Asn–Glu–COOH
●–Asp–Leu–Ile–Ala–Tyr–Leu–Lys–Lys–Ala–Thr–●–Glu–COOH
●–●–Leu–●–Ala–Tyr–Leu–Lys–●–●–●–●–●–COOH
●–●–●–●–●–●–●–●–●–●–●–●–●–COOH
●–●–●–●–●–●–●–●–●–●–●–●–●–COOH

Figure 2-12   Increasing specificity of cytochrome *c* genes with increasingly restricted taxonomic categories. The top sequence is that coded for by the wild-type human gene. The second sequence specifies only those residues that are identical in all mammals so far examined. The third sequence stipulates only those residues held in common by all vertebrates examined so far. The fourth sequence specifies only those residues common to all eucaryotes examined so far. The last sequence indicates only those residues held in common by the eucaryotes and the cytochrome *c₂* of the procaryote *Rhodospirillum rubrum*. The unspecified spaces can be occupied by any of several amino acids. Since the last sequence has so many of these, it approaches being a statistical protein.

mined by the availability of its energy source—was the most successful and ultimately gave rise to the first eobionts. Two such systems could, of course, combine if one were transported to the vicinity of the other. This would be a chance event and, therefore, probably unique or at least uncommon. The greater the number of such fusions one visualizes as being necessary in order to reach the level of eobionts, the greater the probability that life is a rare phenomenon in the universe. Aside from fusion, there is no known means of incorporating new catalytic activity into such systems in the absence of genetic code; it is, therefore, necessary to visualize at least a few such fusions prior to the origin of a genetic code if the metabolic system antedates the code. For a discussion of the further evolution of biochemical pathways, see later in this chapter and the beginning of Chapter 3.

ORIGIN OF THE GENETIC CODE

Perhaps the most difficult process of biogenesis to visualize is the origin and evolution of the complex system of control over protein synthesis, involving a DNA genetic code and its replication, a system of transcription into RNA, and a system of translation into protein. Replication of the information stored in the genome is associated with the enzyme DNA polymerase (replicase). Transcription is accomplished with the aid of the enzyme RNA polymerase (transcriptase); this enzyme is responsible for producing transfer RNA (tRNA), or soluble RNA (sRNA) and ribosomal RNA (rRNA), as well as the message itself, messenger RNA (mRNA). Transfer RNA—a different one for each codon—functions as an adaptor, bringing the amino acid into appropriate proximity to the message and also activating it so that it can become a part of the growing protein. The amino acids are attached to tRNA by means of the enzymes amino acyl synthetases—a different one for each amino acid. The mRNA becomes attached to the ribosomes, composed of rRNA and proteins, as does the tRNA with its amino acid, which also must contact the mRNA transcription of the code. The tRNA is released as the peptide bond is formed between the amino acid and the growing polypeptide with the aid of an amino acid polymerase enzyme, making way for the next one, carrying the next amino acid. Thus, we have four different kinds of enzymes involved, one in replication and three in protein synthesis. In addition, there is at least one kind of repairase that is capable of repairing broken DNA. It is certain that this complexity is not a primitive but rather a highly evolved system.

In the absence of a genetic code and its control over protein synthesis, metabolic systems must be either very stable or very general and simple. In view of the fact that fats and proteins would be degraded abiotically in only a few hundred years, stability in the sense of the long duration of single special accidental enzymes is a possibility that we may discard. In the absence of both stability and a system for increasing the numbers of specific enzymes, the metabolic systems had to have been limited in both their versatility and their efficiency, being restricted to "enzymes" with limited structural specificities and, therefore, with limited kinetic capabilities. These "enzymes" might do a number of things, as suggested above in connection with iron-containing proteins, but would do none of them so efficiently as do modern enzymes. Another problem with such non-hereditary systems is that if they contain more than one kind of catalytic activity—for instance, if they are the products of fusions of different, simpler systems—then any pieces detached from the parent system as a consequence of growth may not contain all the necessary components. This would make reproduction a very chancy matter indeed, suggesting that such systems could not come to occupy much of the earth's surface.

As indicated above, there is no clear answer available as to which was primal, the genetic code or metabolism. Because both proteins and nucleic acids are informational macromolecules—the specific sequence of different kinds of monomers is not predetermined by thermodynamic forces and is, therefore, either random or open to arrangement in serial patterns that can constitute information—it seems simplest to conceive of the earliest systems as being composed only of proteins with a protein code. For this to work, the sequence of amino acids in one polypeptide would have to be able to influence the amino acid sequences of newly forming polypeptides. This has never been demonstrated, and there is no known mechanism for it. In any case, even if there was a time when a protein code was being used by the ancestors of present-day living systems, the problem at hand is to visualize the origin of the nucleic acid code. It may be noted that Carl R. Woese, of the University of Illinois, has suggested that the earliest systems contained both polypeptides and polynucleotides, the one acting as a template or catalyst in the synthesis of the other in reciprocal fashion. He sees the major problem to be the origin of translation. This is discussed below.

Another question of primacy is whether DNA or RNA can be considered to have appeared first. The information available favors RNA primacy. First, the internal structure of the coding-translation system is such that the system would be intact if DNA were conceptually deleted; the mRNA could take its place. One problem here is that mRNA is very unstable as a rule; the DNA bonds are much more stable. (More stable mRNA is known, however, and, indeed, some viruses have an RNA code storage rather than one involving DNA.) From this point of view the invention of DNA could be seen as a result of selection for greater stability in the stored code. This question of stability bears on the time of origin of the code as well. If the code originated before there was sufficient oxygen in the atmosphere to generate a protective ozone screen (see Chapter

3), it would have to have been in connection with a benthic system at least 10 meters below the surface of the ocean in order to be protected from excessive "mutations" caused by UV radiation. In such a "hot" environment selection might overwhelmingly favor systems that incorporated DNA as a code storage, and so this may have happened early. On the other hand, the earliest codes were probably not very precise (see below), and so the stability required of modern code storages would have been irrelevant until a higher degree of precision had been achieved. Another clue here is that deoxyribotides are synthetically derived from ribotides in living systems, and so, by using the principles of biosynthetic pathway evolution discussed above, we should have to conclude that the latter were used by these systems first.

The complexity of the genetic code–protein synthesis system is such that it is difficult to proceed further looking at the system entire. In order to facilitate discussion, three aspects of the code (universality, particularity and degeneracy) will be dealt with separately, followed by an attempt to envisage the development of the complexity of the translating mechanism.

UNIVERSALITY OF THE CODE

Although there are a number of differences in chromosome structure, in the ratios of various nucleotide bases in the genomes, and in the UV wavelengths giving maximal mutation rates among various organisms—particularly large differences are found between procaryotes and eucaryotes—it seems that the specific genetic code used by all these systems is the same; that is, the code seems to be universal. There are two major classes of explanation for this universality. One is a deterministic model in which it is assumed that there is, or was at one time, some steric relationship between an amino acid and either the codon(s) or anticodon(s) associated with it. It has been pointed out by F. H. C. Crick, of the Medical Research Council of England, that a steric relationship between an amino acid and its codon is irrelevant, because the amino acids do not interact with their codons in existing systems. He feels that there might be a greater possibility of some past steric relationship between the amino acid and its anticodon, because it is the adaptor (tRNA) that the amino acid interacts with, not with the message. The idea is basically that the three-dimensional structure of the amino acid selected the only thermodynamically possible combination of nucleotide bases during a time when the nucleic acids functioned in close apposition to the polypeptides. An idea of this kind is capable of explaining universality in the simplest possible way: There is only ône thermodynamically feasible way to code for each amino acid, and any system developing a nucleic acid code for proteins would have to do it the same way; that is, the code is predetermined in organic chemistry. Selective coupling of amino acids to various polynucleotides has been demonstrated, and the study of molecular models has shown the possibility of selective steric interactions between codons and amino acids and also between amino acids and both codons and anticodons simultaneously. Crick has suggested that there might have originally been some steric interactions between a few amino acids and the code at a time before the final structure of the code was established but that such relationships need not have existed between the code and all the amino acids now coded for.

The other model that has been proposed to explain universality is that all living systems are descended from a single ancestral system after the code had been originated, regardless of the means of its origination. In the absence of a compelling demonstration of a necessary relationship between some components of the code and the amino acids, this is the most plausible idea (see below for further data on this point).

PARTICULARITY OF THE CODE

The arguments concerning the particularity of the code (why should UUU and UUC, in mRNA, code for phenylalanine?) are very similar. One explanation is the same deterministic steric-fit model just explored. The only alternative is that the association of certain codons with certain amino acids was fundamentally a matter of chance. Crick has referred to this as the "frozen accident" theory. Again, at present this seems to be the only viable alternative, with the refinements to be pointed out below in connnection with degeneracy.

DEGENERACY OF THE CODE

By far the most interesting aspects of the code are the internal patterns that can be discovered in it, including the pattern of redundancy—degeneracy, or more than one codon per amino acid—in the sense that historical accident cannot explain them. Nine amino acids have two codons, five have four, three have six, one has three, and only two have unique codons—methionine and tryptophan. The codons for a given amino acid are related. Thus, ACU, ACC, ACA, and ACG code for threonine; that is, ACX codes for this acid. Another pattern is where the third base can be either purine or a pyrimidine. Thus AApyrimidine codes for asparagine, AApurine for lysine. Still another fascinating pattern is that physicochemically related amino acids tend to be coded for by similar codons, so that any single-base mutation is most likely to lead to the replacement of an amino acid by one that is chemically related if, indeed, it does not lead to simply another codon for the original acid because of the high degree of degeneracy. These facts have led to a number of speculations concerning the origin and evolution of the code.

First, it is clear that not every codon assignment can have been due solely to chance. Thus, although it may have been a chance event that associated ACX with threonine, once the first triplet, say, ACC, had become associated with this acid, chance alone could not ensure that the other ACX triplets became associated with the same acid. Indeed, the steric-fit hypothesis does not account for this degeneracy either, because the surface configurations of all the ACX triplets are different. Again, however, some kind of steric interaction in the past, rather than chance, may have associated threonine with ACC, but this mechanism is not capable of associating all the synonymous codons with this acid. All workers have agreed that natural selection had to have been involved in generating both degeneracy and an internal pattern favoring conservative amino acid substitutions.

Tracy M. Sonneborn, of Indiana University, was the first to clearly suggest this in connection with degeneracy. With completely random codon assignments, any mutation is likely to lead to an amino acid substitution and, therefore, quite likely to some undesirable phenotypic response, reducing the chances for the system involved to continue existing or to reproduce. Under this circumstance, systems will be most successful that gradually alter their codon assignments to minimize the potential phenotypic effects of base substitutions. This could be achieved by either altering existing assignments or adding new codons in an appropriate pattern. The system that survives is the one that achieves the most optimally degenerate code. A corollary of this argument is that all living forms must be considered to be descended from a single ancestral system, because two separate systems independently evolving degeneracy in this way would most probably evolve two different patterns, while the code seems to be the same in all extant forms. The particular pattern of degeneracy in this model is seen to be due to chance alone. One piece of evidence possibly favoring this idea is that amino acids that are most frequently found in proteins show the most degeneracy in their code assignments and that rarer acids show less degeneracy, although this pattern can be interpreted in other ways. Woese has suggested a refinement of this theory to the following effect: Because of a poor (inaccurate) translation mechanism in the earliest systems, some codons are more easily mistaken for each other than are others. These would be those which are similar, of course, such as GAA and GAG. These codons selection gradually guides to the same amino acid, minimizing the phenotypic effects of translation errors.

A completely different way of thinking about degeneracy has been explored by a number of people, including Crick and Woese. Examining the code as it is now known (see Figure 2-13b), it is clear that in most cases the operational coding is inherent in the first two bases, the third being relatively less important. Thus, CCX codes for proline and GApyrimidine for aspartic acid. It has been suggested that perhaps at first the code was a doublet code, and only when more amino acids were incorporated did it change to a triplet code, at which point degeneracy ap-

	U	C	Purine
U	Leu	Ser	? Chain terminator
C	Leu	Pro	? Unspecified
Purine	Val	Ala*	Glu* Gly* Asp* Ser

(a)

	U	C	A	G	
U	Phe	Ser	Tyr	Cys	U C
	Leu		Term	Term Try	A G
C	Leu	Pro	His Gln	Arg	U C A G
A	Ile Met	Thr	Asn Lys	Ser Arg	U C A G
G	Val	Ala	Asp Glu	Gly	U C A G

(b)

Figure 2-13 (a) Postulated primitive amino acid code involving only very stable amino acids and those (*) formed in relatively large quantities in abiotic synthesis. The code is conceived of as a doublet code, albeit physically involving three bases, since only the first two convey information. (b) The modern amino acid code. This is a triplet code; in many cases information is conveyed by all three bases in the triplet.

peared, because there were more triplets possible than amino acids. Crick has added an important modification. Noting that a change from a doublet to a triplet code would be a revolutionary event at the molecular level, he has suggested that the transition would have been less difficult if there were triplets of bases from the first, but with only the first two being read; the triplet pattern may have been imposed by the dimensions of the nucleic acid helix. In any case, degeneracy is seen in this view as a result of increasing the size and versatility of the code, not as a result of having been selected for as such. It would not have been selected against because of its beneficial buffering action against the phenotypic expression of mutations. In Chapters 8 and 12 there is further discussion of traits that are beneficial to individuals or species that need not be considered to have been selected for specifically but can be seen to arise out of selection pressures operating in other directions. Such traits, although not brought into being by natural selection as such, will certainly be preserved by this force if they are of advantage to the organisms having them.

From the present point of view it is of some interest that (1) the amino acids formed in the largest amounts spontaneously in abiotic experiments are coded for by doublets beginning with G (alanine, GCX; aspartic acid, GApyrimidine; glutamic acid, GApurine; glycine, GGX) and (2) that amino acids that have the greatest stability and would, therefore, be the ones to accumulate in the prebiotic world include the same ones plus valine (GUX) and three others, each with at least four codons—leucine, proline, and serine. From this information, one can suggest that at an early stage the only amino acids incorporated into proteins were these and that they were coded for by a doublet code such as suggested in Figure 2-13a. This array would allow for enough variety in amino acids to form proteins almost as variable in structure as those found today. It would not, however, specify every amino acid residue, and the proteins would still be quasi-statistical in nature. From this doublet code the existing triplet code could be derived, as study of the figure will show.

From a beginning of this kind, new amino acids would become incorporated into the proteins of living systems as the need for greater variety in protein structure increased and as the precision of translation increased. Crick has suggested that the new amino acids selected to become part of the system would be only those which, although they brought greater variety of structure, were not physicochemically so different from the acids already incorporated as to disrupt the existing system. He has also suggested that the number of amino acids incorporated into the system would decrease as the system became more complex, stopping at a point where there were so many proteins coded for that the addition of a new acid would be bound, by chance, to damage the function of at least one of them; this point was historically reached after 20 amino acids out of the thousands ultimately available had become incorporated. The restriction of proteins in all living systems to basically 20 amino acids is thus seen to be a matter of chance; the presence of the *same* 20 amino acids in all living systems can best be explained by common ancestry (except for the eight "primitive" amino acids referred to above, which would be expected to become incorporated into any biogenic process under similar conditions).

To finish this section on the evolution of the genetic code, one possible hypothesis concerning the origin of the complexity of the present system will be presented, drawn principally from the ideas of Seymour Cohen, Woese, and Crick. The earliest system can be seen as involving no code and no translation but being made up of the ancestor of rRNA and a mass of statistical proteins capable of working as RNA polymerase. The proto-rRNA functions catalytically to insert certain active groups—possibly amino acids—into the proteins, thereby converting them into RNA polymerases. Thus, the proteins synthesize RNA, which in turn converts more proteins into RNA polymerase. Any increase in one component of this system results in an increase in the other component, and so the system is capable of growth. A plausible next step is the incorporation into the system of a proto-tRNA as an activator of the amino acids that proto-rRNA attaches onto the statistical proteins, as a kind of coenzyme. The selective advantage for this can be seen to have been an increased efficiency of converting statistical proteins into RNA polymerase.

The source of tRNA is problematical. Sooner or later it must become a product of the system; perhaps the generalized RNA polymerase envisioned above would be capable of catalyzing its production as well. The system, however, is capable of producing only a single kind of enzyme. If it is to come to produce more than one kind, it must incorporate some kind of code, and this, in turn, necessitates the invention of some kind of translating mechanism. The two can best be visualized as originating and evolving together. The earliest code can be seen as the ancestor of mRNA, while the translation mechanism made use of the preexisting tRNA. The proteins coded for in this way are best seen as classes of proteins not very precisely defined, and limited to a few kinds. Increased precision of translation alone would allow more diverse and efficient proteins and would be accompanied by the continued evolution of the code proper, as discussed previously. Eventually some selection pressures arise favoring the development of a nonfunctional code storage, perhaps, as Cohen has suggested, the need for quantitative control over the synthesis of the different enzymes. This need would favor mechanisms for the degradation of mRNA, and, without storage elsewhere, this would destroy the code. Thus, only systems that evolved some code storage could control the amounts of different proteins synthesized, and only these could give rise to higher systems.

The genetic code affords an excellent example of the concept of levels of organization. From a thermodynamic point of view the sequence of amino acids in a polynucleotide can be completely random. There is no geochemical force capable of ordering these bases into patterns. It is clear that the forces ordering the bases in the code do not arise directly from the kinetic interactions of molecules but are imposed on the code from higher levels of organization. A good analogy would be a sentence: The vocabulary is analogous to the nucleic acid bases; the rules of grammar are analogous to chemical reactions allowable in replication. Both of these set limits to the kinds of things that may happen, but neither can determine the sense or message of a full sentence. This comes about only when an intelligence arising out of a higher level of organization orders the words in a pattern consistent with the rules of grammar. The analog of intelligence in the genetic code is the process of natural selection. Genetic mutations are constantly working to disorder the code, while selection is constantly acting to maintain order in it. As George Wald, of Harvard University, has so aptly put it, "We are the products of editing rather than authorship."

With the acquisition of a genetic code, biological growth—an increase in the amount of the specific kind of living material that causes the increase—became possible. Replication of a storage code (DNA) made reproduction possible. Each piece budded off from the parental system, however, must be assured of getting a full copy of the genetic information. This necessitated either a mitotic system comparable with what is found today, with the genetic material gathered into chromosomes, or, more probably, simply a vast reduplication of the stored information so that at least one copy would wind up in each of the daughter systems.

We will examine briefly the selection pressures that may have operated to elicit reproduction in living systems. Immortal systems would probably not experience selection favoring the origin of reproduction. If any immortal living systems ever originated on earth, each became extinct the first time that some drastic accident befell it, crushing it, say, under a landslide. (Incidentally, any immortal system that did by chance evolve reproduction would soon destroy its environment through overuse and would cease to function, becoming extinct.) Thus, the only systems that survived the ravages of chance accidents were those which had evolved reproduction. For systems of this kind there is nothing to be gained by evolving, or preserving, immortality. Therefore, the same condition of environmental change that favors reproduction would not work to preserve immortality, and it would undoubtedly be lost or not evolved. The strategy that has survived until today is the sacrifice of existing parts of the system (individuals) after they have reproduced, placing new copies of the system into new environments in which the drastic accident of the moment may not have any effect. Only systems that could keep moving around by the dispersal of budded-off pieces would survive. Under these conditions reproduction and even mitosis could evolve without any serious directional change in the overall environment. Reproduction is simply a means of moving around at this level (it also provides this function in modern plants). In order to survive continued drastic change in the environment overall, there would have to be some source of genetic change—mutation—by which the encoded information could become altered, possibly resulting in a phenotype favorable under the new conditions. Mutation—or miscopying—was, of course, built into the code from the start. Under its pressure different daughter systems would gradually become genetically different, resulting in variability between systems. This variability is the basis of natural selection (see Chapter 8).

EARLY EVOLUTIONARY TRENDS IN PROCARYOTES

With the acquisition of an orderly means of distributing the hereditary information to the daughter cells when the system divides, we have made the transition from eobiont to what is fundamentally a procaryote. This had to have happened by at least 3.2 aeons years ago. The characteristics of procaryotes are as follows: They are genetically individual single-celled individuals bounded by a typical cell membrane. They are haploids, and the chromosome is circular and attached to the cell membrane. Reproduction is by mitosis, followed by cell division; parasexual mechanisms for the exchange of genetic material have been demonstrated in the laboratory in a few. Most are saprophytic heterotrophs, absorbing food materials through the cell membrane; others are photosynthetic autotrophs (photosynthetic bacteria and blue-green algae) or chemosynthetic autotrophs (some bacteria) capable of building up carbon compounds from simpler compounds, such as carbon dioxide (CO_2). Ecologically, the living forms cover an entire spectrum, ranging from autotrophs capable of living with or without oxygen (facultative aerobes) to specialized parasitic heterotrophs found only in specific higher organisms and from thermophiles living in hot springs to photosynthetic forms found on moist tree trunks in cool rain forests. Typically, their generation time is quite short—up to 200,000 generations per year—but they tend to be evanescent in any specific location, with the population increasing rapidly in size, say, on a carcass, and then dying back as waste products accumulate or food becomes used up, so that the standing crop at any time in a total ecosystem contains very little procaryotic biomass compared with other kinds of organisms, although the numbers of individuals may be comparatively large.

It is not clear what characteristics of the living procaryotes can be considered to be primitive (characteristic of the earliest ones to evolve) other than the overall architecture of the cell described above. Some interest has centered on the variety of chemosynthetic and photosynthetic systems found among them, and we may look at this material and associated hypotheses briefly.

The original eobionts and the first procaryotes are considered to have been heterotrophs, deriving energy from the breakdown by anaerobic fermentation of complex carbon compounds in the dilute, anoxic "soup." Some procaryotes are living this kind of life today, such as some spirochetes and lactic acid bacteria—organisms that show no trace of more complex assimilative pathways. Whether these are secondarily simplified organisms or survivors, as a group, from the remotest past is not clear, but most workers in this area prefer the latter idea. Continued fermentation of the finite supplies of organic compounds in the seas placed selective value on the invention of ways to become less dependent on these compounds. The earliest invention, considered thus because it is found in all organisms other than those few bacteria just referred to, seems to have been CO_2 fixation; this compound is used to build up organic material. What is needed is some way to reduce the CO_2, so that it condenses with other carbon compounds. Sources of reducing power (hydrogen) are needed; in many modern systems these are reduced nucleotide coenzymes, but in the earliest systems they may have been inorganic reducing agents, such as ferrous iron (Fe^{++}) or reduced sulfur compounds. Some members of the modern genus *Clostridium* show this ability, using hydrogen gas (H_2) as the hydrogen source by way of reactions mediated by the enzyme hydrogenase, and, indeed, we can think of the key invention involved at this stage as the invention of this enzyme by some lineages of procaryotes.

Organisms at this level of complexity are, however, still dependent on carbon compounds more complex than CO_2 for their energy sources. The next step seems to have been the incorporation of the ability to use inorganic energy sources. This can be done by using reduced inorganic compounds as sources of energy as well as of hydrogen (chemosynthetic autotrophs) or using light energy (photosynthetic autotrophs). It is not clear whether one of these ways preceded the other or whether different procaryotic lineages happened on different methods, some of which still survive today. All modern procaryotic autotrophs use hydrogenase to obtain hydrogen for reducing power from inorganic sources. The chemosynthetic ones derive energy as well from the same inorganic sources. They do so by capturing the electrons produced during the oxidation of the reduced substrate. This ability involved the invention of ferredoxin, the key electron acceptor—an iron-containing enzyme.

Ferredoxin is used also in bacterial photosynthesis, which, however, involves still another enzyme, bacteriochlorophyll. Here reduced inorganic compounds still serve as hydrogen sources, but the bacteriochlorophyll traps energy from light, passing it on to ferredoxin (see Figure 2-14). Probably because bacterial photosynthesis seems conceptually to be a further elaboration of bacterial chemosynthesis, depending as it does on ferredoxin, some authors think that photosynthesis came later, with the invention of bacteriochlorophyll by some chemosynthetic autotroph.

Both chemosynthesis and bacterial-style photosynthesis depend for hydrogen on external sources of reduced compounds and, as such, are restricted to anoxic regions in which such deposits as ferrous iron or sulfur are present. Today such habitats are not particularly common, but apparently this was not always so. From before 3 billion years to about 1.8 billion years ago there were extensive deposits of ferrosulfides, ferrocarbonates, and ferrosilicates, pointing to the fact that, in the absence of oxygen, ferrous iron and other reduced compounds were common solutes. All the fossil procaryotes found in beds older than 1.8 billion years had to have been anaerobes, and the metabolism of modern anaerobes, as described so far, probably reflects the kinds of metabolism present during

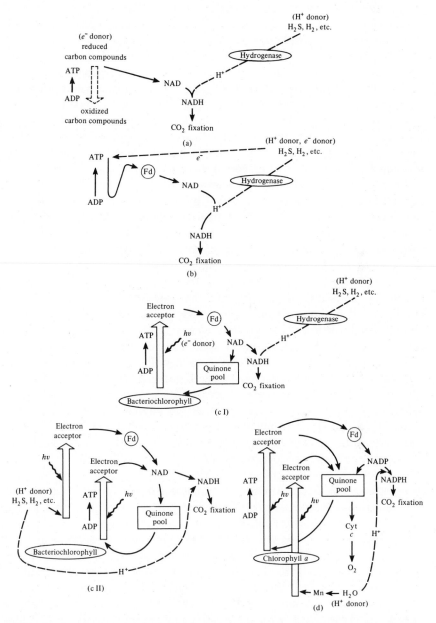

Figure 2-14   Evolution of photosynthesis. (a) Some *Clostridium;* (b) chemo-synthetic bacteria; (c, I and II) two different sorts of photosynthetic bacteria; (d) blue-green algae and eucaryotes. A progression is presumed from (a) to (d). See text for further explanation.

those early times. Probably some time between 2.7 and 2.5 billion years ago the ancestors of the blue-green algae invented chlorophyll a, developing algal-style photosynthesis.

A problem exists here in that chlorophyll a seems to be one of the intermediates in the synthesis of bacteriochlorophyll, suggesting that the former had evolved before the latter. For this reason some workers consider that bacteriochlorophyll was a special invention by some procaryotes allowing them to absorb light more in the violet and orange bands than in the blue band used by chlorophyll a as a response to living in deeper water (orange) or in response to competition with other, overshadowing systems for light. It is then suggested that chlorophyll a was originally used by certain heterotrophs for the photoassimilation of organic compounds from the sea soup, as is done today by some nonsulfur purple bacteria and even some green algae. It is then felt that photosynthesis based on chlorophyll a was a later use of this enzyme, becoming possible with the invention of ferredoxin. In this case, chlorophyll a is seen to have been preadaptive for photosynthesis in shallow water.

Chlorophyll a has the ability to exist in a number of photoreactive forms. In one form it performs exactly the same function as bacteriochlorophyll, supplying energy from light by way of ferredoxin to reduce CO_2. In another form it can absorb light energy at shorter wavelengths, which in algal photosynthesis is used to split water into hydrogen ions (to be used for reducing power) and into oxygen, which is given off into the environment. In this form of photosynthesis the organisms are no longer dependent on external sources of hydrogen ions, because they are capable of deriving these from endogenous (and, in any case, ubiquitous) water. Exactly why this form of photosynthesis became so successful is not clear, but perhaps the answer lies partly in the extensive environmental pollution caused by the waste product oxygen (see below).

By 1.8 aeons ago the ferrous iron exposed to the atmosphere and hydrosphere began being oxidized into ferric iron (Fe^{+++}), forming rust-colored red beds. Increasing amounts of carbonate rocks, such as dolomite, began being deposited as well. It may be estimated that by about 1.2 aeons ago the amount of oxygen in the atmosphere had reached 1 percent, the concentration at which living anaerobic procaryotes die (the Pasteur point). In short, the surface of the earth was gradually becoming oxidized. The only geochemical source of free oxygen is the photolytic splitting of water by ultraviolet radiation. It has been calculated, however, that this source could not account for the massive increase in free oxygen that took place at the earth's surface before a billion years ago. As often pointed out by Harold Urey and George Wald, the only plausible source of this oxygen is algal-style photosynthesis. Calculations place the beginning of this process at least 2.7 aeons ago.

ORIGIN OF EUCARYOTES

If we focus on the period around 1.2 aeons ago, we can see that organisms that were obligate anaerobes would begin gradually to become extinct, unless they began to exploit some more specialized (less common) habitats like those which are anoxic today or unless they could evolve aerobiosis. Organisms that had, for one reason or another, evolved means to detoxify and handle oxygen, on the other hand, would not have faced extinction from this source. Oxygen is a universal poison, largely because it very rapidly forms peroxides, which are intensely strong oxidoreductants and can easily denature proteins. Enzymes, such as catalase and other peroxidases, are necessary for cells to survive aerobic conditions. Among the kinds of systems that had evolved ways to detoxify and get rid of oxygen by 1.2 aeons ago were the very forms that were producing so much of it as a waste product. In order for algal-style photosynthesis to proceed, protective measures had to be present in these systems; they were, therefore, preadapted to survive an increasing amount of environmental pollution by this gas. As their competitors became extinct or became relegated to smaller parts of the biosphere, aerobic photosynthesizers increased in numbers, and, of course, the amounts of environmental oxygen continually increased, ultimately stopping at about 21 percent. The end of continued increases in environmental oxygen was reached when organisms using aerobic respiration had increased in numbers to the point at which the oxygen produced by photosynthesis was balanced by the amount used in respiration (see Chapter 3).

A number of different lineages of procaryotes evolved aerobic respiration. The adaptive value of this beginning some 1.2 billion years ago is obvious, and the origin of the eucaryotes seems to be somehow involved in this, apparently sometime around 1.2 to 1.4 aeons ago. The eucaryotes, with their mitochondrial oxidative phosphorolysis, are as a group aerobic. They have a basic anaerobic fermentative metabolism (see Chapter 3), but in all of them aerobic metabolism surmounts this basis. They differ from procaryotes in a number of ways, for example, in having a nucleus, mitochondria, flagellar-mitotic apparatus, and so on. There are two different major viewpoints about their origin. One view has it that they are descended from a group of photosynthetic procaryotes, with the invention of nuclear membranes, mitochondria and plastids, and flagellar-mitotic apparatus. The other viewpoint sees them as symbiotic systems of different procaryotic lineages that became fused together and no longer live independently. Specifically, the centrioles—involved with cilia, flagellae, and the mitotic apparatus—the mitochondria, and the chloroplasts are seen to be essentially degenerate procaryotes, all inhabiting the inside of still a fourth, the cell proper. Using the example of the mitochondrion, some of the circumstantial evidence for viewing it as a procaryote are the presence in it of ringlets of DNA, much like the bacterial chromosome; the presence of ribosomes of the same size as bacterial ribosomes and smaller than those found in the endoplasmic reticulum of the cell; independent replication of the mitochondria; the presence of a membrane more like the bacterial cell membrane in size and structure than like the eucaryotic cell membrane—and this membrane is not fused with the others in the eucaryotic cell. Another erstwhile possibility, that the procaryotes and eucaryotes are descendants of different origins of life, is no longer felt to be reasonable in view of the large number of biochemical similarities between the two and, specifically, of the sequence homology between eucaryotic cytochrome *c* and that of *Rhodospirillum rubrum* (see Chapter 3), to say nothing of the fact that both use the same amino acid code.

Whatever the merits of these positions may be, it is clear that the appearance on the scene of eucaryotes was a tremendous event in the history of biological systems on the earth. Of major importance was the invention of sexual reproduction, thereby allowing genetic recombination. Procaryotes typically have short generation times and in very short order produce tremendous numbers of individuals clonally. Each generation is subjected to mutation pressure, so that over a period of time, such as a year, many different genotypes are expressed and subjected to natural selection. As will become apparent in later chapters, the amount of genetic variability presented to the environment per unit time to be acted on by natural selection is important to the success of lineages. Procaryotes generate genetic variability largely by having many generations per unit time, indeed, many of them are restricted to this strategy. Sexual reproduction is primarily an adaptation allowing the production of large amounts of genetic variability in each generation (see Chapter 7). Once this has been acquired, generation times can become extended, and, more important, more elaborate physical organization is possible. It takes actual time to construct a mechanism, such as the body of a crab—by whatever means, including biological ones. There would not be time to accomplish this feat if generation times had to be short in order to produce enough genetic variability to remain adapted to the environment. Also, the shortest generation times are possible only by using simple fission as the mode of reproduction. A body like that of a crab cannot be reproduced by fission. Thus, the invention of sexual reproduction allows the invention of multicellular (and, therefore, more complex) organization.

It has long been considered rather mysterious that suddenly, at the end of the Precambrian period, many phyla of invertebrates appear more or less fully developed in the fossil record, without any indication of prior ancestral systems. This is not due to a particularly faulty record before this time. The deposits are there, and they contain algal and bacterial remains but no complex organisms. It has been suggested by several authors that these first invertebrates were actually produced during an extraordinarily rapid evolutionary episode resulting from the acquisition of the eucaryotic grade of organization—what we later refer to as an adaptive radiation (see Chapter 4), that is, a sudden burst of evolutionary activity sometimes triggered by the acquisition of some new key adaptation. We may suggest that the key adaptation in this episode was multicellular organization, allowing more variety in physical adaptations, and that this was made possible by sexual reproduction. On the other hand, the more classical explanation of the sudden appearance of the earliest invertebrates is that their ancestors were present (1) in somewhat less fossilizable form (without skeletal parts) and (2) during the period just before the late Precambrian, when some kind of major catastrophic geophysical event(s) took place—possibly severe worldwide glaciations—eliminating what few remains there might have been from the fossil record. At present it is difficult to choose between these alternatives. We will not pursue in a descriptive way the subsequent history of living systems on the earth; the reader is referred to the time line in Chapter 4.

LIFE ELSEWHERE IN THE UNIVERSE

We end our discussion of the origin and early evolution of life on the earth with some consideration of the possibility of life on planets elsewhere in the universe. We have already noted that one's philosophical position concerning the number of contingencies involved in the origin of living systems has implications for the possible presence of extraterrestrial life. If life is an automatic result of the interactions of complex chemical systems, it is probably present on every planet within the limits of the most extreme allowable environmental conditions. If, on the other hand, very many very improbable occurrences are necessary in order to initiate them, living systems would be far fewer than the number of planetary bodies capable of supporting them. Even in this circumstance, if the universe is effectively infinite in size—usually it is not so considered—then there would be an infinite number of biopoeses that have taken place, are taking place, and will take place.

There are a few shreds of hard evidence bearing on this question. Spectroscopic analyses of light coming from various places in the detectable universe frequently indicate the presence of carbon compounds, including ethane, acetylene, ethylene, and the important cyanide. The basic raw materials of life as we know it are, therefore, not uncommon in the universe. Other and more controversial sources of evidence on this point are the carbonaceous meteorites. The problems associated with the meteorites are numerous, but involve most importantly the possibility of contamination by terrestrial materials in the atmosphere or on impact. There are two classes of materials found on these meteorites—chemicals and formed structures. Many of the formed structures were shown to be materials like ragweed pollen and fungal spores, but apparently not all of them resemble known living forms. A number of them look like some of the simpler Precambrian fossils, and still others bear resemblance to Fox's microspheres. It is possible that the latter were formed during the extreme dry heat of the impact as in one of Fox's experiments from either terrestrial or meteoritic organic material, and so may not be proper fossils.

The organic material in the carbonaceous chondrites is equally equivocal. There seem to be two classes of organic material: one resembling terrestrial material isotopically, the other not. The part with isotope composition similar to terrestrial patterns is that containing compounds associated with earthly life—levo amino acids, purines, and so on. The part with isotopic composition different from terrestrial sources is composed of a mixed bag of carbon compounds, many of which are not associated with our living system. The fact that the amino acids characteristic of terrestrial life are all levo is felt very clearly to indicate contamination.

Assuming that life has originated somewhere else in the universe, to what degree might we expect it to resemble our system? The first thing to consider is whether life may be based on a different kind of chemical background from that on the earth. The general shape of the current answer to this is that, given the conditions of temperature and planetary size (gravity) as they were on the earth during biogenesis, it can be shown in detail why precisely the elements C, O, N, S, and P were incorporated. For example, all organic matter is formed fundamentally by carbon-carbon bonds. It was at one time suggested that, because silicon has a somewhat similar structure, it might form the chemical backbone of a living system somewhere. The chemistry of silicon, however, is very different from that of carbon; CO_2 is a gas, SiO_2 crystallizes to form quartz rock; carbon-carbon bonds are more stable than silicon-silicon bonds; carbon chains readily incorporate elements other than H into the chain, silicon chains do not; carbon easily forms double bonds, silicon does not, and this directly leads silicon to form huge, inert covalently bonded polymers; and so on. Similar arguments can be made for all elements and even the major important compounds (ATP, amino acids, and the like), leading to the conclusion that if life exists elsewhere in the universe, it probably is based on a familiar chemistry.

Put another way, what this really says is that given planets geochemically like the earth (of which there are estimated about a thousand in our galaxy alone and perhaps 10^8 in the total visible universe), any living system forming on them will be of the same type as found on the earth. But what of planets very different from the earth? Thus, on a very cold planet NH_4 is a liquid and could possibly serve in the capacity of H_2O on our planet. As George Wald has pointed out, however, chemical reactions take longer in colder temperatures, and at the temperature of liquid ammonia, it would take, for example, some 64 billion years to achieve biogenesis, which on earth took only 1 or 2 billion years. It is not known whether any planet would even last this long. Furthermore, N_2 instead of O_2 would be formed on dissociating ammonia, and this is a very inert gas and a poor oxidizing agent. Arguments of this kind have convinced most people that living systems probably would not arise on planets very different from ours, and, therefore, any life in the universe must be based on a chemical background similar to that on the earth, or, what is the same thing, that there is only one form of life. This conclusion seems to be the only one possible using logical reasoning and the data obtained by the methods of science on the earth.

If this conclusion is correct, we still need to assess whether or not the forms taken by other living systems would be very similar to ours or quite different. G. G. Simpson, of Harvard University, has suggested that they might by very different indeed on the grounds that current evolutionary theory assigns a large role to chance events—fusions of prebiotic systems, mutations, genetic drift, recombination, preadaptation, and the like. On the other hand, determinism looms equally large in current theory (see the discussion of parallel evolution in Chapter 4). For example, in a system developing on a planet like the earth the early need for chemical reducing power is bound to result in the incorporation of iron into enzymes. Or we may predict that photosynthesizing autotrophs will evolve to achieve a large surface area exposed to the light from the nearest star; they will be flat or have flat appendages (leaves). Or we may predict that active heterotrophs will have a front end and a rear end, with food detecting and capturing mechanisms at the front end. Or we may predict that active organisms in water will show some degree of streamlining and that floating organisms will have large surface areas. These features are imposed on living systems of whatever kind by the nature of the physical world. The question is to what degree of precision such predictions can reasonably be made. For example, most certainly we could not predict the presence of cats, and species identities are out of the question. We might, however, predict the presence of some large terrestrial carnivores at some stage of development of any living system.

SUMMARY

In summary, then, the chemical preconditions for life as we know it are not uncommon in the known universe. It has been possible to demonstrate in the laboratory the abiotic formation of complex organic compounds and their polymers, some of which are catalytically active. Further, it has been possible to demonstrate the formation of supramolecular structure in the absence of a genetic code (liquid crystals). Concerning the origins of membrane boundaries, enzymes, biochemical pathways, and the genetic code we know next to nothing for certain; these areas are still at the level of sophisticated speculation. It must, therefore, be concluded that nothing is really known concerning details of the origin of life or whether this process is rare or common in the universe. The transition from pro-caryotes to the eucaryotic grade of organization is also shrouded in mystery.

REFERENCES

Brancazio, P. J., and A. G. W. Cameron, eds., *The Origin and Evolution of Atmospheres and Oceans.* Wiley, New York, 1963, 450 pp.

Bernal, J. D., *The Origin of Life.* World, Cleveland, 1967, 345 pp.

Bryson, V., and H. J. Vogel, eds., *Evolving Genes and Proteins.* Academic, New York, 1965, 629 pp.

Calvin, M., *Chemical Evolution.* Oxford University Press, Fair Lawn, N.J., 1969, 278 pp.

Cloud, P. E., Jr., "Pre-Metazoan Evolution and the Origins of the Metazoa, in E. T. Drake, ed., *Evolution and Environment.* Yale, New Haven, Conn., 1968, pp. 1–72.

Crick, F. H. C., "The Origin of the Genetic Code," *Journal of Molecular Biology,* **23:**367–379, 1968.

Dauvillier, A., *The Photochemical Origin of Life.* Academic, New York, 1965, 193 pp.

Degani, C., and M. Halmann, "Chemical Evolution of Carbohydrate Metabolism," *Nature,* **216:**1207, 1965.

Ehrensvard, C., *Life: Origin and Development.* The University of Chicago Press, Chicago, 1960, 164 pp.

Fox, S. W., ed., *The Origins of Prebiological Systems.* Academic, New York, 1964, 450 pp.

——— and G. Krampitz, "Catalytic Decomposition of Glucose in Aqueous Solution by Thermal Proteinoids," *Nature,* **203:**1362–1364, 1964.

Glaesner, M. F., "Pre-Cambrian Animals," *Scientific American,* March 1961, 72–78.

Hanson, E. D., "Evolution of the Cell from Primordial Living Systems," *The Quarterly Review of Biology,* **41:**1–12, 1966.

Keosian, J., *The Origin of Life,* 2d ed. Reinhold, New York, 1968, 120 pp.

Margulis, L., "Evolutionary Criteria in Thallophytes: A Radical Alternative," *Science,* **161:**1020–1022, 1968.

Miller, S. L., and H. C. Urey, "Organic Compound Synthesis on the Primitive Earth," *Science,* **130:**245–251, 1959.

Oparin, A. I., *The Origin of Life on the Earth,* 3d ed. Oliver & Boyd, London, 1957, 495 pp.

Shneour, E. A., and E. A. Ottesen, eds., *Extraterrestrial Life: An Anthology and Bibliography.* National Academy of Sciences National Research Council, Washington, 1965, 478 pp.

Simpson, G. G., "The Nonprevalence of Humanoids," *Science,* **143:**769–775, 1964.

Urey, H. C., "Biological Material in Meteorites: A Review," *Science,* **151:**157–166, 1966.

Wald, G., "The Origins of Life," *Proceedings of the National Academy of Sciences,* **52:**595–611, 1964.

Woese, C. R., *The Genetic Code,* Harper & Row, New York, 1967, 200 pp.

CHAPTER 3

Economies

COEVOLUTION OF BIOSPHERE AND ATMOSPHERE

Until only yesterday people in industrial cultures have been taking for granted the air they breathe. Now, the growing threat of dangerous contamination from industrial wastes has resulted in an unprecedented popular awareness of this particular mixture of gases, which includes the all important CO_2 (0.03 percent) and O_2 (21 percent). Scientists, of course, have been studying this gaseous blanket covering the earth, the atmosphere, for some time, and the more historically oriented have asked questions concerning its origins. One curious and, for us, important fact uncovered by these workers is that there is no known geochemical process that can account, either now or in the past, for the amount of oxygen contained in the atmosphere. It has been calculated that the photolytic splitting of water into oxygen and hydrogen, which occurs today in the ionosphere and occurred in the past even at the earth's surface, could not have produced the large amounts of oxygen now found in the atmosphere. At present the overwhelmingly largest source of this oxygen is the process of photosynthesis as carried out in green plants. The marine members of the group of algae known as diatoms alone contribute some 70 percent of the currently produced oxygen.

The photosynthetic process as it occurs in the green plants can be described in chemical shorthand as follows:

$$6 \ CO_2 + 12 \ H_2O \xrightarrow[\text{chlorophyll}]{\text{light energy}} C_6H_{12}O_6 + 6 \ H_2O + 6 \ O_2\uparrow$$

The six-carbon structure on the right is glucose; the arrow after the oxygen indicates that it is produced as a free gas by this series of reactions. Several people, including most notably George Wald and Harold Urey, have developed the rather bold notion that perhaps oxygen was originally put into the atmosphere by photosynthetic processes (see Chapter 2). Because there is no simpler explanation available, this remains the only plausible hypothesis, and its conceptual importance cannot be overrated. It casts in very dramatic terms the idea that organisms do not so much evolve *on* the earth as *with* it. The system to be considered is not a living one operating on the surface of the earth (the lithosphere) or in the water (the hydrosphere) but the entire earth and even the planetary system in that light energy from the sun, at least, has to be considered to be part of the system. Some astronomers predict that in some millions of years in the future the sun will radiate more heat than it now does, making life as we know it impossible on the earth. This, too, in a rather different but equally dramatic way, makes the same point that the living and inanimate are intimately connected.

If free oxygen originally appeared as a result of photosynthetic processes, it seems necessary to postulate that there were rather large amounts of CO_2 in the atmosphere before such processes could function, because this is the carbon source in all known photosynthetic processes, not only the one utilized by green plants. Do we have any evidence that CO_2 can have been present in the atmosphere before the generation of O_2 by photosynthesis? The photosynthetic production of oxygen is associated with the complex subcellular structures known as plastids (chloroplasts), and the use of oxygen during aerobic metabolism is associated with the equally complex subcellular structures, the mitochondria. Plastids and mitochondria are more than superficially alike in structure and are produced by similar ontogenetic processes. In this way we can recognize an aerobic cycle associated with special, complex subcellular structures. On the other hand, anaerobic pathways are typically found to be "soluble"; that is, presumably they occur in the cell in general and not in special centers. Many of these anaerobic pathways as we find them today generate CO_2. Two of them can be summarized in chemical shorthand as follows:

$$C_6H_{12}O_6 \rightarrow 2\ C_2H_5OH + 2\ CO_2\uparrow + 2 \sim P \qquad \text{alcohol fermentation}$$
$$6\ C_6H_{12}O_6 + 6\ H_2O + 12 \sim P$$
$$\rightarrow 24\ H + 5\ C_6H_{12}O_6 + 6\ CO_2\uparrow \quad \text{hexosemonophosphate shunt}$$

Each of these pathways is composed of several steps, many of which are oxidations that occur, not by adding oxygen but by removing hydrogen (an important class of enzymes, the dehydrogenases, mediate these steps).

Now, one aspect of the structure of living systems is that the aerobic pathways tend to exist on an anaerobic foundation. It is rare to find an aerobic pathway in living systems that does not depend on some product(s) of anaerobic metabolism, but the reverse is quite commonly found. This fact, and the association of aerobic processes with complex structures, which presumably were evolved only gradually after generalized eobionts had appeared, suggests an initial period in the history of these eobionts when their metabolism was wholly anaerobic. During this period it is visualized that CO_2 was an important metabolic by-product and that it gradually accumulated in the atmosphere; the substrates during this time are thought to have been the inorganically formed organic constituents of the "cool, dilute soup" discussed in Chapter 2.

During this period, intersystem natural selection (differential survival or endurance) is visualized as having placed a premium on the "invention" or incorporation of a pathway or series of them for (1) using the increasing prevalent CO_2 and (2) using any means to escape from the fundamentally self-destructive heterotrophic way of life. The various forms of photosynthesis can be seen as solutions to both challenges, and presumably developed by mechanisms such as those discussed in Chapter 2. Once the photosynthetic process characteristic of green plants became operative, oxygen could begin to accumulate in the air, after much of the surface of the lithosphere had become oxidized. The oldest rocks on the earth show evidence of having been formed and weathered in a reducing atmosphere. Geochemical evidence, such as the presence of "red beds" (ferric oxide, which could have formed only under oxidizing conditions), also suggests that by 1.2 aeons ago the amount of oxygen in the atmosphere reached about 1 percent. This, the Pasteur point, is considered to be the point at which intersystem selection would begin to favor the invention of ways to handle free oxygen. Here we should recall that the fossil record bears out the idea that living systems were present before the Pasteur point; bacteria-like and algalike fossils have been found in association with amino acids and sugars in various cherts dating from 2 to 3.2 aeons ago (see date line in Chapter 4). Calculations concerning the rate of production of oxygen based on various assumptions indicate that at least the oldest of these (from 2.7 to 3.2 billion years old) had to have been anaerobic organisms, because these calculations estimate the origin of photosynthesis at some 2.5 aeons ago.

Although the origin of aerobic metabolism can be seen to result from an opportunistic realization of an existing potentiality (the presence of free oxygen), its continued success is perhaps best explained from the point of view of efficiency. The chemical shorthand summary

$$C_6H_{12}O_6 + 6\ H_2O + 6\ O_2 \rightarrow 12\ H_2O + 6\ CO_2\uparrow + 30\text{--}40 \sim P$$

shows that this process is some 20 times more efficient in the production of high-energy phosphate bonds than the typical example of an anaerobic system given above, fermentation. Another interesting aspect of this process is that it, too, generates CO_2, and in fact it seems to be very nearly the reverse of green plant photosynthesis. Interestingly, the first definite eucaryotes (all of which today have oxidative phosphorylation) appear in the fossil record at about 1.3 aeons ago, or just after the estimated time when the concentration of oxygen reached the Pasteur point at 1 percent.

At a time when there was some 20 percent oxygen in the air there seems to have been produced a stable relationship between the ability of green plant autotrophs to use the sun's energy to produce complex carbon compounds from CO_2 and the ability of aerobic animal heterotrophs to disrupt these compounds in the process of extracting the energy stored in them into the more versatile form of high-energy phosphates, thereby producing CO_2 as a by-product. Actually, plants do the latter as well, and, indeed, the animal heterotrophs probably arose from some of them by loss of the photosynthetic capability, becoming, as it were, parasitic on the autotrophs. This new heterotrophic way of life was far more stable than the old one in that the animals fed on other living systems, which were not destroyed *as systems* by being fed upon, because they could reproduce. The old, eobiont heterotrophic way was simply to use up an inorganically produced source of energy faster than it could be regenerated and then to cease functioning and become dispersed. Today the use of CO_2 and the production of O_2 by plants is just balanced by the use of O_2 and the production of CO_2 by animals and plants. Whether or not there has been any long-term trend away from stability in this system is not known.

This generalized description of certain ecological "functions" of plants as producers and animals as consumers and the scheme in which both are seen as specialized descendants of a more generalized ancestor can be seen as a paradigm of the topics covered in the remainder of this chapter.

ECOLOGICAL NICHE

It is possible to discover by means of painstaking observations and carefully planned experiments that every kind of living organism enters into many ecological processes, both on the level of the total biosphere and in a specific, local ecosystem—the community—as well. For instance, if we observe members of a species of deer, we find that they feed on the leaves and buds of many kinds of plants. In doing so, they shape to a certain extent the contours of their immediate environment. They could affect the rate of advancement of forest into a grassland by cropping young trees, perhaps slowing their growth. Populations of grasses could be affected by this and in turn so could animals that depend on these for shelter, grazing, or seed fodder. In the meantime the deer, like other organisms of all kinds around them, are producing CO_2 and various nitrogenous wastes by excretion. (The carbon cycle we have already discussed in essence; nitrogen cycles back to plants in a more complex manner, involving some six different kinds of bacteria.) In addition, one could discover these deer also effect various minor and transitory modifications of their environment, such as the transfer of water from place to place (breathing, urine) and the transfer of heat (say, to the sod or stones on which they sleep). Sooner or later each deer falls prey to some predator, and the species in question might make up an important part of the diet of a species of lynx or wolf or puma (or man) to which some 10 percent of the energy stored in their bodies can be transferred. Their putrefying remains (and also, previously, their feces) serve as important microhabitats and energy sources for many other kinds of organisms (fungi, flies, and so on).

In this example we find a number of general ecological functions shared by many kinds of organisms (the cycling and transfer of CO_2, water, and the like) and some very specific functions are not shared by most organisms in the same community. The latter usually involve such factors as what the organism in question feeds on, what in turn feeds on it, where it sleeps or constructs nests or sets up territories. Some organisms modify their environment in very specific but important ways. Thus, the tadpoles of some kinds of frogs, living crowded together in a small, shallow body of water, may actually heat the water as energy absorbed from the sun's rays by their dark bodies is transmitted into it. Species of bats that roost in caves often cover the floor of parts of a cave with their feces, and this material serves as the foundation for the ecology of the entire cave. Another obvious example is man; even primitive cultures modify their environments in important ways.

All the relationships that a type of organism has with other organisms and with the inanimate part of its ecosystem are considered to constitute its ecological niche, which is, then, not a thing, not only a place, but a process in a specific place. Inasmuch as it is impossible to conceive of a species without thinking about aspects of its ecological niche, we can say that these processes are as much a part of the species as its average speed of swimming, its mean length, and its maximum rate of heartbeat. This is why a museum exhibit showing an organism in action in some part of its habitat is superior to a stuffed specimen in a bare case, and a film would be even better. The stuffed specimen, or even a statistically satisfactory collection of bones, represents only a tiny fraction of what the species of organism is or was. Some of the more general kinds of things that a species does, like contributing CO_2 to the general pool of that substance, are not usually considered in any attempt to describe its ecological niche. This concept is usually restricted to aspects of the life processes of a species that differ from homologous processes in other species. Indeed, in order to make the concept more immediately useful, it is almost always restricted to functions and interactions that are characteristic of only a small, restricted group of species. For instance, there is almost nothing to be gained by comparing in detail the ecological niches of an elephant and a species of squirrel. After some very general observations one has already appreciated whatever fundamental differences there are between elephants and squirrels. There is no point in considering further the differences between the African elephant (*Loxodonta africana*) and the eastern gray squirrel of North America (*Sciurus carolinensis*); differences of interest would be found between, say, the gray squirrel and red squirrel.

In actual practice some ecologists further restrict their view of the ecological niche to only the food relationships that a species enters into—to questions of who eats how much of whom when. Although this is done mainly in order to simplify the observational field, occasionally the niche is conceived of in this restricted sense, and this is tied in with the notion of competitive exclusion. Food is one of the resources in the environment that is commonly, though not always, in a condition of potential shortage. Although there usually are mechanisms regulating the population size of a species at or below the carrying capacity of a given habitat, problems of food shortage could, in principle at least, arise if more than a limited number of species used the same food organism as a major part of their diets. Of course, many other resources in an environment could potentially become saturated or used up and may also become important in interspecies competition. This concept of interspecies competition is a special case of the intersystem competition that we considered in Chapter 2 and has some problems associated with it.

COMPETITIVE EXCLUSION—INTERSPECIFIC COMPETITION

Even a superficial examination of most ecosystems would show that the habitat is complex and composed of many microhabitats and that there are diverse organisms in it and making it, each with its own peculiar preferences and activities. Using the most inclusive concept of the ecological niche, that is, everything a species does and all the relationships it enters into with the rest of the biotic world and with its inanimate surroundings, one might deduce that no two species would have exactly identical ecological niches; indeed, in this case they would be the same species. This would be a good guess, and one could consider it a reasonable hypothesis that, however, must be tested. An opportunity to do so could be presented by a careful New England bird watcher. He could report frequently seeing, feeding together in one forest or even in a single spruce tree, four or five species of wood warblers of the genus *Dendroica*. They seem certainly to be sharing the same food insects at the same time in the same place, but ornithologists do consider them to be "good" species. Robert H. MacArthur, then at Yale University working on his Ph.D. thesis, made a series of detailed observations of this system. He found that three of the species tended statistically to occupy different parts of a tree; thus, the myrtle warbler tended to feed in the lower, lichen-covered limbs and near the trunk, the Blackburnian warbler tended to feed on the outer parts of the uppermost limbs, and the black-throated green warbler fed mostly on branches of intermediate height (see Figure 3-1). The two other species, the Cape May warbler and the bay-breasted warbler, also fed in different parts of a tree but overlapped quite considerably with one or more of the other three species. Further analysis showed that the Cape May and to some extent also the bay-breasted tended to occur in appreciable numbers only during outbreaks of spruce budworms, indicating that they may be what have been called fugitive species, living a nomadic existence, essentially following their food supply around the New England area. Thus, all five species occur together only when the carrying capacity of the environment for warblers is increased by the presence of the budworms. At other times only three species are present and feeding on three separate food supplies, because different kinds of insects tend to occupy different parts of a tree. The distribution of insects is connected with the quite different microhabitats present in a complex substratum like a tree; the lower branches are characterized by less light, less wind, higher relative humidity, and so forth, in comparison with the airier and lighter upper branches. These three warblers, then, can be seen as occupying different ecological niches insofar as food supply represents an important aspect of the niche. If the Cape May and bay-breasted feed mostly on spruce budworms, their

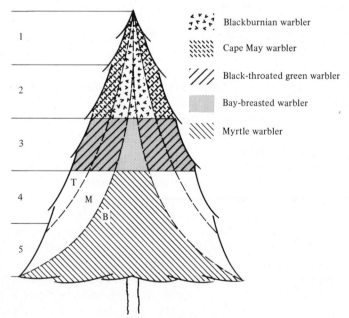

Blackburnian warbler

Cape May warbler

Black-throated green warbler

Bay-breasted warbler

Myrtle warbler

Figure 3-1 Differences in location of the most frequented regions in spruce trees for five species of wood warblers. B, bare or lichen-covered bases of branches of spruce trees; M, middle zone of old needles; T, terminal zone of new needles and buds. (From data of MacArthur, 1958.)

feeding in the same part of the same tree as one of the other birds need not indicate anything beyond a slight overlap in habitat. On the other hand, if all the birds feed on budworms, there is an overlap of ecological niches—that part of the niches involved with food relationships—which, however, does not imply that the entire ecological niche of any of these birds is identical with that of another. In this case identity is not present even in food relations, because, besides budworms, each species finds the insects characteristic of the part of the tree in which its members tend to feed.

Detailed observations like these cannot be cited in very great numbers, but enough have been made so that we may feel confident that the structure of the biosphere is one in which each species occupies a niche such that competition with other species is minimized or avoided altogether. If this is so, we may ask whether the concept of interspecific competition has any objective reality, and if so, in what sense it is real. Joseph H. Connell, working in Scotland with species of barnacles from different genera (*Balanus* and *Chthamalus*) occupying the same shore, was able to observe in just what manner they competed with each other. When both are found together, *Chthamalus* occupies the zone highest up on the rocks (above neap tide); *Balanus* occupies all the deeper regions where barnacles may live. The swimming larvae of both kinds settle down and transform into barnacles in a zone of broad overlap. In areas where *Balanus* is absent, *Chthamalus* can and does live in the part of the *Balanus* zone above the mean tide line. Where both kinds of larvae attach in this zone, *Balanus* appears to crowd out *Chthamalus,* because comparatively larger numbers of its larvae settle in this zone and because its growth rate is greater than that of *Chthamalus.* Thus, we have in this case what amounts to an actual physical struggle between individuals of two species for space, and we might certainly wish to call this competition. But in this case *Balanus* always wins, and it never wins in the typical *Chthamalus* zone, where some of its larvae do settle. Apparently *Balanus* does not have the ability to resist desiccation that *Chthamalus* has and so cannot survive so high up on the rocks. Thus the struggle has an entirely predictable, necessary, and, we might say, almost conventional end. Once we are familiar with the system, there can be no suspense, as it were, about how the two kinds of barnacles will align themselves on the shore (see Figure 3-2). Similar but less elegant studies have been done on other sessile organisms, such as plants, and the same sense of determinism is present in them.

The barnacle story is elegant partly because it suggests further that each species is adapted to survive in certain ranges of environmental conditions and that in sympatric forms these ranges tend to differ. Under optimum environmental conditions, then, a species tends to win a competition with another species whose environmental optima are not identical.

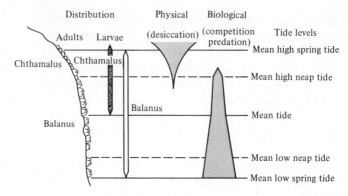

Figure 3-2   Factors controlling the distribution of two species of barnacles in an intertidal gradient. The young of both species settle over a wide range but survive only within more restricted ranges. Physical factors such as desiccation control the upward limits of *Balanus* while biological factors such as competition control the downward distribution of *Chthamalus*. (Redrawn from Connell, 1961.)

Thus, not only does each species have a unique ecological niche, it is also the species that is best adapted to this niche. (We note, however, that probably not every, or even possibly any, population of a species lives in a habitat to which the species is optimally adapted. This is primarily because different components of the environment—soil type, temperature, pH, other species, and the like—have partly independent distributions. To some extent chance is involved here too. Thus, *Artemia* brine shrimps live in barren, saline inland waters but do better physiologically in seawater. It is not known, however, that they could breed successfully in an ocean environment, even without the increased number of competitors to be found there.)

If we could define completely the niches of a group of closely related organisms, we should be always able to predict the outcome of interspecific competition. The experimental work of Thomas Park, of the University of Chicago, with flour beetles of the genus *Tribolium* is perhaps the best attempt to approximate this level of knowledge. These animals can live in extremely simplified experimental environments in which flour is both their medium and their food. By growing cultures of two species in different conditions of temperature and moisture, it was possible to define some important aspects of the ecological niche of *T. castaneum* as warm and moist and that of *T. confusum* as cool and dry. We should then predict that if the two species were reared together under different conditions, as the medium used was cooler, *T. confusum* would win the competition, and that if the medium were moister, *T. castaneum* would win. A series of experiments showed that in fact there was a direct relationship between the optimal environmental conditions for a species living by itself and those under which it tended to win in competition with another species (see Figure 3-3).

Thus, a species is adapted to living in a certain place in a certain way under certain specific climatophysical conditions, and this total adaptation defines its ecological niche for which it is the best-adapted organism in the sense that it outcompetes all comers that challenge it. All comers, that is, that it is likely to meet in its own community, whose members can be seen to form a kind of total mechanism in which each part is adjusted to the presence and requirements of each other part. If a community were in unchanging equilibrium—if there were no long-term trend in weather, if no organisms could arrive from outside the community— there would in fact be no place for the concept of interspecific competition, although some individuals might find themselves physically competing with those of another species for a short time before the inevitable result. If *Balanus* disappeared from Scotland, *Chthamalus* could maintain a larger total population and widen its ecological niche, both of which results could serve as a hedge against some disaster that could readily appear at a time so full of disaster that *Balanus* has disappeared. In other words, the moment that some element of change appears beyond the limits usually experienced in cyclical and homeostatic fluxes such as seasonal weather changes and internal regulation of population numbers, the entire community system loses its stability because each unit is balanced against each other unit.

Under conditions of directional change, species that had not previously had contact may come into a competition whose results are not predictable even in principle because, once change is present somewhere, it

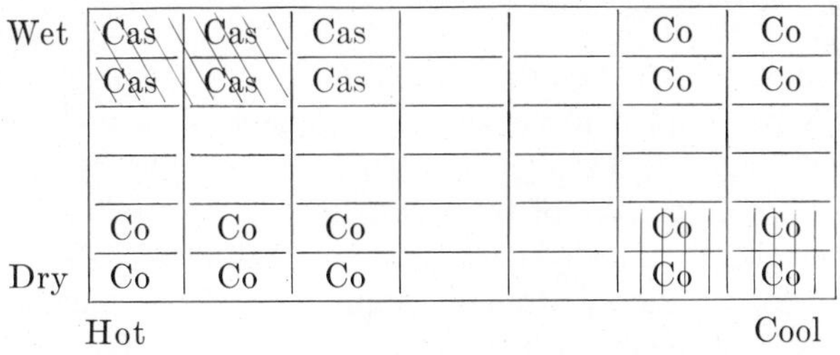

Wet	Cas	Cas	Cas			Co	Co
	Cas	Cas	Cas			Co	Co
	Co	Co	Co			Co	Co
Dry	Co	Co	Co			Co	Co

Hot   Cool

Figure 3-3   The outcome of competition experiments with *Tribolium castaneum* (Cas) and *Tribolium confusum* (Co). Each box represents a bag of flour kept at different conditions of temperature and moisture. The species indicated in some boxes represents the one that always wins the competition under those conditions. Boxes without species indicated represent flour bags where the outcome of competition was indeterminate. The blocks shaded with slant lines indicate conditions under which *T. castaneum* was found to be optimally adapted in prior experiments. The blocks shaded with vertical lines indicate conditions in which *T. confusum* was found to be optimally adapted in prior experiments. (Data from various papers of Thomas Park.)

is potentially present everywhere, so that conditions at any moment are importantly different from those at any other moment. It is the nature of the experiments and observations referred to above to be begun and finished in what amounts to a tiny fraction of a geological moment, and because of this they must always convey this sense of equilibrium and poise; the situation at any moment becomes projected into the future in the literature and maintains a kind of life of its own. The environment along the coast of Scotland may not have changed in any important way since the 1950s, but it will. A glance at historical geology shows us that conditions on the earth are in continual major change in periods longer than a few of our generations.

ECOLOGICAL CATASTROPHE

An example of competition between two species as a result of ecosystem change that we can observe is our own competition with various insects and molds for food plants. We can observe this real competition because of the speed with which we ourselves carry out the primary change, that of putting more large areas under cultivation as our population grows. An interesting aspect of this example is the distant relatedness of the competitors, showing that intersystem competition can occur between any two species of any kind provided that the conditions are right. Other striking examples of interspecies competition that have been observed, although not carefully, are those resulting from the invasion of a community by a foreign species. Although no two species in the world have exactly the same total ecological niche, species living on different continents can evolve niches that overlap to a considerable extent—holarctic weasels, paleotropical civets, new world orchids, old world orchids, nearctic minnows, neotropical characins (see Figure 3-4 and Chapters 4 and 5). Nor will two such species be adapted to coexist (by coadaptation) with each other, so that if they come into ecological contact, interspecific competition would probably occur. Examples of the results of such competition might be the virtual extinction of the thylacine or marsupial "wolf" after the introduction of the dog to the Australian Continent and the increasing rarity of the North American bluebird after the introduction of the house sparrow to this continent. These invasions are examples of what might be called an ecological catastrophe, which can be defined as any change(s) in the system of a community whose effects are felt faster than any species can adjust to them or evolve. True interspecific competition, with unpredictable outcome, can be seen as one of the processes that take place during the transient period when an ecosystem seeks a new equilibrium after the old one was disturbed by an event that impinged catastrophically on at least one member of the ecosystem. Interspecific competition with unpredictable outcome can be postulated to exist generally and therefore to function as

Figure 3-4   Organisms with similar ecological niches from different continents. (a) a neotropical orchid, *Epidendrum* sp.; (b) a tropical Asian orchid, *Dendrobium* sp.; (c) an African civet; (d) a weasel; (e) a tropical Asian cyprinid, *Rasbora* sp.; (f) an African characin, *Neolobias* sp.

one of the evolutionary forces especially important in dividing up a community into many niches because the earth is continuously undergoing major changes, so that no ecosystem remains in the equilibria detectable by ecological studies for very long. These ecological observations are like stills of a motion picture and can show some of the modes by which true interspecific competition can occur but do not usually show the phenomenon itself. Furthermore, their predictive value concerning the outcome of interspecific competition is virtually limited to the geological moment during which the observations were made.

Thus, a competitive exclusion principle can be defined that does two things. First it describes an existing situation that is everywhere present, that is, that no two species operate exactly the same ecological niche in a single community. It can also be invoked to predict that if two species with very similar niches (or important niche overlaps) come into contact, one or the other or both will adjust to a slightly different niche (by a mechanism—character displacement—to be described in Chapter 6), in that way eventually avoiding interspecific competition, which will thus exist only until the area of niche overlap is reduced, or one of them will become extinct in the region where they meet (where they are sympatric). In the latter sense competitive exclusion is a process rather than a description of existing structure.

DIVERGENT EVOLUTION

The process of competitive exclusion is immensely important in producing evolutionary change in organisms and communities. It is a fundamental process involved in divergent evolution, which is in part a dividing or partitioning of ecosystems into ever more precisely delimited and probably narrower ecological niches. A simple descriptive model of this process would be one in which an ancestor feeds on vegetation in general and gives rise to descendants of two kinds: one of which is exclusively a grazer (feeds on grasses), the other of which is a browser (feeds on broad-leafed plant leaves). The descendants are more specialized than the ancestor, and the community system of the descendants' biome is more complex, unless one of the descendants becomes extinct. We now look at the concept of divergent evolution.

If we compare the skeletal structure of the forelimbs of several vertebrates (see Figure 3-5), we are almost certain to be struck by some startling similarities. The whale's flipper, man's arm and hand, the bird's wing—there is no reasoning by which we can conceive a necessity for these disparately functioning structures to be built on an identical basic design, as indeed they are. The different parts of these structures are said to be homologous; there are carpal bones and a humerus, and the like, in each one of them. These similarities need to be explained.

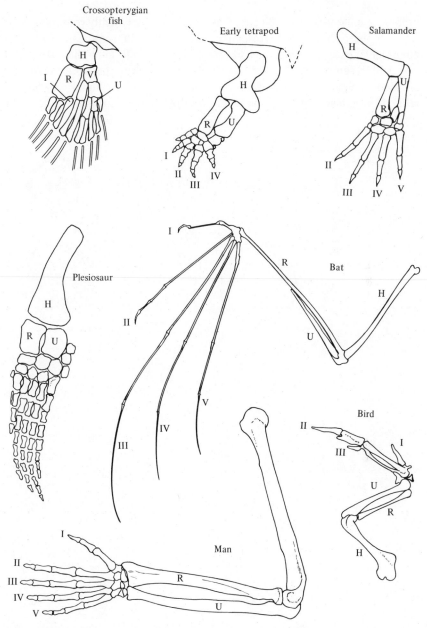

Figure 3-5   Comparison of the forelimbs of various vertebrates. R, radius; U, ulna; H, humerus. The digits are given their conventional numbers.

If we compare the embryonic stages of a series of animals, we are certain to be struck again by some strange similarities. Thus, until the blastula stage there is no fundamental difference among embryos of a sea urchin, a salamander, a pig, and a man. The early embryonic stages of the last three continue to resemble one another in significant ways—all three have external gills, for example. During still later stages the last two still resemble each other much more than do the adult organisms of these species (see Figure 3-6). Thus, the more similar or "closely related" organisms seem to us to be, the longer they continue to resemble

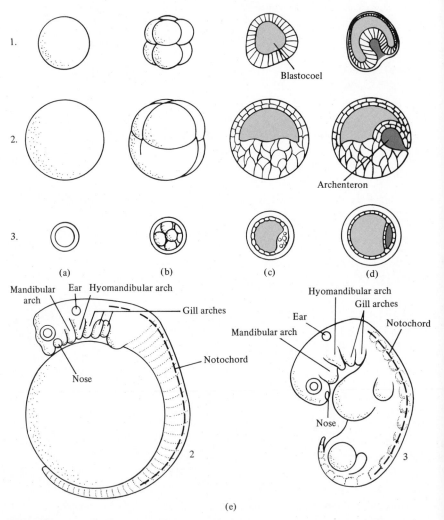

Figure 3-6 Comparisons of three different animals during different embryonic developmental stages: (1) sea urchin; (2) salamander; (3) man. (a) ovum; (b) 8-cell stage; (c) blastula; (d) early gastrula; (e) gill-bud stage (for salamander and man only).

one another during development. These similarities need also to be explained.

If we examine closely the amino acid sequences of homologous proteins from a number of different organisms, a situation similar to the last appears. The cytochromes c of man and a macaque differ by only a single amino acid out of about 100 or so (the alpha and beta chains of the hemoglobins of man and gorilla differ by a single amino acid out of some 145; these proteins are identical between man and chimpanzee, as are the cytochromes c of these last two species and also the fibrinopeptides). The cytochromes c of man and pig differ by 10 out of 100, and of course these organisms are less similar in obvious ways than are man and monkey. The difference between man's cytochrome c and that of a moth is 31 per 100; the difference between human cytochrome c and that of the black bread mold is 48 per 100 (see Figure 2-12). Thus, as organisms remind us less and less of ourselves, their primary gene products become increasingly different from ours. Nevertheless, human and black bread mold cytochromes c are identical in some 50 percent of their total sequence. These similarities too need to be explained.

Logically all these kinds of similarities can be explained in two different ways. We shall see that these ways are not necessarily mutually exclusive, but they are conceptually very different. First, we can take a completely functional, or teleological, approach and begin with the assumption that all biological structure—macro, micro, and molecular—is the way it is because it must be this way in order for the system of which it is a part (the organism) to function. Thus, all cytochromes c have certain structural identities, because, given the intracellular conditions in which they function, they must be made in just this way and no other or they would be functionally inferior and the systems of which they are a part would not be able to compete with more competent systems and would become extinct. Implicit in this approach is the idea that biological structure can be explained completely in terms of its function; that is, functional considerations are both necessary and sufficient to explain any structural feature of an organism. The second approach can be called the heredity approach. From this point of view all the cytochromes c bear certain clear resemblances to one another because they are descended from a single ancestral form of the enzyme. It is considered that, as copies of the cytochrome c gene were passed on to descendant populations, they became slowly modified from the ancestral structure by an accumulation of mutations. Two different descendents would end up with two somewhat different cytochromes c partly because their lineages had been separate for an appreciable length of time so that different random mutations occurring in both lines would have accumulated in them (see Figure 3-7). It is true that the mutations that accumulate in a lineage do so in part because they are functionally useful (see below and Chapter 6), but what is being

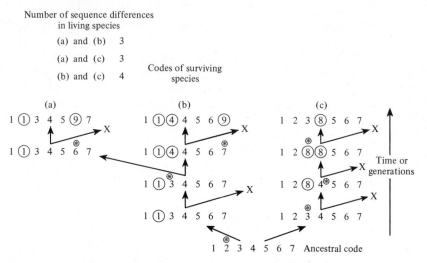

Number of sequence differences
in living species

(a) and (b)   3

(a) and (c)   3

(b) and (c)   4   Codes of surviving
                              species

Figure 3-7 The accumulation of sequence differences in a gene in a given lineage. ⊛ represents a mutational event; X indicates a sublineage that has not survived until today; circled numbers indicate codons (or nucleic acid bases, or bits of information) that have been altered from the ancestral condition. The last mutation indicated in lineage c is one that returns the coding unit to the ancestral condition, and can be considered either an example of convergence or a type of back mutation.

stressed here is that the pattern of accumulation would be different in different lineages regardless of whether the mutations accumulate at random or under the aegis of selection. This is so because the chances of the same mutation occurring at exactly the right moment in two lineages are small and because even if this did happen, selection would not be likely to favor the same mutation in lineages occupying different niches. The heredity approach stresses that, although it is necessary for a biological structure to be functionally adequate, this consideration is not sufficient by itself to explain the form of any structure.

If we visualized evolution occurring from the purely functional point of view, we should see the random origin of many quite different eobiont systems followed by the gradual elimination of all those not maximally fit to operate under the conditions present on the earth. We must also include in this vision the possibility of the fusion of two or more systems to construct more complex ones. Ultimately only very few of these systems remain, having constructed the requisite kinds of molecules, pathways, and structures. Thus, mammals could be one such system, birds another, bacteria still another. The other point of view begins with a single origin of life, whose units become separated from one another time and again, gradually becoming more and more different, ultimately producing such quite different systems as mammals and bacteria as elaborations of the basic (original) plan of organization.

Now, in favor of the first approach is the obvious fact that biological structure at any level shows unmistakable signs of being functionally useful (adapted) in a way that the inorganic world is not. Also, we can note that when it becomes functionally necessary, two quite different-seeming systems can construct, for instance, a complex structure like the vertebrate eye. An eye based on essentially the same principles and constructed in much the same way was developed by the cephalopod mollusks. The same visual pigment, 11-*cis*-vitamin A aldehyde, is used in both systems, and the kinds of specialized proteins found in the transparent lens are the same. The latter form in both cases a series of proteins of different molecular weights—α-crystallins, β-crystallins, γ-crystallins, and so on. Clearly, only certain kinds of proteins will do for this specialized functional task.

There are, however, some problems with this purely functional approach. First, if it were true in the form stated here, all organisms would have only a single kind of, say, cytochrome c; there would be only one best way of performing its function of electron transport, given the conditions on the earth. (One cannot think about macrostructure in this way, because not all organisms have, for instance, eyes. We need structures possessed by all organisms in order to test this functional necessity hypothesis, and these would have to be primary gene products.) It turns out that procaryotes have a significantly different kind of cytochrome c than that present in eucaryotes. Although they have the same configuration around the heme part of the active site, the procaryote cytochromes c are quite different in almost every other residue (see Figure 2-12). Now, cytochrome oxidase is necessary for the functioning of cytochrome c, and the one enzyme from any eucaryote functions in a test tube with the other from any other eucaryote. However, mixtures of procaryotic and eucaryotic cytochrome enzymes do not function as well *in vitro,* and in some cases hardly at all. Thus, one cannot argue that only a single kind of cytochrome c is functional on the earth and that this is why all organisms eventually came to have it.

Another conceptual problem with the functional necessity approach for explaining widespread similarities in the biotic world is that it does not fit well with our knowledge of the ever-changing nature of conditions on the earth. Consider again the cytochrome c molecule. It is unlikely that, with all the possible forms that a sequence of 104 amino acids could take (20^{104}), any two separately originated living systems would have had a similar sequence. Therefore, the functional necessity approach must assume a gradual transformation (by accumulation of mutations) of quite dissimilar molecules into cytochromes c by convergent evolution. Such a transformation would have to have taken quite a long time in order to arrive at molecules as similar as human and bread mold cytochromes c.

This being so, we must visualize a very long period of time during which one kind of possible cytochrome c (the kind all the others in different lineages were converging toward) remained the best possible molecule of this kind, and this implies that conditions on the earth would have to have not changed very much during this time. The data of historical geology tell us that this was never so.

This brings us to the fossil record, which provides the most dramatic refutation of this approach, in which many different lineages are visualized as being continually weeded out until those remaining are the ones we find today. The first fossils are found some 3.2 aeons ago. From that point until 1.3 billion years ago we find evidence only of what are probably procaryotes. By 1 billion years ago fungi are added to the list of forms present on the earth. By 600 million years there were many forms of invertebrates as well. By 450 million years we find evidence of vertebrates (see dateline in Chapter 4). The point of this recital is that the fossil record shows, despite plentiful extinctions, that with time the biosphere has increased in complexity—in the numbers of different kinds of organisms—and not decreased. A decrease is necessitated by a thoroughgoing functional necessity argument.

The heredity explanation has none of these problems associated with it but, as stated above, is incomplete, primarily because it does not take into account the functional adaptiveness of much of biological structure. By adding to this concept that of negative selection (weeding out inadaptive mutations), we can increase its power to explain how descendant organisms become different from one another. We can visualize two such descendants living in different climates—one in the tropics, one in the arctic—so that in each case the mutations weeded out are different, because the adaptive requirements of communities in these regions are quite different. Thus, descendant organisms differ from one another and from the ancestor, because (1) mutation is random and would not be identical in each lineage and (2) different mutations are selectively favored in different environments after they arise (or different ones are weeded out).

CONVERGENCE

The concept of divergent evolution (divergence) is consistent with the known fossil record and presents no problems in explaining why organisms of different kinds are as similar as they are. It is, furthermore, consistent with what is known about reproduction and heredity at the cellular and molecular level. It can be justly called the major mode by which evolution occurs—in Darwin's phrase, "descent with modification." Examples such as that of the vertebrate and squid eye given above are not in conflict with this formulation. These examples are often cited as examples of convergence, by which is meant that distantly related organisms produce similar solutions (adaptations) to similar challenges from the environment by modifications of, by definition, quite different genomes. Thus, by testing squid α-crystallin against an antibody prepared using beef α-crystallin, it was shown that the detailed chemical structure (and probably the entire amino acid sequence) of these proteins was very different—there was no cross reaction. For the present we should note that, although convergence does occur here and there and is by its nature a fascinating phenomenon, its evolutionary role has been small and peripheral. N. W. Pirie, of the Rothamsted Experimental Station in England, has pointed out that a kind of convergence may have had a larger role during prebiotic evolution, before the chemical system characteristic of living organisms as we now know them became the dominant or only system to go on to organic evolution. The above description of evolution based on the functional necessity argument could validly pertain to some aspects of prebiotic evolution. This is because almost certainly metabolic systems arose that were destined to be short-lived because they were based on some rare or eventually rare substance (see Chapter 2). One can visualize many such systems coming to a halt at various times, while the eventual winner in this field was using water and CO_2 and other plentiful substances. This process could be considered a kind of converging of preliving metabolic systems on the best possible chemical base, given the conditions of the earth in a most general sense (see Figure 3-8). Also, the physical fusion of two prebiotic systems to form a larger, more versatile one can perhaps also be seen as a kind of convergence of separate units upon the one superunit that was to survive.

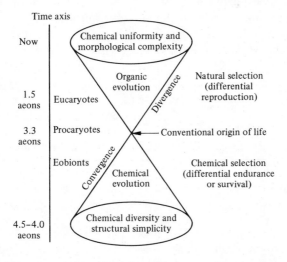

Figure 3-8   The different major modes or stages of evolution through time. (Adapted from a basic idea of N. W. Pirie, 1959.)

ECOLOGICAL LIMITATIONS ON DIVERSITY

We launched into a consideration of divergent evolution by noting that the process of competitive exclusion could produce it. The model is that of an ecologically more generalized ancestor that gives rise to several more specialized descendants, and then several, narrower, ecological niches are present where originally there was only broad one. If the biosphere has increased in complexity with time, something like this must actually have occurred overall, dividing up ecosystems into ever more complex interactions of ever-increasing numbers of niches. Actually one feels intuitively that such a process cannot go on indefinitely, and as an overall phenomenon perhaps has already stopped some time in the past. No community or even microhabitat can be said to be fully represented in any fossil digs, and so it is virtually impossible to fully analyze a past ecosystem. A rough comparison can be made, however, between a fossil aquatic assemblage and an existing lake. A comparison of two modern lakes with two fossil fresh water assemblages in terms of numbers of genera of each kind of consumer (based on sizes of the organisms—see Figure 3-9) gives one the feeling that the Pennsylvanian waters from which the fossil samples derive were not terribly different in terms of numbers of niches from today's lakes, some 300 million years later. At what stages various communities become saturated with niches is not known or even whether they typically are saturated. This is a difficult question; because the earth's climate is in continual flux, even today's communities are probably not all saturated. Perhaps only in the relatively stable tropics is this condition approached. The more boreal regions may still be "recovering" from the latest glaciation. Perhaps some overall maximum number of niches became established in the past when the total biomass of living systems reached its maximum. We do not know when this was (see Chapter 4), but we do know that

	Pennsylvanian		*Modern*	
	Linton, Ohio	*Mazon Creek, Illinois*	*Cassadaga Lake, New York*	*Lake Champlain, New York*
Tertiary consumer	7	3	4	10
Secondary consumer	10	5	8	11
Primary consumer	14	11	21	34

Figure 3-9 Comparison of two fossil aquatic vertebrate faunas with two modern ones. The fossil assemblages are composed only of fishes and amphibians, the modern ones of fishes, amphibians, turtles, snakes, birds, and mammals. (The fossil data are from various sources, those for the modern lakes are mainly from a New York State Biological Survey.)

there must exist some upper limit on total biomass determined by limited amounts of space and raw materials. Because we visualize a beginning to organic evolution, there must have been an increase in the size of the biosphere from this beginning. We can see the functioning of the limitation on biomass today, perhaps, when we note that as man increases his biomass, that of many other systems decreases. At least in some respects, organic evolution may alreay have stopped.

SPECIALIZATION

A process by which fewer kinds of ancestors produce more numerous kinds of descendants is one that would generate increasingly more specialized kinds of organisms, even if most of the descendants of an ancestor did not survive until the present. Divergent evolution is at least in part a continued partitioning of ecosystems into ever more specialized niches. In large part this process can be seen as one in which interspecific competition favors increased efficiency in aggrandizing raw materials. An organism that feeds on all vegetable matter cannot possibly feed on any one kind as well or quickly as a specialist that feeds only on roots, grasses, leaves, or buds. This is all a matter of the amount of energy expended to obtain the food versus that obtained from it. The important ratio here is energy obtained/energy expended; we should expect this ratio to be maximized by selection. For a cat to live on Indian nuts or an elephant on bird's eggs would be very inefficient in this sense. The ultimate result of divergent evolution in this respect should be the production of very highly specialized organisms, such as the Florida kite. This hawk has specialized in feeding on a certain kind of snail found in mangrove swamps. Its behavior and beak are finely adjusted to the task of removing the snails from their shells, which it can do with great speed. This organism is feeding on only a single other kind of organism, which by any standards is specialized indeed. In this sense the Florida kite is much like a parasite, which could perhaps be defined as an organism that feeds on only a single other species of organism at any stage in its life cycle. The koala bear is another very highly specialized species, feeding, as it does, on the leaves of only certain species of eucalyptus trees. It is clear, however, that such highly specialized organisms are rare in nature (they may be continually produced, however, making up a large part of the lineages that become extinct). More commonly a species feeds on a number of different organisms. A typical example could be the ruffed grouse, which in the fall feeds on the fruits and buds of some 30 different plants in any one area. Indeed, one has less trouble finding examples of omnivores—for example, bears, raccoons, opossums, many primates—than finding examples of very specialized feeders.

From these considerations it is clear that evolution has not in fact produced a biosphere composed of uniformly very highly specialized niches. There are at least two reasons for this. First, even slight reflection allows us to see that a high degree of trophic specialization is a danger akin to "putting all one's eggs in a single basket." If an ecological catastrophe overtakes the single food organism, the specialized consumer is also endangered. Probably most organisms (perhaps excepting what are usually termed "parasites") that have taken up the strategy of extreme trophic specialist became extinct on the average sooner than nonspecialists. Parasites of the usual kinds may have escaped this restriction on maximizing efficiency by having extremely high reproductive rates. A perhaps more important factor that has operated to prevent wholesale maximum trophic specialization may have been the fact that in order for an organism to continue specializing (evolving toward a specialization), the environment must not change radically very fast. But the data from historical geology suggest that this is just what has been happening most of the time on much of the earth. As a general rule there may simply never have been enough time to specialize very far in some direction before the environment had become radically changed. Perhaps this is why many more examples of extreme specialization are found today in the tropics, which have undergone much less radical climatic change in the last 50 million years than have other regions of the earth. Nevertheless, all organisms are in one way or another and to some degree specialized. This is because they are specifically adapted to perform adequately in their various ecological niches.

ADAPTATION

Most structures possessed by an organism can be considered to be adaptive, that is, formed so as to perform some task(s) that increase the probability of their possessors' reproducing. Thus, the teeth of horses are well-suited to chewing grasses, the "wings" of penguins to swimming, the antlers of deer for combat. Some structures are adapted to being adaptable, for instance, man's hand. It can be a club, a pincers, a hook, a tickling device, a drumstick, a symbolizer, and the like. In contrast, a horse's hoof is highly specialized indeed in its roles as foot and spade. The last example gives us a good way to define "specialized" for a structure; that structure which can assume many different forms (and therefore functions) is less specialized (or more generalized) than one that is restricted to fewer forms (see Figure 3-10).

All the adaptations and adaptive features of organisms mentioned so far function in reference to the external environment—to such processes as obtaining food and mates or to avoiding being eaten by a predator. It is usually structures that are involved in these pursuits that are studied in connection with adaptation, including behavioral studies on animals.

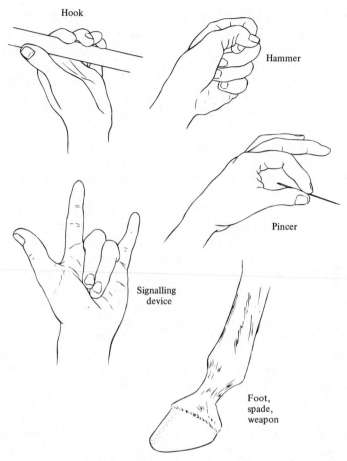

Hook

Hammer

Pincer

Signalling device

Foot, spade, weapon

Figure 3-10   Comparison of a generalized structure (man's hand) with a more specialized structure (horse's hoof), showing the plasticity of form and function of the more generalized forelimb.

These features are usually specialized in some way as specific adaptations but typically retain some evolutionary potential and often have a small range of adaptability to more than one use. It is these features that determine and operate in the ecological niche, which is a system of relationships with those aspects of the external environment to which these features are specifically adapted.

The general functional adaptedness of any kind of organism is not usually studied in terms of adaptation. Thus, an orchid has a photosynthetic apparatus that is like that of other plants and so is not thought of as being specialized for the photosynthetic use of CO_2. Many orchids are truly specialized for life in trees, with thick, grasping, water-absorbing roots and specialized water-storage organs. In addition each orchid species is adapted to being pollinated by a specific insect through the structure of its flower. To say that the vascular system of a plant is an adaptation to keeping the plant alive is a statement of the same kind that says that the atmosphere on the earth is adapted to being used by living organisms. Such statements (or truisms) are not only not instructive but are somewhat misleading. Because living things as we know them evolved in connection with a certain kind of atmosphere, they had to be "adapted" to this atmosphere or they could not have evolved. Adaptation to this atmosphere would be a precondition for evolution, not a specific adaptation. The kinds of adaptations that we are usually interested in as evolutionists are those which are related to specific aspects of the external environment and which are built on top of a general physiological adaptedness that we take for granted if an organism is to exist in the first place. These specific adaptations determine, not whether an organism is capable of distributing glucose to various parts of its body and metabolizing it, but the efficiency with which an organism exploits its food supplies or nesting materials, and so forth. Usually these specific adaptations are considered to be identically characteristic of members of a species and are not used to discriminate among them. Rather they are used to distinguish between members of different species. Thus, one species of orchid is seen to be adapted to being pollinated by insect A, while another orchid in the same biome is adapted to being pollinated by insect B. In actual fact no two orchid flowers even in one population of a species are exactly identical, and so one of them may be better able to attract the insect than another in the same species. Although this is true (and forms the basis for natural selection in this example), it is nevertheless also true that both the successful and the unsuccessful versions were both, identically, adaptations to being pollinated by the same kind of insect.

Specific adaptations, then, are in reference to the external environment and are characteristic of species as a whole. They are the attributes that determine the ecological niches of species and are also the means by which species avoid competing with each other. The myrtle warbler and the Blackburnian warbler avoid competing with each other by being adapted behaviorally to feeding in different parts of a tree and therefore on a different food supply. Every member of a population of each of these species competes with every other member (except sometimes its mate), but the members of a population of one species do not compete with members of the other sympatric species because they are adapted to feeding in different parts of a tree (they are not syntopic). A specific adaptation is a feature present in all members of a species, although individuals may vary in their competence in respect to the adaptation in question. An adaptation can be, for example, a structure, the ability of a structure to assume a given form, or a behavior. Most usually adaptations involve a certain degree of specialization, restricting the abilities of organisms to the functions that they at the same time facilitate. Thus, the peculiar mouth of an anteater makes it easy to eat ants and at the same time makes it difficult to eat anything else. Adaptations are rare that expand the ability of a species to function in many areas. This is not to say that populations and species do not sometimes expand their niches—by being initially able to muddle through in a slightly less than optimal environment. But structures and behaviors that facilitate this are not "adaptations" as such, because organisms cannot foresee the future; that is, they cannot become adapted to an environment they have not experienced. The ability to expand the niche is a function of preadaptation. (See Chapter 4.) Man's hand with its many forms and functions is perhaps an example of a specific adaptation that resulted in the expansion of the ability of a species to operate in many habitats. Perhaps the elephant's trunk would be another such example. Man's brain might be discussed in this context as well, because it has allowed him to function in almost every environment present on the earth. Because man is not specifically adapted to many of these environments—for instance, the polar regions—he lives there only with a difficulty not experienced by penguins or polar bears, which have specialized to live in these cold places at the expense of not being able to live elsewhere. In these respects, as in many others, man seems to be unique.

ADAPTABILITY

A few words should be said about adaptability, an important part of physiological adaptation. The typical kinds of evidence that we have in this area consist of studies showing, for example, that within some taxon individuals of species living in colder regions can survive colder experimental temperatures than can individuals from species adapted to living in warmer regions, and vice versa. Similar studies are available for the ability to withstand desiccation or different degrees of salinity. In the aggregate, these studies show that individuals of a species can adapt themselves by means of physiological regulating mechanisms to living under conditions not typical for the species. They can live longer and more normal lives if the conditions are not extremely different from their "normal" ones. This is because each species has only limited capabilities of regulating, and the limitations are linked to the species adaptations in specific ecological niches. In order to live well in arid regions, cacti have evolved certain structures and physiological responses that by their very presence eliminate the possibility of individual cacti living in swamps. Some organisms have comparatively greater powers of adjusting to different environments than do others, for example, man and various organisms associated with him, such as the house sparrow and the Norway rat. Also in this category would be organisms that live in estuarine and tidal regions, which commonly experience rather large changes in salinity, temperature, or moisture. In the latter cases, we can certainly consider the ability to function over a wide range of environmental conditions to be part of a series of specific adaptations to an unstable environment. But the ability to regulate physiologically in response to the challenge of a somewhat abnormal environment seems to be quite generally present in all living systems. It is only the unusual ability to regulate over rather large vicissitudes that need be considered a special adaptation. Organisms with these unusual abilities have been termed regulators, as opposed to the usual kinds of organisms, termed conformers. Thus, extreme ability to regulate, as found in regulators, can be considered a specific adaptation; adaptability in general is not. Needless to say, one organism might be a regulator in some regards and a conformer in others.

THE OPPORTUNISTIC NATURE OF ADAPTATION

Specific adaptations, and particularly the more strikingly complex ones, such as the vertebrate eye, have been taken by the supernaturally or religiously oriented to suggest the presence of a designer external to the biological systems themselves. Such adaptations seem to exist "in order to" perform this or that function; they seem to exist for a "purpose." Thus, the purpose of the shearing carnassial teeth of carnivores is seen to be to cut meat. The implication is that the idea of cutting meat was present before the ability to do so. To the evolutionist the latter statement is not acceptable; a structure cannot exist without its function, but neither can function exist without a structure. Simply, fur would not be invented until it is cold, except by chance. But without cold, chance alone could not be responsible for the continued development of fur. Fur would be invented in the cold because those individuals without fur would die off, while those who by chance had a tendency toward developing fur would tend to leave relatively more offspring than other members of their species. Other responses to the problem of conserving body heat would appear with equal probability as an environment became cold—for instance, feathers or larger body size. Which response to cold is exploited depends on what raw materials, in a genetic sense, are at hand. This is the sense in which organic evolution is opportunistic; the first response of a system to altered conditions that allows that system to continue to function adequately is preserved and perhaps subsequently refined. It is the process of subsequent refinement, step by slow step, that is visualized as the means by which a complex mechanism, such as the vertebrate eye, with its many cooperating parts, has evolved.

A crucial point in this argument is that at any given time in the past, the stage reached in the development of an adaptation was functional and so not weeded out by selection. Some persons have expressed doubt that any less refined version of the vertebrate eye than is now present could have been functionally good enough not to be selected against. The simplest answer to this is that, although the stage reached by this adaptation at any time in the past would not have survived if it had to compete with today's version, at that time it was the best version available and therefore the best from this point of view of selection (as will become clear in later chapters, fitness is a relative quality). Each further refinement (or improvement) would in turn be added to the adaptive system, because it increased the probability of the reproduction of the organism of which it was a part over that of other members of its species. This is the rough shape of the mechanistic argument to be developed in this book.

COEVOLUTION AND COMMUNITY STRUCTURE

It should be clear that, although many adaptations are responses to environmental changes in the inanimate world, equally as many are responses to the presence of other organisms. In a very general way aerobic metabolism can even be seen as an adaptive response to the production of oxygen by plants and so an adaptive response to the presence of plants. There are many more specific adaptations of a species to the presence of other species. Thus, the various adaptations of a carnivore—claws, teeth, speed—are adjusted to the kinds of prey species making up most of its diet. The prey species, in turn, evolve adaptations—speed, protective coloration, chemical-warfare devices—that reflect the modes by which their main predators detect and catch them. In addition, the reproductive rates of both predator and prey may be adjusted to each other. Hummingbirds are adapted to feed on nectar in certain kinds of flowers by the ability to hover in the air next to the flower, by having long beaks to reach the hidden nectaries in the flowers, by being small, and so on. In turn some kinds of plants are adapted to being pollinated by feeding hummingbirds, by having tubular flowers with nectaries hidden deep in the funnel, stiff floral construction so that the hard beaks will not damage them, red colors to take advantage of the fact that birds tend to be most responsive to these wavelengths, and so on. Many plants attract ants by having external nectaries located here and there on their surface. These ants, in turn, tend to keep other insects off the plants in question. Some of the tropical American *Acacias* have evolved very complex specific adaptations along these lines, providing not only external nectaries but also little food packets high in protein (Beltian bodies) and expanded thorns in which the ants can build nests. The ants need not leave the tree at all. In turn they instantly attack with pincers and stingers any organism of any kind that alights on or touches the *Acacia,* and they also clear away all competing plants in the area by destroying them.

Symbiotic relationships like those just described are often referred to by using the terms commensalism (one partner in a relationship derives benefit; the other is unaffected), predator-prey relationships, parasitism (one partner in a relationship injures the other), and mutualism (both partners derive benefit from the relationship). Community structure can be seen to be held together by an intricate skein of these relationships. In this sense each organism is coadapted in special ways to function with many others. As with adaptation, some of these coadaptations are so general and fundamental that one need not consider them specifically— for instance, the coadaptation of plants and animals in respect to CO_2 and O_2. The Beltian body of some Acacias, however, is a specific coadaptation in relation to only certain kinds of stinging ants.

Another large category of adaptations is that made up of those in respect to other individuals of the same species. These include, for instance, various sexual attracting devices and courtship performances, dominance systems, territoriality and associated behaviors, some kinds of responses to overcrowding, and so on. We will discuss these extensively in later chapters.

SUMMARY

Organic evolution, then, is the continuing process of adaptation of organisms to changing geoclimatic factors and to the presence of other kinds of organisms. It proceeds overall by divergence from fewer, more generalized kinds of ancestors to a greater number of kinds of more efficient, more specialized descendants. Various specific adaptations define the ecological niches of organisms, no two of which are identical or even alike in the same community. Each kind of organism is adapted (and adapting) to its own ecological niche, the limits of which are set by the niches of other organisms and by the inanimate environment. Organisms and the inanimate environment evolve together as one system. The system of relationships among different species in a given community is also evolving in response to changing environmental conditions.

REFERENCES

Allee, W. C., A. E. Emerson, O. Park, T. Park, and K. P. Schmidt, *Principles of Animal Ecology*. Saunders, Philadelphia, 1949, 837 pp.
Andrewartha, H. G., *Introduction to the Study of Animal Populations*. The University of Chicago Press, Chicago, 1961, 281 pp.
Berkner, L. V., and L. C. Marshall, *The Origin and Evolution of Atmospheres and Oceans*. Wiley, New York, 1964, 300 pp.
Cold Spring Harbor Symposia on Quantitative Biology, vol. 22, 1957. Population Studies: Animal Ecology and Demography.
Connell, J. H., "The Influence of Interspecific Competition and Other Factors

on the Distribution of the Barnacle *Chthamalus stellatus,*" *Ecology,* **42:**710–723, 1961.

Elton, C., *Animal Ecology.* Sidgwick & Jackson, London, 1927, 203 pp.

——, *The Ecology of Invasions by Animals and Plants.* Methuen, London, 1956, 181 pp.

Jukes, T. H., *Molecules and Evolution.* Columbia, New York, 1966, 285 pp.

MacArthur, R. H., "Population Ecology of Some Warblers of Northeastern Coniferous Forests," *Ecology,* **39:**599–619, 1958.

Odum, E. P., *Fundamentals of Ecology.* Saunders, Philadelphia, 1971, 400 pp.

Park, T., "Beetles, Competition, and Populations," *Science,* **138:**1369–1375, 1962.

Simpson, G. G., *The Meaning of Evolution.* Yale, New Haven, Conn., 1949, 365 pp.

——, *Principles of Animal Taxonomy.* Columbia, New York, 1961, 247 pp.

——, *This View of Life.* Harcourt, Brace & World, New York, 1964, 308 pp.

Wald, G., "The Origins of Life," *Proceedings of the National Academy of Sciences,* **52:**595–611, 1964.

CHAPTER 4

The fossil record: I

In this and the following chapter, the discussion will center on observations about and concepts derived from the study of the fossil record. However, these will serve as focal points from which the discussion will frequently radiate in many directions; the fossil record, or the viewpoint of the paleontologist, will serve as a catalyst for thinking about organic evolution.

It was pointed out in Chapter 1 that the fossil record is the only hard evidence we have that organic evolution has in fact occurred. In order to deny the reality of this evidence, it would be necessary to deny not only our understanding of nuclear physics and radioactive decay in particular but our entire approach to physical science as embodied in geophysics and historical geology. This realization, coupled with the fact that we can hold in our hand a piece of undeniable stone curiously and intricately sculptured into the shape of some part of an organism that is not found living on the earth today, forces us to think of organic evolution as a fact (see Chapter 1).

Taken as a whole, the fossil record, some of whose high points are summarized in Figure 4-1, suggests three important and related generalities: (1) there has been overall an increase in the complexity of the biosphere; (2) there has probably been an increase in the area or volume of the biosphere; (3) there probably has been an increase in the total biomass of the biosphere.

INCREASING COMPLEXITY OF THE BIOSPHERE

The first generality is apparent in Figure 4-1. There was a time when only procaryotes were fossilized. At a later time certain of the simpler eucaryotes joined them, while at a still later time some kinds of invertebrates begin to appear in the record. At this time there were no mammals, and these appeared only later, after other kinds of vertebrates had been fossilized. The general impression is that of new groups continually being added to an existing biota while old groups remain. A closer inspection of the fossil record reveals that this is an oversimplification. There are groups that appear in the fossil record for longer or shorter periods and then disappear (seed ferns, trilobites, bony placoderm fishes), but at the level of the highest taxonomic categories (say, phyla) the process of organic evolution seems to have been predominantly additive.

The trouble with this observation is that it depends on what we wish to call a higher category. For instance, in the late pre-Cambrian invertebrate assemblage found in the Ediacara Hills of Australia there are a number of forms that do not clearly belong to one or another major living invertebrate taxon. Do they represent early members of some living group that had not yet evolved the distinguishing characteristics of these groups as we know them, or do they represent potentially major phyletic lineages (evolving gene pools) that subsequently become extinct? Were the graptolites a major group of invertebrates that became extinct in the Carboniferous, or were they simply a specialized lineage of hemichordates that became extinct at that time? We might almost say that of all the major taxa that we recognize as living today not one has become extinct. This is because the only major taxa that we recognize are those which we know to have produced numerous different forms still alive today. Most of the fossilized forms found in the earth have been assigned to one or another major taxon of living forms, and extinctions are considered to have taken place within major taxa (phyla), while the latter themselves have not, as a rule, become extinct.

Do the major taxa represent lineages of organisms with distinct enough adaptive zones so that none competes (or competed in the past) with any of the others? Were there other major taxa that became extinct partly because of some important overlap in adaptive zones, leaving the field open for the forms found today? Actually, the competition would have to have been between species (in the sense developed in Chapter 3) at a time when both groups were small as groups; that is, the competition was not between higher taxa but between species that were capable of giving rise to a multitude of different descendant species that we could subsequently group together as a major taxon. If this is true, the only major taxa that ever existed are in fact those which we find today, because the species that became extinct did not actually give rise to any group of de-

EVOLUTIONARY DATELINE

Years Ago	Events	Relative Time
5×10^9–4.5×10^9	Origin of the earth	Jan. 1
4.6×10^9	Oldest meteorites and terrestrial lead	
3.3×10^9	Oldest crystalline rocks—volcanic in origin and apparently cooled under water	
3.2×10^9	Algalike microfossils in Onverwacht sediments, South Africa	
3.1×10^9	Algalike and bacterialike microfossils in association with amino acids in Fig-Tree Formation, Barberton, South Africa	
2.7×10^9	Presence of alkanes in rocks of the Soudan Formation, Minnesota, and possible bacterial microfossils in Southern Cross Formation, Australia. In Rhodesia, Limestone and dolomite apparently deposited by lime-secreting organisms	
	Estimated origin of oxygen-evolving photosynthesis	
2.6×10^9	Fossils of blue-green algae (stromatolites–lime-secreting intertidal forms) in the Bulawayan Formation, southern Rhodesia	May 25
2.5×10^9	Stromatolites in Steeprock Lake, Ontario	
2.1×10^9	Amino acids and hexose sugars in rocks of Witwatersrand, South Africa	
2.0×10^9	Blue-green algae and bacteria in the Gunflint chert of Ontario. Weathering of minerals left them unoxidized, and so no appreciable oxygen in the atmosphere. Stromatolites in South Africa	
1.8×10^9	Red beds (ferric iron, Fe^{+++}) beginning to be deposited instead of banded iron formations (ferrous iron, Fe^{++}), indicating substantial amounts of oxygen in water	
1.3×10^9	Green-algae and golden algae (first eucaryotes) in Beck Spring Dolomite, California	
1.2×10^9	Estimate about 1 percent oxygen in the atmosphere (Pasteur point) Fossil actinomycetes in Canada	
600×10^6	Late Pre-Cambrian, jellyfish, soft corals, segmented worms, and several previously unknown (to man) forms of invertebrates in the Ediacara Hills, Australia	Nov. 17

EVOLUTIONARY DATELINE

Years Ago	Events	Relative Time
450 × 10⁶	Oldest known vertebrates (ostracoderms) in what appear to be marine deposits	Nov. 28
400 × 10⁶	Beginnings of land colonization—horsetails and clubmosses (Pteridophytes—first vascular plants) in Europe and Australia; scorpions and millipedes in Scotland (mid to upper Silurian)	Dec. 1, evening
350 × 10⁶	Late Devonian. First land vertebrates (ichthyostegids) of amphibian grade. First seeds evolve in plants	Dec. 5, noon
320 × 10⁶	Late Mississipian. Reptile grade achieved; origin of the cleidoic egg	Dec. 7, late morning
250 × 10⁶	Mid-Permian. Earliest coniferous plants. Beginning of separation of original land mass (Pangaea) into a northern continent (Laurasia) and a southern one (Gondwanaland) by continental drift	
200 × 10⁶	Late Triassic. Earliest mammals. Separation of North America from Eurasia	Dec. 17, 10 A.M.
180 × 10⁶	Jurassic. Earliest known birds, lizards, and frogs. Earliest angiosperm plants. Beginning of breakup of Gondwanaland into the southern continents, South America first	
80 × 10⁶	Late Cretaceous. Extinctions of many groups, including the dinosaurs. Mammals increasing in importance. Spread of the angiosperm flora	Dec. 26, noon
60 × 10⁶	Mid-Paleocene. Earliest primate adaptive radiation	Dec. 27, noon
35 × 10⁶	Early Oligocene. Earliest anthropoids	Dec. 29, morning
27 × 10⁶	Early Miocene. Origin of hominoids	Dec. 31, morning
14 × 10⁶	End Miocene. Oldest known hominid forms(men)	Dec. 31, 8 P.M.
575 × 10³	Mid-Pleistocene. Homo erectus	Dec. 31, 11 P.M.
250 × 10³	Earliest Homo sapiens	Dec. 31, 11:30 P.M.
10 × 10³	Neolithic revolution—switch from being a secondary to a primary consumer by man	Dec. 31, 11:59 P.M.

Figure 4-1 Dateline for the evolution of the earth, emphasizing organic evolution. The dates on the right are included so that the student may at least partially comprehend the enormous amounts of time involved—the entire span of time is contracted into the period of one year.

scendant species large enough to group as a major taxon. Any species could become the ancestor of an entire phylum if a very great many of its descendants continued to survive. Because the environment on the earth is heterogeneous and because there is a limited carrying capacity for any species in each environment, a species can leave a very great many descendants only if the latter become members of a number of different species (that is, if they come to have different ecological niches).

To sum up, the known major phyla of organisms are the only major phyla—give or take a few smaller potential possibilities in the pre-Cambrian—to have existed on the earth. They do not all first appear in the fossil record simultaneously but come in sequentially, with the higher plants being the last major group to appear, and, just before this, the vertebrates. Thus, there has been an increase in the overall complexity of the biosphere after the initial biogenesis. In Chapter 3 the problem of whether the complexity of particular habitats has increased in time was briefly touched on. It is clear that before the appearance of invertebrates, the biota of a lake was not so complex as it was after there were invertebrates present. The general increase in complexity of the biosphere at the ecological level may have stopped a long time ago if scored as the number of different forms living in some given major habitat—say, in the oceans or on land. The continual flux of the earth's climates, however, results in the process leading to habitat saturation having to be repeated periodically in any particular geographical region. For example, glaciations sweep the biosphere from the rocks of boreal land masses, while subsequent interglacial periods witness the recolonization of these areas, one group at a time, by a process partially analogous to the increase in complexity that must have taken place in the past as the major phyla made their appearances one by one.

It is a very long time since a new phylum last arrived on the scene—perhaps 350 million years. Yet any living species could be the ancestor of a future phylum. Because the carrying capacities of environments are in principle limited, such an event would probably depend on some catastrophic elimination of whole groups of the living organisms now present. An example from the past might be the replacement of most procaryotes by eucaryotes, accompanying the increased concentration of oxygen in the atmosphere after about 1.2 aeons ago. Again, man is now altering the earth's environment at a rate so fast and to so large a magnitude (for example, in the last 50 years the CO_2 content of the air has increased by 10 percent because of the burning of fossil fuels) that we may be on the threshold of another episode of dramatic evolutionary activity. It is clear that man, like glaciers, has the ability to radically simplify the biosphere in large areas on the earth. Much of the Mediterranean region is desert or semidesert partly because of the long-continued activity of man in this area; the wheat fields of the Russian steppes or the prairies of North America are about as simple as ecosystems can be and would probably spontaneously become more complex if man did not prevent that from happening.

The spontaneous increase in complexity of biological systems with time can be seen to be a kind of acquisition of stability analogous to the attainment of a chemical equilibrium and, therefore, to be highly probable occurrence. Thus, beginning with glucose and water, at a pH of 8.6 the system increases in complexity of molecular species even while it decreases in total free energy and, therefore, assumes a more probable configuration (see Figure 4-2). This proposition can be illustrated with some ecological examples. Most food relationships in nature are food webs, rather than simple linear food chains. Consider why this is so. If an organism feeds on 12 different kinds of prey species, the extinction of one of them in its community will have less effect on it than if the disappearing prey species is the only one it feeds on. Therefore, while selection itself has generally favored fairly specialized trophic relationships, differential extinction has tended to leave only those forms that were not too specialized in the field. Once these relationships are set up, a typical community has achieved the kind of stability in which a single species can be lost without affecting the major characteristics of the system. Experiments carried out by C. B. Huffaker, at Berkeley, California, on the cyclamen mite that attacks straw-

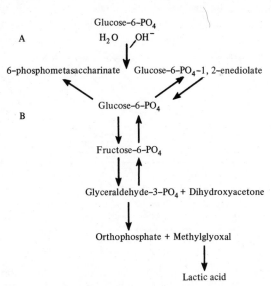

Figure 4-2   The spontaneous change of a molecular system from a less probable configuration containing higher free energy to a more probable, but more complex, one with less free energy. (a) beginning configuration at pH 8.6; (b) the final state of the molecular system. (From the data of Degani and Halmann, 1967.)

berries afford an interesting example in which complexity confers stability. Two experimental plots of strawberries are set up. In one, cyclamen mites are released and increase in number. At some point their numbers exceed the carrying capacity of the plot, and they begin to starve. Eggs deposited before starvation began keep hatching and adding to the problem. The food plant cannot grow fast enough to replace eaten leaves. This is followed by a precipitous decrease in the numbers of mites. When the number of mites goes below a certain level, the strawberry roots send out new runners and the plants can again grow. This revives the mites' food supply, and they in turn increase in numbers, and the cycle is repeated over and over. Such a system is characterized by wide fluctuations in the numbers of producers and consumers—in short, by instability. In a second plot a species of predatory mite is introduced along with the cyclamen mite. In this case the latter's population never reaches the carrying capacity of the strawberry patch, being kept in check by predation. The addition of another species makes the system more stable than it was when only two kinds of organisms were present (see Figure 4-3). Given enough time without major climatic catastrophes (such as glaciations), it is probable that biological systems will become increasingly complex (by divergent evolution) up to a point of optimal complexity, (determined by the population sizes of organisms, which cannot be less than what is required to generate an adequate amount of genetic variability per year) because once a sufficient degree of complexity is attained, further major change is inhibited.

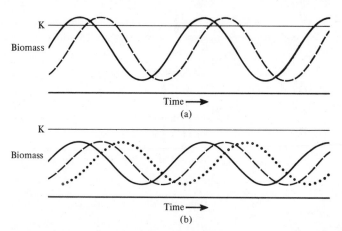

Figure 4-3   Greatly simplified presentation of data from the works of C. B. Huffaker. (a) the oscillations of biomass of strawberry plants (solid line) and strawberry mites (dashed line); (b) the oscillations in biomass after the addition of a third organism to the experimental ecosystem; solid line, strawberry plants; dashed line, strawberry mites; open circles, predatory mites. In both (a) and (b) K indicates the carrying capacity. The system in (b) is relatively more stable than is the system in (a).

Before this point, further change promotes increased complexity. C. H. Waddington, of the University of Edinburgh, has suggested that the mere presence of one kind of organism makes possible adaptive zones not possible without them. This then results in the evolution of newer sorts of organisms exploiting the first kinds, and the presence of these, in turn, allows for still other adaptive zones (previously not possible) which are again realized by still further evolutionary activity. This process leads to adding to the existing ecological complexity and does not result in replacing the older kinds of organisms, which remain as integral parts of the developing biosphere. Actually, in most parts of the world long periods of time without major ecological catastrophes and the resulting local simplifications of biosphere complexity, are a rarity.

INCREASING SIZE OF THE BIOSPHERE

Along with an increase in the complexity of the biosphere there has occurred an increase in the area or volume occupied by it. At first, life was present only in waters—perhaps at the very first only in certain waters; later, exposed lands were colonized—first by plants, then invertebrates, then vertebrates. The atmosphere does not seem to have been colonized as such. Birds fly through it, as do insects, seeds, pollen, and resistant spores reside in it temporarily, blown by winds; but nothing seems to have taken up permanent residence there. The problem has perhaps to do with the low concentration of nutrients in the air. In order to use these, one would have to be anchored instead of planktonic, because in the latter case one would have to be small enough to be blown along with the food particles and could, therefore, not come in contact with them. Some plants—for example, the Spanish moss and other members of the genus *Tillandisia*—obtain their mineral and nitrogen nutrients solely from wind-borne dust particles, but they need to be anchored to do so. Attached to branches, they are as terrestrial as are barnacles and other littoral-zone–attached organisms that strain food particles from their turbulent medium. Water, unlike air, is dense enough so that a planktonic organism can be large enough to move slowly through it toward food particles.

In order to colonize successively more and more of the earth's surface, the biosphere must have had to have increased its biomass from the presumably small amount that was involved in biopoesis. When the average maximum biomass was reached is not known. Possibly it was at the point at which the volume of the biosphere reached its maximum. Relevant here is the geochemical observation that the amounts of phosphorus and nitrogen dissolved in Cambrian seas seem to have been only about half of what they are today. The concentrations of molecules involving these atoms can limit photosynthesis, and so we may infer that this process was operating at as little as half the rate at which it is today. This in turn would narrow

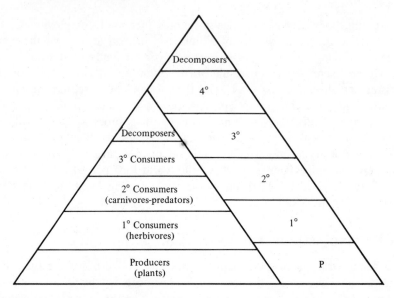

Figure 4-4   Pyramid of biomass, showing how the biomass of various levels in the food chains is limited by that of the producers. The more productive ecosystem (with the broader base) has an added link in its food chains, the quaternary (4°) consumers.

the base of the triangle of biomass (see Figure 4-4), so that we may induce that the total Cambrian biomass (all of it aquatic) was as little as half of what we have today. If there was a time when a maximum total earth biomass was achieved, that would have been the end of a growth phase of the evolution of the biosphere. Continued evolution past this point was then partly a matter of dividing the biosphere into smaller and smaller units (ecological niches; see Chapters 4 and 8). How far this process can continue is not known, or whether there has ever been a stable enough environment for it to have continued to "completion."

EXTINCTION

The variety of living systems, however, does not only continue to increase. An equally important factor in organic evolution is extinction. G. G. Simpson, of Harvard University, has supplied us with the following interesting statistic. Roughly 2500 families of animals with a family longevity of about 75 million years have left a fossil record; of these, about a third are still with us. Two thirds of them became extinct. Clearly, a family of animals is more likely to become extinct than to continue existing. Of course, new families are continually appearing as well, as the continuingly successful descendants of some successful ancestral lineage multiply and

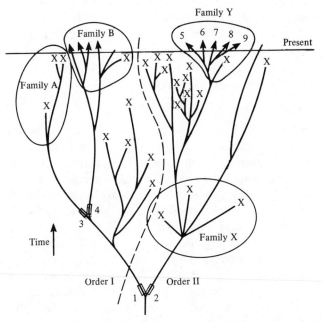

Figure 4-5 A phyletic dendrogram. 1 and 2 are species that gave rise to two different orders; 3 and 4 are species that gave rise to two different families (A and B). The lineage descending from 3 is extinct. Some families are extinct (family X), some are alive today (family Y is composed of species 5–9).

become numerous enough for them to be grouped by us into a family. Both processes can be detected in the fossil record (see Figure 4-5).

EXTINCTION AS SUCCESS

In a sense, every species becomes extinct when it evolves into something else. This would be extinction as a necessary concomitant of success. In a world that is constantly changing, no gene pool can remain adequate for very long; that is, no population of organisms can continue to exist and not evolve. After a gene pool has become significantly modified from what it was at an earlier time, the chances are that the populations at the two times being compared would be considered to be different species if they were to be found existing together at a single time and place (see Chapter 14). Therefore, a species can never be successful in the sense of avoiding extinction. What can be successful in this sense is the lineage. The lineage eventually branches (speciation during divergent evolution) and produces more lineages, each of which may branch again and again. Some of the branches may continue for very long periods of time without producing many new lineages. Others may quickly be superseded by many descendant lineages, becoming successfully extinct as distinct entities. Still others may become extinct as lineages, having failed to evolve and leaving no descendants. The latter phenomenon has especially interested evolutionary biologists and has provided tinder for many theoretical flames. It is really this which is referred to in discussions of extinction, and we too use the term extinction to refer to the unsuccessful evolutionary event—the disappearance of a lineage from the record.

EXTINCTION AS FAILURE

As mentioned above, no major phylum seems to have become extinct. Species always become extinct (in the special sense discussed above). Therefore, we are concerned with taxonomic categories ranging from genus to class when discussing extinction. The major problem here is to understand why some lineages do not continue to evolve and others do, or why some species cannot respond adequately, by evolving, to the challenge of an ecological catastrophe (see Chapter 3). This involves factors both intrinsic and extrinsic to the organisms involved. For example, if the ecological catastrophe is a geophysical event of large magnitude that occurs at an extremely fast rate, that would be an extrinsic causative factor. The event must be of large magnitude from the point of view of the organism in question. Thus, if the temperature on an elephant's back changes from 90 to 70°F in the period of an hour after sunset, that would be a rapid change, but its magnitude is not sufficient to affect the elephant or its species. On the other hand, the change has to occur too fast for the organisms to respond by evolving. If the mean temperature of an environment changes significantly during 500 generations, there may be adequate time to adapt. If the same magnitude of mean temperature change occurs in five generations, there may not be enough time to adapt (time for sufficient numbers of new genetic recombinations to be produced and to be tested by the environment; see Chapters 6 and 7). Needless to say, the effect must also cover sufficient area, so that organisms that can be adversely affected by it cannot simply migrate by gradual change of the species ranges. If an event of this kind is of sufficient magnitude, as, for instance, a glaciation would be, the time involved is not critical if the organisms cannot disperse from the region being affected. An archetypal catastrophic event of large magnitude and fast rate would be the volcanic explosion that blew the biosphere off the island of Krakatau in 1883. Nothing could evolve fast enough to keep pace with the environmental change in this case, and some endemic species became extinct.

INTRINSIC FACTORS IN EXTINCTION

During catastrophic environmental challenges of lesser intensity than that which occurred on Krakatau, factors intrinsic to the organisms play greater roles in extinction. The rate at which an ecological catastrophe occurs in terms of generation time can be increased without causing extinction if the number of females produced per female per unit time (intrinsic rate of natural increase, r) is increased; that is, other factors being equal, species that produce more offspring per unit time can survive faster rates of catastrophic change in the environment than can those producing fewer offspring per unit time. This is so because they can generate more new genetic recombinants per unit time (see Chapters 6 and 7). r can be increased, for example, by increasing clutch size or by decreasing generation time. r is only one factor that determines whether an environmental change is or is not too fast. As mentioned above, the ability to disperse is critical in escaping extinction in organisms that cannot adapt to new conditions. The seeds of plants frequently show adaptations related to dispersion—burrs, wings, parachutes, and the like. The larval forms of sessile animals often show similar adaptations to being carried about in water currents. Other kinds of animals, of course, are mobile. But a population of water fleas, for example, could probably not escape a catastrophic environmental event by voluntary movements, because they are small and physically restricted to a given lake or watershed.

Ecological catastrophes involving rapid changes in the biotic (rather than in the climatic-physiographic) environment, such as the invasion of a community by a species from another biome, bring into play various intrinsic factors that could cause extinction. The latter may result if any of the species present in the community cannot adapt to the presence of the new species (for instance, by undergoing character displacement; see Chapters 3 and 6). This topic can be taken further if we now consider some specific examples of extinction.

Man's rise to ecological dominance in most parts of the world has resulted, and will probably continue to result, in the extinction of numerous species of plants and animals. These are of interest here in that we know more of the details of their extinctions than of those which occurred in past eras. A convenient example is that of the heath hen, a subspecies of the prairie chicken of North America. This was a very common bird in shrubby blueberry barrens in New England when the first Europeans arrived. It is said that they were the original Thanksgiving turkey. They were hunted extensively throughout the seventeenth and eighteenth centuries, and their habitat was destroyed by cultivation throughout most of their range. In the 1880s the government of Massachusetts, seeing that the species was imminently threatened with extinction, began protecting the few left on the island of Martha's Vineyard. Their numbers were originally small, but they began to increase and apparently built up a population close to the carrying capacity of their new environment. Then a severe fire during the nesting season of 1916 decimated their numbers, followed by an unusual influx of goshawks the next season, after which there were fewer than 100 left. The few that survived these catastrophes were unable to rebuild the population, and the last one was seen in 1932.

The primary and fundamental (ultimate) ecological catastrophe in this case was the arrival of the Europeans. The most important aspect of this event was probably not the increased predation on the species by the Europeans. Most organisms produce more offspring than can survive, given the finite carrying capacities of all environments. Natural selection weeds out those that are least fit to survive in an environment until the total number of survivors equals the carrying capacity. Increased predation simply means that more offspring per generation have a chance to survive until they too are harvested by a predator or destroyed by other agencies; that is, all other factors being equal, a chick entering a community with its conspecific population at only half of its carrying capacity has twice as much chance to survive (reproduce) as one entering a community saturated with conspecific individuals. In the former case there is less direct intraspecific competition and also probably fewer kinds of predators other than man because it is less probable that a nonparasitic predator will hunt and uncommon prey than a common one. The result of increased predation, after an initial decrease in the number of adults followed by an increase back to carrying capacity, would be to increase the turnover of individuals, and to decrease selection based on factors other than the predation in question. These are the results if the new predation is not of such extreme magnitude as to destroy more individuals than can be replaced each generation. This actually happened to the passenger pigeon. This bird was hunted on so large a scale that special railroad trains were enlisted to follow its breeding season northward each year in the Mississippi Valley, carrying the hunters and their families. Some kinds of animals might have been able to respond to this situation by evolving greater r's, but in birds this is difficult because the parents have to be able to feed their young as well as simply producing the eggs (see Chapters 6 and 12). This point emphasizes an intrinsic factor that in birds may be relevant to extinction. Probably the most important aspect of the arrival of the Europeans on the ecology of the heath hen was the widespread destruction of its habitat and food sources. This factor seems to be mainly responsible the world over for extinctions caused by man's activity. Cultivation, urbanization, and environmental pollution are the three major means by which habitats are destroyed by man.

The example of the heath hen has one further point to contribute to the discussion of extinction. That is the concept of critical population size. After the birds were protected on Martha's Vineyard, they did quite well as long as no major disaster threatened them. The first bout of real adversity wiped them out. It is inconceivable that the species had not met this kind of event repeatedly during its history on the mainland, because its major habitat was a disclimax caused mainly by fire. After all, no year is average. Every year is unusual in some respect, and many years are bad ones for any given species—too dry, too cold in spring, too much snow, and the like. On the mainland the birds were able to survive periodic catastrophes and yet they could not do so on Martha's Vineyard. Because they were native there and were able to increase in number rather rapidly, we cannot assume that in some way Martha's Vineyard was a community to which they were not adequately adapted. During catastrophes and bad seasons for any species anywhere, an unusually large number of individuals is killed, and this represents a passage of very intense selection, because, usually, or at least often, some individuals are better able to respond to the relevant selective agents than are others in the population (see Chapter 6). In order for populations to recover from these episodes of very intense selection, a certain minimum population is required. This critical size is a function of r. For species with a small r a larger number of individuals is required to regain carrying capacity numbers before a new catastrophe overtakes the population, causing a further large reduction in number, than would be required of species with a high r. There probably was simply not a large enough carrying capacity for heath hens on Martha's Vineyard to allow the population to survive two bad years in a row. They might have survived one such year but did not survive two. This is simply a matter of having enough reserve individuals to obtain a sufficient increase in number when the population can begin breeding again. Perhaps very few species have population sizes that would allow them to survive five to ten severe years in a row. Fortunately this contingency is undoubtedly very rare.

The actual agents of destruction of the few remaining survivors of an overly severe catastrophe are not important. There are probably a certain number of accidental deaths every year in a population. Add to this the continuing predation from predators that have not been affected by the climatic catastrophe, and the probability of survival becomes very low. The probability of survival of any given individual is less than that of one in a group of 10 individuals, and this is less than that of any 1 out of 100. When the probability of survival of a species is reduced to the probability of survival of certain specific few individuals, the probability of extinction becomes very high.

Small populations have other problems. They cannot generate much genetic variability per unit time by sexual recombination, because there are so few individuals available to contribute their genes to the gene pool of the next generation. The fewer individuals there are, the fewer different alleles there would be given at any genetic locus (see Chapters 7 and 13). In the case of the heath hen, man's agricultural activity reduced the population below the critical size. After this it was simply a matter of time before extinction overtook them. What remain of interest are the factors that contribute to the reduction of populations below critical size—in other words, the ultimate, rather than the proximate, reasons for extinction—that is, the ecological catastrophes themselves.

EXTRINSIC FACTORS IN EXTINCTION

A great deal of thought has been devoted to the possibility that there may be periodic occurrences of single factors that result in widespread extinctions among many different groups of organisms. This possibility arises from the appearance in the fossil record of a few relatively short-term episodes of widespread extinction. One such time was at the end of the Cambrian period, about 500 million years ago, when about two thirds of all the trilobite families in existence till then disappeared relatively suddenly from the record. During the Permian period, some 260 to 230 million years ago, there was another such episode. The last of the trilobites disappeared then, as well as many other families of marine organisms, including many corals, brachiopods, and vertebrate fishes. Many kinds of reptiles and amphibians became extinct on land during the same time as well. At the end of the Cretaceous period, about 80 million years ago, we find the famous extinction of the dinosaurs, accompanied by the extinctions of many kinds of fishes and ocean-living invertebrates, such as sponges.

Because extinction does occur all through the fossil record as well, it is worthwhile questioning the validity of these data. For instance, if the upper Cretaceous is defined as that place in the fossil record where dinosaurs are last found, it need not represent the same actual time in the past at different localities. Thus, the dinosaurs could have lasted millions of more years on one continent than another, and we might not detect this lack of correlation between the sediments of the two continents. It is reassuring to know that paleontologists themselves worry about these questions and that the information presented does seem to be valid. This has been arrived at from two approaches. First, each paleontologist is a specialist in a certain group. Each makes his own correlation between his special fossils and geologic time. When later put together, the data cluster, as indicated above, giving three major periods of extinction. Second, radioactive dating has done much to increase our confidence in correlations between strata found on different continents. Thus, there do seem to have been periods of widespread simultaneous extinctions in the past.

The question then arises as to whether simultaneous extinctions could have had some single ultimate cause. A number of possibilities have been suggested, such as radiation storms from solar flares or other galactic sources; widespread relatively drastic changes in climate, as in glaciations; widespread orogenic activity that could simultaneously build new mountain chains and eliminate shallow seas by making them deeper. None of these, however, is capable of explaining all the extinctions of any particular episode. Nor is any of them, unaided by further forces, capable of causing the extinction of even one species. Two questions always arise in connection with such sweeping explanations. First, why did not *all* living forms become extinct during an episode of, say, radiation storms? In other words, we still have to explain why genus X and not genus Y became extinct at a particular time. Why did some lineages of sponges survive into the Cenozoic era while others did not? Second, in the face of some large ecological catastrophe, why was genus X unable to evolve (adapt)?

The example of the ruling reptiles, including the dinosaurs, is interesting. They were a varied group, ranging in size from that of a chicken to that of a small whale; they ranged ecologically from specialized herbivores to carnivores, from fully aquatic to terrestrial; some were quadrupeds, others bipeds; some were slow and armored, others naked and swift; some glided in the air. They had entered and long occupied a large number of adaptive zones. This varied group became extinct during a period of about 10 million years. Put another way, they left no descendants in the Cenozoic regardless of how long their extinction took; their lineages stop in the upper Cretaceous. A number of possibilities have been suggested to account for their demise as a group—mammals increasingly preying on their eggs, disease, change in climate, change in vegetation, and the like. None of these specific agents, however, could have plausibly caused the extinction of so many varied forms. Presumably an ichthyosaur in the ocean could not be attacked by the same disease organism that attacked a carnosaur; climatic changes would not be equally drastic everywhere; live bearers would not be subject to egg predation; and so on.

The upper Cretaceous was a period of high orogenic activity. The Rocky Mountains, the Andes, and the Himalayas date from this time. As the mountains rose, the floors of the oceans sank, eliminating most of the shallow seas. Large areas of the terrestrial regions of the earth became both colder and drier. In response to this, the vegetation changed drastically by the spread of a more xeric angiosperm flora, with a concomitant decrease in the importance of gymnosperms and ferns. Any of these simultaneous events could have influenced the extinction of a given species, and some of them may have had more widespread effects. For example, living reptiles are not homeotherms; that is, they regulate their temperature behaviorally by moving into desirable ambient temperatures, not by conserving metabolic heat as do birds and mammals. If dinosaurs were also heterotherms—we do not know—then changes in temperatures would have affected them all. Some of them, however, could simply have migrated away from regions of uplift and remained on the tropical coasts. In a particular case we might be able to explain why such dispersion was not possible, but it is not likely to be the same explanation for each one of the varied dinosaurs. Why did some not evolve homeothermy? Again, the change in vegetation would have had important effects on the herbivorous forms—but why did they not evolve new adaptations to cope with coarser food?

Thus, there is the possibility that widespread climatic changes were the general ultimate cause of the extinction of the dinosaurs as a group, even though the spectrum of proximate stresses on each species was not the same. In one case temperature changes coupled with low dispersability may have been the major cause of failure; in another the changed vegetation; in still another the loss of large schools of shallow-water food fishes may have been crucial. If one focuses on specific proximate causes—unknown for the most part—the dinosaurs as a group did not become extinct, rather many separate species that happened to be dinosaurs became extinct. On the other hand, it may be equally true at a more abstract level that the rapid climatic changes during the late Cretaceous, spurred largely by orogenic activity, were a sufficient general ultimate cause for the destruction of the dinosaurs as a group.

But, again, why was none of them able to adapt to the changed conditions? For example, one group of large reptiles that became extinct shortly after the dinosaurs was the crocodilelike (or, more precisely, gavial-like) champsosaur lineage. Why should they not have survived, although the crocodiles did? Indeed, why were the crocodiles able to adapt while the semiaquatic dinosaurs (in quite different adaptive zones, it must be said) were not? Clearly the problems were not generalized reptile problems but were rather special problems faced by many different dinosaurs and champsosaurs. If this is so, some of the possible causes listed above, such as heterothermy in connection with cooling climates, could not have been more than ancillary factors, because all reptiles would have faced the same problems at that time. Although the dinosaurs were simultaneously faced with a whole spectrum of different challenges, so were other reptiles. But for the dinosaurs and champsosaurs these added up to too much environmental change too fast; for the crocodiles, turtles, and lizards this was not so. (It should be pointed out that some species and other taxa of surviving reptilian groups did become extinct in the late Cretaceous, such as the giant marine lizards, the mosasaurs. But all the species of reptiles that became extinct during the late Cretaceous include in their number *all* the then living lineages of dinosaurs.)

An interesting extinction problem is presented by biotal replacements, in which whole groups replace other whole groups in the fossil record. One such case has been presented by G. G. Simpson, that of the virtual replacement of the perissodactyls (living forms—horses, tapirs, and rhinoceri) and other ungulates by the ruminant artiodactyls (living forms—deer, antelopes, giraffes, camels and llamas, pronghorns, goats, sheep, and cattle). There is considerable overlap in the adaptive zones (generalized ecological niches) involved in this comparison. The adaptive zone of horses, for example, is very close to that of deer, antelopes, and cattle. One could characterize in a very oversimplified way an adaptive zone including all these forms as follows: swift, herd-living grazers or browsers (primary consumers) of small to intermediate size, with precocious young, that form the staple food of the larger kinds of cats and dogs. The adaptive zones of horses and cattle are similar enough so that one could imagine some species of each kind competing with species of the other (because they had significant overlap in their ecological niches) should an ecological catastrophe bring them into contact. Figure 4-6 shows how the number of different kinds (genera) of the artiodactyls increased gradually while the number of genera of perissodactyls decreased. Did artiodactyl species on the average outcompete perissodactyl species more often after a number of ecological catastrophes? Or did most of the perissodactyls become extinct for various other reasons, primarily, leaving vacant adaptive zones in which the artiodactyls were quick to construct new ecological niches? There is in fact no way to choose between these alternatives.

In either case, the artiodactyls would be considered somehow adaptively superior to the perissodactyls. It is difficult to see which of the seemingly minor differences between them might have been crucial. Perhaps the cloven hoof of the artiodactyls has more evolutionary potential or can assume more forms and functions than can the large middle hoof of the perissodactyls (see Figure 4-7). For instance, the cloven hoof must be important in obtaining purchase in climbing in such forms as mountain goats; a single middle toe would probably not function well in this situation. In addition, the astragalus (a bone in the ankle) of artiodactyls can function as a double pulley, being attached, not only to the tibia (as in perissodactyls), but to other ankle bones as well, thereby allowing a greater degree of flexion and extension in the hind limb than is found in perissodactyls, allowing in turn remarkable abilities for leaping. The artiodactyls other than the pigs (that is, the ruminants) also have the ability to

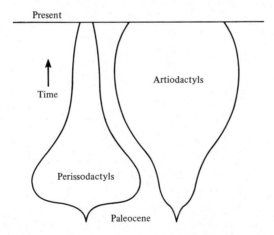

Figure 4-6   The faunal replacement of the perissodactyls by the artiodactyls. The width of each figure roughly represents the number of genera present at a given horizon in time. (Redrawn from Simpson, 1953.)

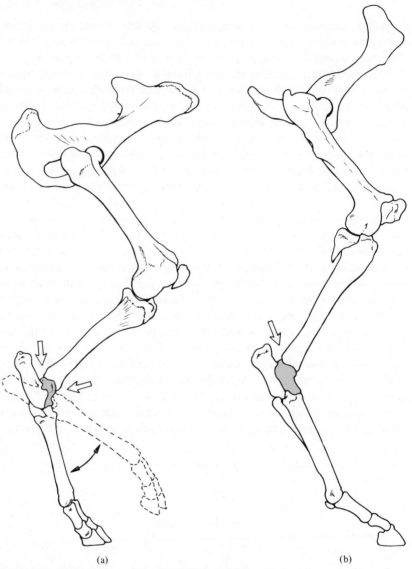

(a)   (b)

Figure 4-7 Comparison of the hind-limbs of (a) artiodactyls, and (b) perissodactyls. The astragalus is shaded. The arrows point to moveable joints associated
with the astragalus (two in artiodactyl, one in perissodactyls). Note the double
hoof of the artiodactyls as opposed to the single one of perissodactyls.

gobble food very rapidly, store it in the rumen, regurgitate it, and chew
it later. Perhaps this gives ruminant individuals a slight edge in that they
can feed away from the food source after having escaped some predator
and can, therefore, remain in hiding more of the time than could a
perissodactyl.

Simpson has discussed still another faunal replacement in which it is perhaps possible to choose between whether one group first became extinct, leaving vacant potential niches, or whether there was more step by step direct interspecific competition. Figure 4-8 shows the geographic origins of the genera of mammals in North and South America before and after the establishment of the land bridge between the two in the Pliocene. North American mammals were more successful at leaving descendants in South America than vice versa. The establishment of the land bridge initiated a classic ecological catastrophe, with different biotas meeting and interacting. It is conceivable in this case that interspecific competition may have played an important role, because there is no indication that families endemic to South America were in any kind of trouble before the establishment of the land bridge.

This example affords an opportunity to point out an important feature of natural selection. The best organism or individual at a particular time is the best in respect to the other organisms or individuals in the same population or community. This is a relative concept. The best is not some designer's absolutely best model; it is the best that happens to be at hand. If something better arrives, the previous best is no longer the best; or if the best is lost by chance, the next best becomes the best. As long as no descendants of North American mammals were at hand, various forms in South America—rhinoceroslike toxodonts, horselike litopterns, elephantlike pyrotheres, and others—were adequately adapted to their environments. Adaptation is completely opportunistic and only produces organisms capable of producing other organisms in a given environment. As the environment changes, a lineage either continues to produce organisms capable of producing more organisms or becomes extinct. If the environmental change is relatively fast (catastrophic), more lineages will not be able to continue producing adequately adapted individuals than when environmental change is slower. If the rapid environmental change involves the invasion of a community by foreign species expanding their range, then

| | South America | | North America | |
	Autoch-thonous	From North America	Holarctic	From South America
Recent	16	14	20	3
Pleistocene	23	13	26	8
Mid-Pliocene	24	1	26	1
Mid-Miocene	23	0	27	0

Figure 4-8 Origins of families of land mammals present in North and South America at various times in the geological record. (Data from Simpson, 1953.)

some of the specific reasons for failure to produce enough adapted individuals on the part of some species will be related to interspecific competition (see Chapter 3).

There are also examples in which it is clear that there was not any interspecific competition involved in the faunal replacement. One group of reptiles became adapted to living in the oceans—the ichthyosaurs. They were fishlike in shape and are known to have been live bearers because a fossil has been found with a smaller one inside, more or less in position for being born. Their tooth row indicates that they were fish eaters, and their streamlined shape indicates that they were fast-moving pelagic forms. These facts, and their general size, indicate that they occupied an adaptive zone close to that of modern toothed whales and porpoises. The ichthyosaurs became extinct with the dinosaurs at the end of the Cretaceous, but the cetaceans did not appear on the scene until some 20 million years later. As another example, during the Triassic there was a dominant group of crocodilelike reptiles, the phytosaurs, that apparently became extinct just before the appearance of crocodiles in the Jurassic (see Figure 4-9). In both of these examples the antecedent group must have vacated its niches,

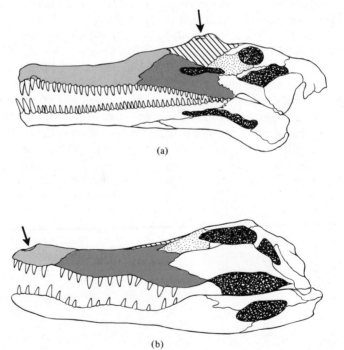

(a)

(b)

Figure 4-9 Comparison of the skulls of (a) a phytosaur, and (b) a crocodile, showing slightly different arrangement of the bones in relation to the different location of the external nares. The postcranial skeletons of these two groups are also very similar.

which then remained more or less unfilled, until by chance some new group evolved characteristics that preadapted them (see below) to enter the vacant adaptive zones in question.

To sum up on extinction, this occurs to a lineage only in the sense that it occurs to each species in it separately. Whole groups become extinct when all the lineages in them separately become extinct. Each lineage or species that becomes extinct must be dealt with on its own terms; general explanations for the extinction of whole groups (or many different groups) are weak in that exceptions can always be found. On the other hand, there are periods of very rapid environmental change that place numerous hurdles simultaneously in the paths of many evolving lineages, so that the number of separate extinctions during these periods is unusually high. These general ultimate causes of extinction operate by generating numbers of more specific proximate causes, some of which operate on some species, others of which operate on others. There is no way to predict which lineage will become extinct and which will continue to survive. This has something to do with intrinsic factors in the gene pools of the species themselves. Those which become extinct do so because their numbers become reduced below some critical level, after which they cannot survive ordinary environmental vicissitudes. Their numbers become reduced below the critical level because for one reason or another they were unable to generate enough genetic variability per unit time so that they could evolve instead of becoming extinct. There is an element of chance involved as well, in that the particular ecological catastrophe that presents itself cannot be anticipated by the gene pool—or else it would not be a catastrophe. The rate of environmental change (relative to generation time, clutch size, and chromosome number of the species involved) is important. There are probably rates of environmental change that would be catastrophic for any species. But rates of environmental change in the past were such as to allow some lineages to survive while others did not. Each specific case is probably forever lost to analysis, because what are needed are details (such as those presented for the barnacles in Chapter 3) that cannot usually be reconstructed with any confidence. Thus, specific examples of extinction usually cannot be explained; for groups it is often possible to suggest some ultimate cause if allowance is made for significant exceptions; in the most general sense, extinction is due to the unpredictability of environmental change.

ECOLOGICAL DETERMINISM

The concepts of vacant adaptive zones and unfilled ecological niches referred to above in connection with ichthyosaurs and cetaceans are peculiar. In a sense, there is no ecological niche without a species occupying it, because the niche *is* the interaction of the members of the species with their environment. It is clear, however, that in most places where there are trees, there are potential niches for squirrels; wherever there are minnows, there are potential niches for pikes; wherever large animals come to drink, an adaptive zone is available for crocodiles. Just as clearly, not every drainage system has either pikelike fishes or crocodilelike tetrapods; not every continent has diurnal squirrel-like mammals. Everything that seems possible does not necessarily happen. More interesting perhaps, many things that seem possible, that is, predictable, do happen.

Perhaps the most dramatic way to demonstrate the possibility of unfilled ecological niches is to compare the biotas of different land masses isolated from each other by large distances of ocean. For example, one can compare the mammalian faunas of Australia and Eurasia. Australia has been isolated from the Eurasian land mass for some 150 million years. During this time mammalian faunas evolved independently on both continents with the exception only of bats, which found their way from Eurasia to Australia after they have evolved flight, as did also a group of rodents (some aquatic). The endemic mammalian fauna of Australia consists (with the exception of these rodents, bats, and monotremes) wholly of marsupials, a group that has evidently become extinct on Eurasia. In order to facilitate a rough comparison of the adaptive zones represented in these two faunas we can refer to those represented on Eurasia and elsewhere as placental mammals, and enquire whether any such zones have evolved on Australia, and if so, which ones.

Looking at the Australian fauna in this way, we find the ecological equivalents of shrews, moles, civets or mustelids (including a wolverine type), a small cat, a dog or wolf, an anteater, a raccoon, a slothlike form, woodchucks, squirrels (but nocturnal instead of diurnal), flying squirrels, nocturnal lemurs, and large to medium sized herd-living herbivores. In some of these cases the anatomical structure is very close to that of the equivalent eutherian, as in the case of the woodchucks and the Australian wombats (see Figure 4-10). In other cases the anatomical differences are striking, for example, those found on comparing ungulate artiodactyls with kangaroos. Missing from the Australian marsupials are a large number of adaptive zones found in Eurasia: for a few examples, pigs; large cats; large thick-skinned herbivores, such as elephants and rhinoceri (although the extinct diprotodonts were rather rhinoceroslike in general habitus); anthropoids; aquatic forms, such as otters and beavers. At the same time there are a few adaptive zones found among the Australian marsupials that are not found elsewhere, such as a flightless flower (nectar, pollen, insect) feeder, the honey possum.

For another example, the Hawaiian Islands have an endemic family of birds, the sicklebills or honeycreepers. These have diverged into a number of different adaptive zones roughly matching those of woodpeckers, finches, thrushes, wrens, parakeets, and nectar feeders. Other kinds of birds

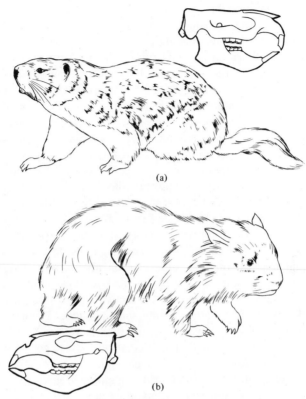

(a)

(b)

Figure 4-10   Comparison of (a) a woodchuck (a sciuromorph rodent, placental mammal), and (b) a wombat (a marsupial mammal).

are present on the islands and fill many other adaptive zones, and so the example is not as clear as the first. Even so, some bird zones are not represented at all on the islands—owls, gallinaceous birds, and others.

Looking at the forms that we do find present on different land masses, we might well be struck by a sense of evolutionary determinism. It seems that given a mammalian gene pool and a great deal of time, a woodchuck-like animal will ultimately evolve with a high probability. It is as if this adaptive zone were somehow implicit in the structure of nature, having been approximated on all continents by marsupials, rodents, hyraxes, and lagomorphs and, in the past, by multituberculates. There seems to be a mammalian anteater on every landmass (in tropical regions only), except the smallest ones. This exception is interesting and may have a great deal to do with the fact that there are no owl-like birds on the Hawaiian Islands. (1) Small land masses do not have as much variety of habitat as do large ones. (2) The total biomass of a small land area may be so little (have such low total productivity) as to limit the length of food chains, thereby eliminating the possibility of tertiary and quaternary consumers. It has been calculated that the amount of energy that it is possible for a consumer to store in its own biomass is about 10 to 15 percent of the energy trapped in its prey. If the total biomass produced by plants is less than a certain amount because there are not enough plants because there is not enough place for them to grow because the land mass is too small, then there cannot be, say, any predators or carnivores. (3) Another consideration is that small land masses frequently are new or recently formed, and so the requisite amount of time has not been available for either colonization or divergent evolution to occur. A related factor for very small islands is the distance from a source of colonizing species—the further away, the smaller the chances of successful rafting. For these three reasons, only the faunas of large continents can be expected as a rule to display the evolutionary determinism referred to here (see Figure 4-11).

We must now look at the other side of the picture. There are no large cat equivalents in Australia. Such animals do occur on all the other large land masses. But the large cats on all the other large land masses are all felids, presumably descendants of a single lineage that originated in Eurasia, and so we cannot test the possibility of the independent origin of catlike forms from other mammalian lineages. Presumably the ability of the cats to disperse was so great that there has been no chance for other catlike forms to originate. We can, however, get an answer here by looking into the fossil record. True cats did not enter South America until the establishment of the land bridge mentioned above. Before this time there were in fact numerous carnivorous animals on this continent—they were all marsupials and they included forms that were catlike. One of these, *Thylacosmilus* (see Figure 4-12), was a very good duplication of

	Africa	*Neotropica*	*Australia*
Small insect eaters and predators	9.1	2.2*	9.3
Fossorial moles	1.7*	Nil	0.3
Specialized ant and termite feeders	0.8	0.4*	1.1
Small to medium-sized omnivores	3.7	3.8	5.5
Rabbit-sized herbivores	4.0*	2.1	1.9
Medium to large terrestrial herbivores	12.6	2.8*	8.8
Arboreal herbivores, omnivores, and insectivores	11.4	13.1	11.0
Carnivores, weasel to fox size	4.8	4.0	1.9*
Large carnivores and scavengers	0.9*	0.04	0.03

Figure 4-11 Comparisons of the mammalian faunas of three continents. The figures are percentages of the total mammalian faunas. Asterisks indicate where a continent is most different from the others. Only those mammals are used which probably evolved their adaptive zones in parallel, independently on each continent. (Data from Keast, 1969.)

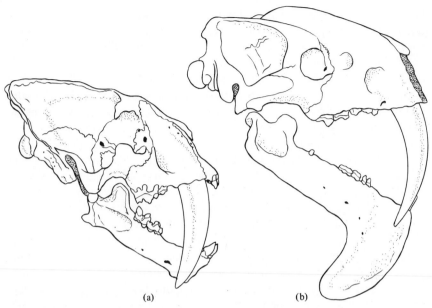

Figure 4-12 Comparison of the skulls of two sabertooth carnivores: (a) *Smilodon,* an Upper Pliocene eutherian from North and South America, and (b) *Thylacosmilus,* a Pliocene marsupial from South America. (Simplified from Romer, 1966)

a sabertooth tiger, and no doubt occupied exactly the same adaptive zone as the latter—that of killing large, thick-skinned herbivores by stabbing. Thus we find the independent origin of big cats on two isolated large continents but not on a third, Australia. If we can explain why this origin did not happen on Australia we can retain our hypothesis of ecological determinism.

Big cats are adapted to hunting big herbivores. The latter were present on all the continents where big cats, felid or otherwise, lived. There are no really large herbivores on Australia. The largest kangaroos weigh about the same as a man or a small antelope; of the 41 living species of kangaroo, only 6 are not smaller than most antelopes. During the Pleistocene, however, there were both giant kangaroos and giant wombats, which, together with the rhinoceros-sized diprotodonts, would have made up an ample large herbivore biomass for large catlike carnivores. The question then shifts to why there is not today present an Australian marsupial pachyderm, or rhinoceros, or even an eland. At this point we can return to the strictures mentioned above concerning the size of the land mass. Australia is something less than half as big as South America (3 million square miles as opposed to 7 million). In addition, about half of this is relatively recent desert, to which the animals have yet to adapt. The extinction of the large Pleistocene herbivores on this continent may well reflect this recent restriction in the size or mass of the total biosphere. Is it possible that there is a relationship between total biomass and the detailed structure of that part of it represented by primary consumers? If the biomass of the producers is less than a certain amount, is it unlikely that a very large herbivore will be able to survive? Perhaps there is some optimum number of types of primary consumers for each level of producer biomass, preventing the occurrence of a large animal at lower levels because too much primary consumer biomass would be tied up in a single species. The giant tortoises found on the Galapagos and on the island of Aldabra near the Seychelles Islands probably are about as large as a primary consumer can be on a small island. It may be that their great size (nowhere near that of a rhinoceros) is possible only because of a relatively low metabolic rate coupled with a low intrinsic rate of natural increase. A more difficult problem is presented by the 10-foot predaceous dragon lizards in the Lesser Sunda Islands in the East Indies. These are certainly relics from a time when all the islands were connected to the mainland. The large animals, including carnivores, that occupied this region then have become extinct on these islands, except for this huge lizard. At present one can only suggest that certain peculiarities of energy flow on some of these smaller islands allow this apparently oversized predator to exist in them. Perhaps it involves the presence of wild pigs introduced by man to these islands. It is clear that we cannot at present answer these questions, but if we could, we could perhaps also deal successfully with each other case in turn. If this were true, the principle of ecological determinism could be established; at present it remains a suspicion, an interesting hypothesis.

Certainly, we must allow some leeway for the operation of chance in these matters. For instance, why did a nonflying flower-feeding mammal evolve on Australia but not elsewhere? In birds, flower feeders are common (see below), and some bats and many insects feed this way. This adaptive zone must be potential everywhere in the tropics. It may not, however, be a very probable direction for the evolution of a mammalian lineage in that, in the absence of flight, feeding in flowers is not as easy or safe, given a mammalian structure, as feeding elsewhere, and so only in the absence of other forms more suited to this way of life do we find a mammal evolving it.

Citing cases of unusual adaptive zones can be entertaining, but the main point is that we can cite them. Are they due to concatenations of rare events or are they better interpreted as having been the inevitable or high-probability products of the peculiar structure of given biomes? Low-probability (or chance) events would play a larger role in small land masses, such as islands, or other isolated habitats, such as some stream drainage systems or mountain lakes. This is so because the choice of colonists has a large random component, so that, given only a few tries, the distribution of colonizing organisms is less likely to reflect the total probability distribution of potential colonizing organisms as well as more tries would; and the smaller the isolated community, the fewer the number of tries. The specific organisms that colonize newly formed islands are determined partly by low-probability events—by chance. Thus, not every island has a large marine lizard, such as that found on the Galapagos; indeed, such an organism is not found anywhere else in the world. This is surely not because there were no potential niches of this kind available elsewhere, or no potentially ancestral lizards. The presence of marine iguanas on the Galapagos is probably best viewed as a chance phenomenon in that (1) an ancestral potentially herbivorous lizard was carried by unpredictable events to the islands, (2) there happened by chance to be available a large biomass of easily accessible seaweed for food, and (3) there happened by chance to be no major shallow-water predators. No doubt a host of other biological and nonbiological factors could be cited in this case. Chance events establish that a species with certain evolutionary potentialities should arrive in a specific habitat that is capable of eliciting certain responses from its gene pool. The actual evolutionary steps involved are not, of course, indeterminate or chance phenomena in that they are susceptible to causal explanation.

There is, then, enough evidence to warrant formulation of the following statement in the form of an hypothesis: If different members of the same basic stock (similar gene pools) arrive in different but similar isolated biomes—or continents with similar biomes—they will, over a long period of independent divergent evolution, produce roughly the same number and kinds of adaptive zones. The continents, or a number of connected biomes, must be about the same size, and the length of time involved must be of the same order of magnitude. Predictability decreases as the size of the unit of connected biomes or the time involved decreases. This we can take as a statement of the hypothesis of ecological determinism. One aspect that we have not yet discussed is the matter of similar gene pools and just how similar they need be (see also the discussion of the evolution of the mammalian jaw joint in Chapter 5).

PARALLEL EVOLUTION: I

For the purposes of this chapter, a gene pool (see Chapter 6) is a collection of hereditary units isolated from other such collections. Each species has its own gene pool. The gene pools of two species of frog are more similar to each other than either is to that of a species of salamander; the gene pools of a species of frog and a species of salamander may be considered more similar to each other overall than either is to that of a teleost fish; the gene pools of teleost, frog, and salamander are more similar than any of them is to that of a lamprey; and so on. All mammals have similar gene pools in reference to a comparison with lizards or birds. An overall similarity of gene pools is taken to be reflected in taxonomic distinctions (see evolutionary rates in Chapter 5).

We may now ask whether in fact it is common for similar gene pools to evolve similar new adaptive zones. We have already explored the eutherian-marsupial situation in which it is clear that two different therian gene pools proved capable of generating similar adaptive zones in response to similar challenges from the environment. These two groups evolved in parallel, and they represent a fine example of parallel evolution. There are in fact many examples of parallel evolution. Two different groups of rodents evolved into porcupines, one in Africa (strictly terrestrial) and the other in the New World (but these are partly arboreal); five different lineages of birds evolved into nectar feeders with long, sickle-shaped bills and similar behavior—the American hummingbirds, the sunbirds of Eurasia, the honeyeaters of the paleotropics and Australia, the asities of Madagascar, and the Hawaiian honeycreepers; three groups of lizards independently evolved a manner of living in trees involving frictional foot pads and other anatomical similarities—the anoles of the New World tropics, the skinks of Oceania, and the Eurasian gekkos; two families of plants independently evolved round, leafless water-storing stems with spines—the euphorbias in South Africa and the cacti in the American deserts; two genera of orchids have evolved independently a whole series of identical body shapes, including long reedlike stems, bulbous water-storing stems (pseudobulbs), and no stems—*Epidendrum* in the New World and the Asian *Dendrobium;* two groups of carnivores evolved into seals. Examples could be multiplied into a huge list. In each of these cases there is no question about the basic similarity of the genome structures of the lineages involved, about the basic similarity of the environmental challenges, or about the fact that these were the results of independent evolutionary events on different land masses.

Do the environmental challenges involved have to occur during the same period of time? How similar could the starting gene pools be if the independent parallel development were started at different times? We have indicated that the crocodiles were preceded in their adaptive zone by the phytosaurs, and, we may add, these by some of the amphibian stereospondyls in the Triassic. The toothed cetaceans were preceded in their adaptive zones by the ichthyosaurs. Among fishes, succeeding dominant groups—paleoniscoids, holosteans, teleosts—each has produced a series of very similar adaptive types—the pikelike form, the flattened panfish or angelfish form, and so on (see Figure 4-13). It is clear that in these examples the evolutionary events occurred independently. There is, furthermore, strong suspicion that the environmental challenges involved were very similar if not identical. Thus, at any geological age a panfish shape is an adaptation to relatively static waters (lakes, lagoons) and to sudden

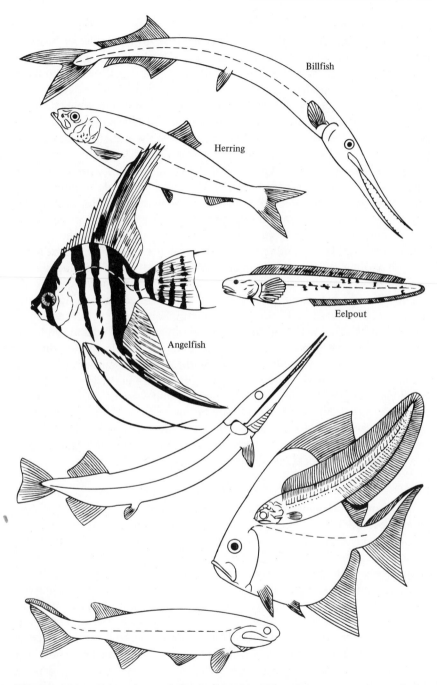

Billfish

Herring

Eelpout

Angelfish

Figure 4-13 Comparison of four adaptive types of teleosts (top) with four similar adaptive types of paleoniscoids. Four similar sorts of holosteans could have been included as well.

spurts of great speed, as when capturing prey or escaping the jaws of larger fish. Given a lineage of fishlike vertebrates living in lentic waters, we may predict that some kind of panfish will evolve. In this sense the gene pools of all "fishes" can be considered to be similar to each other. Mollusks or sponges living in static waters do not evolve panfish shapes. They have genomes that are very different from those of all vertebrate fishes. Looking at the example of the crocodile, it is clear that all tetrapods must be considered to have relatively similar gene pools from the present point of view; that is, they are capable of coding for, for example, a long snout with many conical teeth, short legs, flattened tails, large size, lurking aquatic habits, turreted eyes, raised nostrils, and the like. As a rule, genome structures seem to become dissimilar in our present sense at the level of the phylum or possibly subphylum or superclass, with exceptions, of course.

It may be noted that parallel evolution need not involve the production of exactly identical adaptations. Thus, in the crocodiles the need for breathing while the mouth is under water resulted in the development of a secondary palate connecting the anteriorly placed nostrils with the trachea. In the phytosaurs, no secondary palate formed; instead the nostrils were moved far back and raised up to just in front of the eyes (see Figure 4-9). In both cases selection has produced a crocodilelike form, and both could occupy very similar adaptive zones, but the details of how this was done differ. Again, a few of the New World monkeys adapted to arboreal life partly by developing prehensile tails; this did not occur in tailed forest-living monkeys in the Old World occupying quite similar adaptive zones. Details in parallel lines differ primarily because of the opportunistic nature of natural selection. The first answer to an ecological problem that presents itself will be seized on and worked up; the chances that the mutations in two lineages will be identical are almost nil (see also Chapter 7).

PARALLEL AND CONVERGENT EVOLUTION

We may now make an explicit comparison between parallel and convergent evolution (see Chapter 3). Parallel evolution typically involves the whole organism and its ecological niche; convergence involves restricted organ systems, chemical pathways, end products, specific physiological adaptations, neuronal pathways, and so on. Neither concept necessarily involves temporal simultaneity. Parallel evolution is the result of relatively similar gene pools adapting as a whole to similar entire environments, involving the entire structure, behavior, and ecology of the lineages involved. It should be noted that the similar adaptations involved do not need to be considered to result from the selection of identical alleles at identical genetic loci. As explored in greater detail in Chapters 7 and 8, identical polygenic traits (most continually varying traits) can be coded for by quite different alleles in different lineages; there is no one-to-one correspondence between primary gene products and complex multigenetic traits (or, between genotype and phenotype).

Convergent evolution is the production by very different genomes of some similar circumscribed "structure," often in response to similar challenges from identical environments—the eyes of squids and vertebrates—and almost certainly as a product of allelic combination at predominantly nonhomologous genetic loci. Thus, the lens crystallines of squids and vertebrates, although having similar overall tertiary structures, can be shown to have different primary structures; they are convergent on certain tertiary structures necessary to produce transparency. Convergence is the production of a few very similar adaptations in very distantly related lineages; parallelism is the entry into the same or very similar adaptive zones, with the generation of *many* similar adaptations, by fairly closely related lineages. The definition of convergence at the molecular level is perforce a bit different in detail (see Chapter 3), while parallelism at this level may not be a useful idea.

ADAPTIVE RADIATION

The entry of a lineage into a new adaptive zone has been shown by G. G. Simpson and others to be frequently followed by a tremendous burst of evolutionary activity referred to as an adaptive radiation. Paleontologists can measure this activity by the number of new families or genera produced by a lineage per unit time or by the number of existing families or genera in a lineage at particular periods of time (see Figure 4-14). Characteristically, a group (some species in parallel) acquires some new adaptation that effectively removes it from its ancestral adaptive zone—and also from its previous taxon. Its descendants gradually increase in number and diversity by divergent evolution for a period of, say, about 55 million years (a mean figure derived from Simpson's data in his 1953 book) and then, relatively suddenly, begin increasing in number and diversity at an explosive rate, reaching a crest of evolutionary activity and then subsiding.

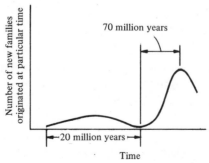

Figure 4-14 One way of plotting a curve showing the rate of evolution of a group of organisms. This curve is for the mammals as a whole; the period of time during which their major adaptive radiation took place was 70 million years. Comparative figures: agnathans—50 million years; placoderms—70 million years; paleoniscoid fishes—80 million years. (Data from Simpson, 1953.)

The explosive phase is characterized by a rapid production of great numbers of very diverse, often highly specialized forms. Examples of groups that have only recently undergone such adaptive radiations, so that the products of the crest are still with us and observable, are the beetles, the spiny-rayed teleost fishes, the characin fishes of South America, the rodents, the orchids, the drosophilid flies, the composites, the finches, the hylid frogs, the gekkonid lizards, and many others. This listing includes groups of different taxonomic rank; the products of an adaptive radiation may form a genus, a family, or a subclass at any moment in time. Over long periods of time some of these radiations may reach the rank of class by having had the numbers and diversity of their descendants further increased. The number of descendants alone is not of importance here; what is important is the number of different kinds of descendants (exploring somewhat different adaptive zones or subzones) in short, the diversity of the lineage.

An adaptive radiation is sometimes held to be a process by which a relatively generalized ancestral group gives rise to many relatively more specialized descendants. Many of the species present at the crest of the curve could fairly be called specialized in that their ecological niches restrict them to diets of single species or to uncommon habitats, such as beer mats or oil pools, with relatively little regulatory powers. These are said to have narrow niche breadth. It is not clear, however, that the ancestral group or species was not itself a very specialized member at the crest of an earlier adaptive radiation. In some cases this must have been true. Consider the ancestral bird; it would probably have been described at the time as a fairly specialized little dinosaur. In terms of anatomy, what might be specialized for a dinosaur might be a general trait for a bird, say, wings. One problem is that, with fossil material, it is almost never possible to discuss the breadth of ecological niches. We have to deal only with anatomical specialization, and this is always a relative concept (see Chapter 3) and need have no relationship to ecological specialization, which is the important factor in the evolutionary potential of the organisms in question. Probably the notion of increasing specialization accompanying an adaptive radiation is of restricted usefulness. It arises directly out of the commonly-used taxonomic definition of a generalized feature as one that is held in common by many or most species or groups in a taxon. In this definition "generalized" is more or less synonymous with the term "primitive."

ENTRY INTO NEW ADAPTIVE ZONES

An adaptive radiation can be considered to be a detailed exploration of different specific ways to engage in a new, previously unavailable mode of existence or major adaptive zone. Sometimes, entry or breakthrough into the new adaptive zone is facilitated or made possible by factors intrinsic to the organisms involved, such as the acquisition of some key trait, as in the development of wings in birds. During the history of the vertebrates there have been a number of such key acquisitions, among them the origin of jaws and paired fins in the Devonian and the origin of the terrestrial egg (amniotic, cleidoic egg) in some reptiles probably in the Permian. Each such acquisition was followed by significant adaptive radiations. Not all adaptive radiations, however, were sparked by the acquisition of single, major, key characteristics. For example the rapid radiation of teleost fishes in the Cretaceous cannot be traced to any single feature. They differ from the antecedent holosteans in a number of features, but only in degree or quantitatively. Thus, they have thinner, less bony scales combined with a more fully ossified backbone, making them much more flexible than the holosteans. The loss of dermal bone was important also in the head. Here, crucial bones in the upper jaw and cheek region were lost or decreased in size, allowing greater motility and flexibility of the jaws. The incredibly diverse jaw mechanism of modern teleosts based on a single basic structural plan demonstrates the evolutionary potential that was generated by this slight change in patterns of ossification (see Figure 4-15). Bobb Schaeffer, of the American Museum of Natural History, has suggested that the combination of a more flexible body and a more flexible mouth allowed the exploration of many new ecological niches unavailable to the holosteans.

Another revealing example is the transition from aquatic to terrestrial life by the vertebrates in the Devonian. In this case it can fairly be said that not a single new feature other than a slight behavioral change was needed to make the initial transition. The ancestral rhipidistian crossopterygians had stiffish fins (lobe fins) on which they probably walked on the bottom in shallow waters. They had fairly rigid backbones, allowing the strong sinusoidal flexation needed in this kind of progression in shallow water, and also allowing a fairly rigid suspension of the body off the ground. They had lungs, internal nostrils, and the habit of breathing air. Many fishes today are capable of breathing air. This is usually associated with living in stagnant waters—lungfish, electric eels—or with excursions out of water to catch insects—mudskippers—or with excursions to new waters under the pressure of population expansion—the walking perch of India. The rhipidistians also had a solid covering of bony scales that would have served to prevent too much water loss during short sojourns on land. The picture that we can conjure up of their lives as follows: They were fairly large predaceous fishes that lived in shallow, stagnant waters, perhaps in swampy regions. The lobe fins were probably adaptations for progression in very shallow waters where fishes of such size could hardly swim. We can picture them lurking about in weedy shallows, lifting their heads occasionally to breathe by lifting their snouts out of water, crawling slowly from

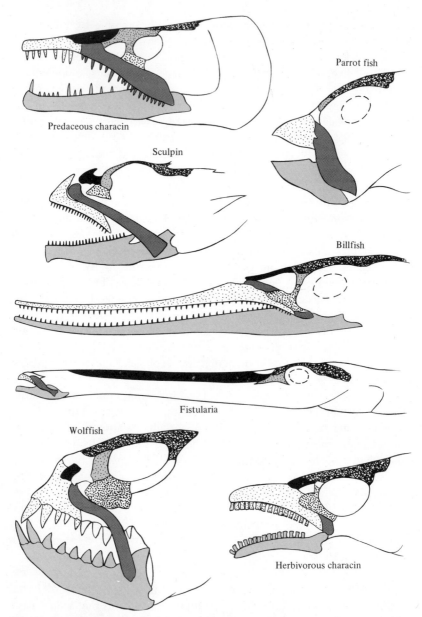

Figure 4-15   Comparisons of several jaw and skull bones of several different sorts of teleost fishes, showing a few of the many different feeding mechanisms evolved by this group. (Redrawn and simplified from Gregory, 1933.)

place to place, and occasionally dashing forward by means of a powerful lunge of the tail to grab some prey organism or slashing their heads sideways to snap up small prey. As youngsters they may even have snapped at large dragonflies perched on aerial leaves. One needs only to imagine some selection pressures driving them to come into shallower and shallower waters. One possibility is that they nested in the shallowest possible water, and finally some of them even in damp sites out of the water, as an adaptation to escape egg predation. In this case, the individuals who placed their eggs farthest out of the reach of aquatic egg predators may have left the most descendants. Whatever specific selection pressures we wish to imagine, the point is that the rhipidistians were already walking around breathing air long before they left the water. All that was needed for the initial move into a habitat associated with a new adaptive zone was a behavioral change. Once this occurred and was successful, selection began to overhaul the anatomy of the animals, converting them from what we call lobe-finned fishes to what we call amphibians.

PREADAPTATION

In the last example, the rhipidistian fishes, in the process of becoming adapted to life in shallow water, had, by chance, become preadapted to life on land. Preadaptation is one of the very important ways in which chance influences evolution. Low-probability events need not be considered to have been of crucial importance in the step-by-step evolution of any adaptation, say, lobe fins. Chance enters the picture when the organisms in question happen to be in a geographical, anatomical, or ecological position to begin in a small way, and probably accidentally at first, exploring a potentially new adaptive zone. One might argue that, because the rhipidistians were becoming adapted to shallow water, it was inevitable that they would encounter the shore, and so on. But the point here is that the particular adaptations developed by these fishes were what allowed them eventually to venture out on shore. Other forms, even other fishes, may have developed adaptations to similar habitats that would not have been ultimately successful in this sense. Thus, lungfishes have been living in shallow waters for as long a time as the rhipidistians (indeed, longer) but did not originate any terrestrial descendants. Furthermore, even if we wanted to consider it very probable that a particular group would originate a particular adaptive zone, we should have to include in this prediction some other prediction concerning the (high) probability that the relevant environments would have a certain structure. This is so because future events or possibilities simply cannot have any effect on the gene pools of organisms (see Chapter 6). Adaptations that turned out to be preadaptive could not have been influenced in their evolution by a future adaptive zone or the habitat associated with it, and so it is by definition unlikely that any particular preadaptation should occur.

As a general rule, entry into new adaptive zones is made possible by the nature of some of the adaptations that were developed in reference to already existing adaptive zones. Sometimes these transitions involve many separate adaptations, as in the example just cited; another such example would be the many different adaptations of desert plants that have allowed some of them to enter the epiphytic adaptive zone—some cacti and bromeliads. Further examples of adaptive traits preadaptive to a new way of life are viviparity in the ancestors of cetaceans and probably ichthyosaurs; bipedalism in the ancestors of the birds; loss of limbs in the burrowing ancestors of the snakes. Actually any trait might turn out to be preadaptive to some new adaptive zone if the opportunity for it to do so presented itself by chance. There is nothing inherently special about traits that we know *post hoc* to have in fact turned out to be preadaptive to a new way of life.

It may be noted that it seems necessary to postulate in every case of transition to a new adaptive zone by animals an initial change in behavior. If the rhipidistians had never ventured out on land, even accidentally, it is unlikely that they would have given rise to a terrestrial lineage. The initial change in behavior is allowed (or even favored?) by certain preadaptive traits. If the behavior is reinforced by selection, it would be in the form of supportive behavior and anatomical or physiological changes that simultaneously make the behavior both more adaptive and more probable. In plants, however, it is not behavior that carries prospective makers of new adaptive zones into new habitats but seed dispersal (this is also true for sessile animals in the form of dispersion of gametes or larvae). Thus, a seed of a desert bromeliad may be dropped by a bird that has eaten the berries onto a tree branch instead of onto a rock. Probably nothing so extreme would work, but this is the general kind of occurrence that is necessary in more gradual form.

The idea of preadaptation is important in showing us how factors intrinsic to the organisms themselves could have an important role to play in the entry of lineages into new adaptive zones. The role that chance plays in this process suggests that extrinsic factors are no less important. For example, the mammals were around for some 90 million years before, suddenly, they began their big explosive radiation in the latter part of the Cretaceous. Both the fact that many of the adaptive zones to be developed by the mammals were not unlike some of those of the earlier dinosaurs, and the fact that the major mammalian radiation did not really get underway until the dinosaurs were on the way to extinction suggest that the two events were not independent. A reasonable interpretation is that the mammalian radiation occurred after the dinosaurs had vacated their adaptive zones. The explosive nature of an adaptive radiation suggests, as well, that the radiations are into unoccupied adaptive zones, meeting essentially no competitive resistance. Indeed, it may be suggested that a precondition for an adaptive radiation is the availability of large numbers of unfilled potential niches. This is also suggested by the fact that not every newly developed adaptive zone has afforded the opportunity for a major adaptive radiation. Thus, the emergence of crustaceans into terrestrial niches—some of the isopods—has not resulted in anything like a major radiation. It may be suggested that the prior presence of arachnids and insects was an important factor here.

G. G. Simpson has suggested another possible reason for the explosive nature of adaptive radiations. He has suggested that once a group begins to explore the possibilities of a new adaptive zone that some preadaptation(s) have allowed them to enter, there should be a period of very rapid evolution, because the organisms are not fully adapted to their new ways of life. Directional selection pressures (see Chapter 10) should be particularly intense in this view because of the need for "postadaptation," bringing the entire organisms into harmonious coadaptation with the preadaptive traits and, obviously, into an adapted relationship with new environmental factors. This process of rapid transition Simpson has termed quantum evolution in reference to its resulting in a large adaptive jump in a short period of time. This is probably a useful concept where the transition to a new adaptive zone involved one or a few key characters. It has the merit of explaining why intermediate "missing links" are so rare in the fossil record: they existed over relatively short spans of time, and so had less chance of becoming fossilized.

CHANCE AND DETERMINISM

The interplay between chance and determinism in the phenomena discussed in this chapter needs explicit emphasis. Extinction, adaptive radiation, and preadaptation all result from chance (or low-probability) interactions between causally determined intrinsic factors—gene pools, phenotypes—and relatively indeterminate extrinsic factors—rapid environmental change, accessibility of unexploited habitats, extinction of or invasion by potential competitors. In these circumstances it is possible to conclude that chance is of such large importance that there should be no predictability at all in evolution. But the phenomenon of parallel evolution (of which further examples are given in the next chapter) is a very forceful suggestion that this is not entirely so. It seems that, for example, given a large enough land mass, a long enough period of time, a forestlike situation, and a mammalian gene pool, a squirrel-like mammal will ultimately evolve. Possibly, if there had been only many small land masses, there would not have occurred any significant parallelisms in terrestrial situations. On the other hand, if there had been only one large connected land mass, the lack of more than a single opportunity for an adaptive radiation at a particular time would leave only sequential opportunities for parallelism to occur; this may be closer to the situation in the oceans.

The implications of these considerations for speculation about extraterrestrial life are of some interest. We have to conclude that, given, say, a mammalian genome on some far-flung planet and a habitat composed in part of plants of some kind, we have to expect the same array of forms that we find on the earth—woodchucks, cats, bats, and the like. The question then becomes one of whether we could expect to find something close to a mammalian genome independently derived or evolved in parallel with that on the earth. This is the same as the question of whether the origins of major phyla or classes on the earth were in any degree predictable. We have no evidence that there was ever a parallel origination of any major phylum. At present, then, we must conclude that, although the general pattern of living systems on different planets can be expected to show similarities (see Chapter 2), the basic structures of the large phyla would be similar only by chance or to the degree demanded by environmental physics (a swift-moving water-living form would have to be streamlined, no matter where in the universe it lived). We may have only the faintest glimmer of knowledge, however, about the constraints placed on living systems by geophysical forces.

REFERENCES

Colbert, E. H., *Evolution of the Vertebrates*. Wiley, New York, 1958, 479 pp.

Darlington, P. J., *Zoogeography: The Geographical Distribution of Animals*. Wiley, New York, 1957, 675 pp.

Drake, E. T., ed., *Evolution and Environment*. Yale, New Haven, Conn., 1968, 470 pp.

Huffaker, C. B., "Experimental Studies on Predators, Dispersion Factors, and Predator-Prey Oscillations." *Hillgardia*, **27**:343–383, 1958.

Jepsen, G. L., G. G. Simpson, and E. Mayr, eds., *Genetics, Paleontology and Evolution*. Princeton, Princeton, N.J., 1949, 474 pp.

Lewontin, R. C., ed., *Population Biology and Evolution*. Syracuse University Press, Syracuse, N.Y., 1968, 205 pp.

MacArthur, R. H., and E. O. Wilson, *The Theory of Island Biogeography*. Princeton University Press, Princeton, N.J., 1967, 203 pp.

Rensch, B., *Evolution above the Species Level*. Columbia, New York, 1960, 419 pp.

Romer, A. S., *Notes and Comments on Vertebrate Paleontology*. The University of Chicago Press, Chicago, 1968, 304 pp.

Simpson, G. G. *The Meaning of Evolution*. Yale, New Haven, Conn., 1949.

————, *The Major Features of Evolution*. Columbia, New York, 1953, 434 pp.

————, *Principles of Animal Taxonomy*. Columbia, New York, 1961, 247 pp.

Stebbins, G. L., Jr., *Variation and Evolution in Plants*. Columbia, New York, 1950, 643 pp.

CHAPTER 5

The fossil record: II

Chapter 4 discussed how certain traits or adaptations could turn out by chance to be preadaptive for some unexplored adaptive zone. This chapter is more concerned with the step-by-step evolution of adaptive traits as such.

GENERAL MODIFICATION OF STRUCTURE

Consider the evolution of a structure such as the bird's wing. Today, near the crest of the adaptive radiation of birds, we find many kinds of wings—long ones for soaring, the flippers of penguins for swimming—used in many different ways—some for hovering, some as sailplanes during running, some for flitting and darting through the air. The earliest known fossil bird, the Jurassic *Archaeopteryx*, was clearly a kind of specialized reptile. It was bipedal, of course, as were many dinosaurs. It had a long reptilian tail, and it had teeth. But it also had feathers, a rather short backbone, a fairly long forelimb with a reduced number of digits, and a comparatively large braincase. It was as nearly intermediate between dinosaur and bird structure as any seeker of "missing links" could wish. The wings were obviously very weak in terms of flying, for there was no expanded sternum for the attachment of powerful flight muscles (the breast muscles of birds). This animal may have glided from high places, as do flying squirrels, and it may have run on the ground much like a chicken, using its "wings" as a sailplane to obtain speed.

It has been suggested that the presence of feathers all over its body indicates that Archaeopteryx was a homeotherm. Indeed, one theory of the origin of feathers is that they were developed from reptilian scales as a means to conserve body heat in forms extending their range into colder climates and that only secondarily did they become involved with gliding and sailplaning, realizing a possibility for which they were preadaptive. These earliest postulated stages have not been found in the fossil record. Selection pressures increasing the size of the feathers to that of those found in *Archaeopteryx* are considered by most workers to have been most plausibly involved with parachuting from high places, perhaps in arboreal forms. There are modern lizards, frogs, snakes, and squirrels that are capable of breaking their fall by parachuting. In some of the arboreal frogs, lizards, and squirrels there are no obvious structural adaptations for this, and apparently they can accomplish the parachuting by the nervous control of the position of the limbs (see Figure 5-1). Frogs and lizards without this control wriggle in panic and fall straight down, heavily. Thus, we can imagine a first stage in a small animal of parachuting by neuromuscular control alone. Some frogs, lizards, and squirrels have gone further and evolved various skin folds and flaps to increase their surface area, becoming efficient gliders (see Figure 5-1).

One group of reptiles in the past, the pterosaurs, apparently explored the gliding way of life more extensively than any other vertebrates. They are always found in marine deposits and have teeth that would have been efficient fish traps. The sternum in many of them had a relatively small keel, so that it has been suggested that they could not actually fly as well as modern birds, and this would be especially true of the largest ones with wing spans in excess of 25 feet. It is generally felt that they lived on cliffs beside lagoons from which they glided down to the water, snatched fish from the surface, as do modern terns, and then glided back to the cliffs. Some recent aerodynamic calculations, however, suggest that they could have been lifted off the ground, like a kite, in only a moderate head wind. The wings of these animals consisted of a single flap of skin, the patagium, extending from the tip of the elongated fourth finger to the hind limb. Bats have a patagium not unlike this, but including more digits, and often the hind limbs and even the tail. They, however, actually fly.

Returning to the evolution of the bird wing, the stage we find in *Archaeopteryx* could have functioned much as the wings of pterosaurs did, for gliding or, as in the modern chicken, for sailplaning. In a sense it is a mistake to call this organ or that of the pterodactyls a wing. It is a structure that we know *post hoc* to have been preadapted to being molded by selection into a wing. We do not meet a wing in this lineage until the Cretaceous. Thus, *Archaeopteryx* did not have a poor or "imperfect" wing; it had a very good gliding organ or sailplane. The difference in emphasis is important in stressing the fact that in the evolution of any structure,

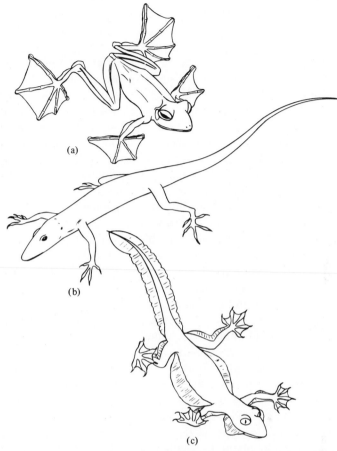

Figure 5-1 Various gliding vertebrates: (a) the Malayan arboreal frog *Rhacophorus pardalis;* (b) the New World arboreal lizard *Anolis;* (c) the Malayan arboreal lizard *Ptychozoon*. Notice the *Anolis* has no obvious flaps or skin folds to increase its surface area, and it cannot actually glide; it breaks its fall by freezing its posture in the position shown. (a) and (c) simplified from Gadow, 1902.

each step or phase along the way must be considered to have been adaptively adequate (and obviously was, because it existed). It is not a question of comparing an earlier stage with what we find present in a descendant form and evaluating the earlier stage in terms applicable only to the later. If a structure had been maladapted, in some way not as good as something possessed by other members of the species, the organism possessing it would have been selectively disfavored; that is, it would have left no descendants.

To consider the *Archaeopteryx* forelimb to be a stage in the evolution of the bird wing is itself a rather questionable procedure, because it overemphasizes conditions and situations as they exist today. In a purely descriptive sense it is true that the *Archaeopteryx* type of forelimb represents such a stage, but this has nothing to do with *Archaeopteryx*, how it used its "wings," or the multitude of selection pressures operative in the Jurassic on this kind of organism. Organic evolution is a completely opportunistic process, works from no plan, and has no mechanism tying it in to some future potentiality of a lineage or organ (see Chapter 3). Existing structures are modified from geological moment to geological moment according to the separate and individual needs of organisms living at those moments. The specific needs of many moments continually integrated with the needs of a continuing present moment happened to result in one lineage in bird wings and bird flight. Selection can have been operating directly on wings and flight only in the moments since flight became possible in this lineage in the Cretaceous.

A structure such as the bird forelimb evolves slowly over a long period of time and changes form only gradually. This is the only way it could change if each unit of change was accomplished separately according to the adaptive needs of the moment. Furthermore, large-scale changes over short periods are less probable in that they are more likely to disrupt existing adaptations—to produce "monsters." In order to explore this idea more fully, it is helpful to introduce some terminology developed by Walter Bock, of Columbia University—that is, the idea that a structure has one or more forms associated with one or more functions. Thus, as pointed out in Chapter 3, a man's hand is capable of assuming many different forms—fists, pincers, and the like—associated with different functions. In the case of forelimbs such as those possessed by *Archaeopteryx*, selection could operate, say, to improve the ability of the organism to flap them, without sacrificing its ability to use them as sailplanes. One form and function can be evolved without necessarily disrupting other forms and functions of the same structure, especially if the changes accumulate slowly and gradually. It often happens that after long periods of time old forms and functions may be lost entirely, transforming the organ or structure into a completely new thing. Examples are the reorganization of the filter-feeding apparatus of cephalochordates into the thyroid gland of vertebrates, of the fins of crossopterygians into tetrapod limbs, of the lungs of primitive vertebrates into the swim bladders of teleosts.

It is obvious, then, that all new structures are developed out of preexisting ones. The nature of the preexisting structure, perforce, sets limits to its evolutionary potential. The wings of vertebrates and those of insects are very different in structure and arose from different preexisting structures. What is needed may be a wing; what is finally evolved is a specific kind of wing, and this specificity is only in part due to somewhat different functions for wings in different groups. Probably the major influence in the specificity of a structure is the nature of the structure from which it was evolved. The wings of vertebrates all have the basic form of a limb; those of insects do not, and this is not surprising, because insect wings were not derived from preexisting limbs.

IRREVERSIBILITY OF EVOLUTION

The evolutionary modification of a structure, then, is generally a very slow, gradual process and is significantly impinged on by limitations set by the preexisting forms of the structure. Furthermore, as an empirical fact derived from the fossil record, once a structure has become so modified as to have lost some or most of its archaic forms, it does not return to them again. Thus, the ancestors of snakes, having lost their limbs during adaptation to fossorial life (as have several other kinds of tetrapods) did not regain these structures when members of this lineage again took to adaptive zones on the surface of the earth. Instead, a totally new method of progression was invented, based, however, on the sinusoidal motion that is implicit in the fundamental segmented structure of vertebrates, remnants of which, at least, are found in them all. Some of the snakes, entering the aquatic environment, became fully aquatic (the sea snakes). These did not reacquire the fins or gills of fishes. The body became laterally compressed, especially the tail, which serves, much as do the caudal fins of fishes, to provide propulsion. The mechanics of swimming in these snakes may be considered convergent on that developed independently by the teleost eels (convergent instead of parallel in that the snakes in an intermediate period had entered an adaptive zone so different from that of the ancestors of the eels as to make them in this sense the adaptive equivalent of very distant relatives, using the terminology of Chapter 4. In the same way, ichthyosaurs and cetaceans became convergent, in general body shape only, on a certain fishlike habitus, but in this case did so in parallel with each other, from similarly structured and adapted ancestors). The structures of sea snakes, ichthyosaurs, or cetaceans are not reconstructions of those of any of their piscine ancestors. There was in no fundamental sense a reversal of evolution in any of these lineages. Returning to aquatic environments was possible only if their bodies were streamlined for swimming. In the case of the sea snakes, their bodies were already preadapted to a certain kind of swimming and already had become streamlined. In the case of cetaceans, streamlining was gradually acquired as an adaptation to living in water, possibly through an intermediate stage, such as that exemplified by seals. Anything that swims fast in the water must be streamlined; this is what a comparison between fishes and other aquatic animals boils down to. In no case was an ancestral structure reconstructed.

Although detailed structure, once lost or changed, is never regained, can the same be said of ecological niches or adaptive zones? Because no two species can occupy the same ecological niche as a matter almost of definition (see Chapter 3), this problem does not arise. There is reason to consider, however, that very similar adaptive zones can be evolved by different groups at different times (see Chapter 4). It seems quite possible that at some time or other a fish has occupied an adaptive zone similar to that of a cetacean or an ichthyosaur, although probably not so similar as those of some of the latter two may have been.

Looking more closely, at smaller details, there do arise a number of possible reversals of evolution. For example, the tooth row of piscivorous cetaceans consists of one after another of simple, conical teeth, not unlike those of predaceous reptiles or even, in a general way, of some fishes. Their ancestors are considered to have had the differentiated tooth row of mammals (see Figure 5-2), which was lost in favor of a tooth row that does

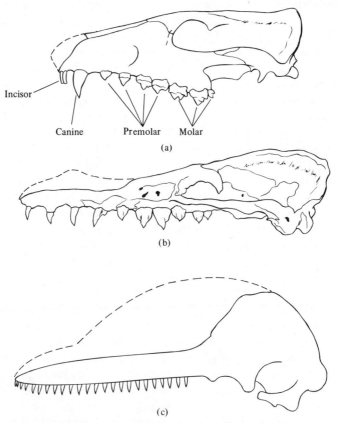

Incisor

Canine   Premolar   Molar

(a)

(b)

(c)

Figure 5-2   The evolution of homodonty from heterodonty in toothed whales (a) a primitive creodont carnivore with typical mammalian tooth row; (b) *Zeuglodon,* an Eocene whale; (c) *Tursiops,* a modern porpoise. Sizes not comparable. The dotted line represents the nose region. (Simplified from Gregory, 1951.)

not differ essentially from that of (ancestral) reptiles. There is a way out of having to consider this to be a true reversal of evolution, which, as we shall see, is well nigh a genetic impossibility. We can handle this approach even better in another example, described by Björn Kurtén, of the Geological Institute of Helsingfors, Finland.

During the evolution of the cats, all the teeth in the lower jaw behind the carnassials (modified first molars) were lost by the end of the Miocene, and in many lineages, including that of the ancestral lynx, several cusps were lost from the carnassial during the Pliocene, making it bicuspid. But beginning in the middle Pleistocene, these cusps began to show up again and in recent lynxes are as fully developed as in the Pliocene forms (see Figure 5-3). Some individuals of the northern lynx are found today that also have a small second molar behind the carnassial, thus reconstituting completely the Miocene condition of some 20 million years ago.

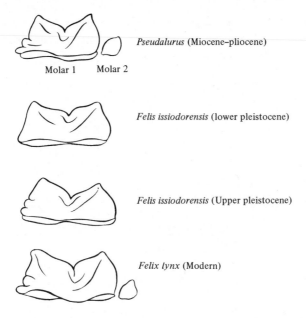

Pseudalurus (Miocene–pliocene)

Molar 1   Molar 2

Felis issiodorensis (lower pleistocene)

Felis issiodorensis (Upper pleistocene)

Felix lynx (Modern)

Figure 5-3 Carnassial cusp evolution in a series of lynxlike felids from Europe. (Modified from Kurtén, 1965.)

We do not have to consider that these changes reflect changes in the gene pools of the populations involved. One way to think of this example is in terms of developmental fields. A developmental field is an influence which is exerted on surrounding cells by a group of developing cells or a developing organ and which decreases in intensity with distance from the influential center. The structure of the field is a gradient with a high point right at the influential center. There is some indirect evidence that such fields may be involved in the development of mammalian teeth. Suppose there were a molarization field located at the proximal ends of the jaws and, possibly, an incisorization field at the distal end (see Figure 5-4). These fields may be established by physical factors, such as temperature or pH, or by chemical factors, such as the accumulation of waste products or the production of some end product. What is needed is that there be an unusual accumulation or depletion of some such factors at a certain place, possibly because the tissues are thicker there or because they happen to abut on some other embryonic structure. These factors, depending on their nature, can have a variety of effects on the production and operation of primary gene products. They can influence the activity of effector molecules (see Chapter 8) and thereby effect the switching on and off of genes; they can influence the activity of other effector molecules (or may themselves be such effectors) acting on the primary gene products themselves, either speeding up or slowing down a particular reaction rate (see Chapter 8); in short, they can in one way or another influence the activity of genes at various critical periods during development but are not themselves wholly genetically determined as such.

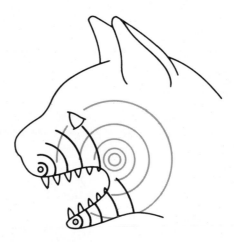

Figure 5-4.   Possible position of the incisorizing field and the molarizing field during tooth development in mammals.

We can then interpret the development of mammalian teeth in the following way: The production of certain dentine-forming and enamel-forming condensations is coded for by a series of genes that are, for some reasons, active only in a certain region of the jaw, in discrete serially arranged rows. Those tooth anlagen which are nearest the front of the jaws are most influenced by the incisorization field and become flattened in a characteristic way. The anlagen further to the rear of the jaw are less and less influenced by this field, but are increasingly under the influence of the molarization field, and develop multiple cusps, rotundness, and other characteristics of molars. Canines could be thought of as molarized incisors, premolars as incisorized molars, and so on.

We can with this model interpret the evolution of cat dentition as having involved a weakening of the molarization field, possibly in relation to a shortening of the jaws. Actually, we cannot and need not stipulate exactly what influences were involved in this; what is important is that they need not, in terms of the model, have been genetic, because the shapes of the teeth in this model are not genetically determined as such.

If the shape of specific mammalian teeth is not genetically or directly determined as such, changes in shape, particularly small changes, in some lineage with time need not be considered to be authentic evolutionary modifications. But with large-scale changes, such as those seen in the cetaceans, it is probably necessary to invoke some genetic changes involved, at least indirectly, in the change in tooth shape. Thus, the upper Eocene *Zeuglodon* is a whale by any definition. But its skull is not yet greatly modified from the primitive mammalian condition, and in particular its jaws are relatively much shorter than those of modern cetaceans (see Figure 5-2). It has 44 teeth of simpler construction than those of typical mammals, but with yet some signs of influence by a molarization field. Suppose that the tendency in this lineage to increasing the length of the jaws was the major factor in eliminating the influence of mammalian tooth fields. Suppose that this lengthening of the jaws was a genuine evolutionary change involving the replacement of many alleles by others in a number of genetic loci. Is homodonty in modern cetaceans then the result of genetic changes? Indirectly, yes. Is it a reversal of evolution to a greater number of identical teeth (homodonty) from fewer, more specialized teeth (heterodonty)? This depends on our point of view. On the face of it, yes, if we define evolutionary change in terms of changes in phenotype. But a consideration of the situation at the genetic level shows why it is probably best not to treat this change as an evolutionary reversal.

Before looking at the molecular level, we can note that regardless of whether tooth shape is directly or indirectly influenced by the genotype, it must be considered to be adaptive. Thus, if in cetaceans increasing the jaw length results secondarily in homodonty, selection could favor alleles that help to produce longer jaws, because individuals with homodont teeth are more successful in catching fish than are those with heterodont teeth. In other words, selection could favor longer jaws, not for themselves, but for the secondary effect of homodonty. The tooth row of piscivorous cetaceans has doubtless been favored by selection, because it makes an adequate fish trap. This is the same ultimate cause of similar tooth rows in other piscivorous vertebrates. Natural selection has favored this kind of tooth row many times in independent lineages; it hardly adds anything to our understanding of the meaning of this kind of tooth row to say that this represents a reversal of evolution in one of these lineages.

As mentioned in Chapter 3, the cytochrome c of man differs from that of the black bread mold by some 50 percent of the amino acid sequence. Therefore, roughly about half of the triplet codons of this genetic locus are different in these two forms. Consider the probability of evolutionary changes occurring so that one such locus begins to return in structure to the condition prevalent in the common ancestor of these forms. What would be involved would be the production by mutation of forms of this locus differing from the "normal" form probably by a single codon. One of these has the new codon the same as in the ancestral form in question. This allele (see Chapter 6) now must increase in the population to become at least a common form involved in a polymorphism with the previously "normal" allele. Now this allele must again mutate in a direction making another codon more like the ancestral from than the previously "normal" form. This new allele now must increase in frequency and become at least a common form, and it must mutate again, and so on. Each new allele, closer in structure to the one possessed by the common ancestor, must be favored by selection over the previously favored allele. This must go on for a long time in order for each requisite allele to become the common one in the population. At the same time any mutations away from the ancestral condition must not become selectively favored. In short, the probability of a reversal of evolution of this kind is so low as to be virtually impossible (see also the discussion of back mutation in Chapter 7).

On the other hand, man and macaque differ by only a single codon at this locus. The macaque locus is more like those of other vertebrates at this codon, coding for threonine instead of, as in man, isoleucine. Thus, a mutation in man going from TAA or TAG to TGA or TGG at this codon would produce a macaque cytochrome c. Actually, because the physicochemical properties of threonine and isoleucine are fairly different, this particular mutation might not be very successful. Even if it did occur as a rare mutant, this would not qualify it as a reversal of evolution, because the total gene pool that must evolve (see Chapters 6 and 7). This mutant would have to become the common one in human populations in order to effect a reversal of evolution. From this point of view, even the reversal of a single codon is not something that we expect to be frequent. Individual humans with monkey cytochrome c or gorilla hemoglobins are no more significant in this sense than would be a few individual lynxes carrying an extra molar, even if this trait were genetically coded as such.

It may be noted that if the presence or absence of this molar is directly genetically determined, it could be in terms of small differences in a codon or so at a single locus. Suppose that the absence of this

tooth derived from the inactivation of a crucial enzyme. This would actually mean that in the environmental conditions present at this particular
tooth anlage, the altered enzyme has an activity so low as to inhibit a crucial pathway (which might remain active at the other anlagen, because
the environments there are not inhibitory to it), so that no extra molar
is formed. In this case, we might have a back mutation, restoring the
original level of enzyme activity, and that would lead to a true evolutionary
reversal if it gradually came to be the wild type for the population.

There is an interesting case studied by M. G. Weigert and A. Garen
involving the alkaline phosphatase of the bacterium *Escherichia coli.* They
discovered a mutation that prevented the production of the enzyme by ter-
minating the amino acid chain before the entire polypeptide was finished
(nonsense mutation; see Chapter 7). Exposing the phosphatase-less bacteria to mutagens resulted in some colonies that again contained the enzyme. These were grown and the enzymes isolated and examined for amino
acid sequence differences. It was found that there were seven different "re-
verted" enzymes, including one that had the original tryptophan at the position in question. The other six each had a different amino acid at this
position, but they all were active to some extent. The enzyme activity was
restored in each case, but in only one of them was this accomplished by
going back to the same codon from which this microevolutionary event
was started. Only this case would qualify as an evolutionary reversal in
our sense. C. A. Yanofsky has done similar work on another enzyme in
E. coli, tryptophan synthetase. In this case inactive enzymes were produced
by mutations going from glycine to glutamic acid or arginine at a particular
triplet. One of the spontaneous revertants that was almost fully active had
alanine at this position instead of the original glycine; some of them went
back to glycine. Another important class of mutations restoring activity
to an enzyme that has lost it by mutation is the suppressor mutation. This
is a mutation at a locus that restores activity to the gene product of another
locus; there are a number of different kinds of these, but we need here
only be aware of them as a class.

The point of this digression is to show that the return of a small morphological trait, such as a molar cusp, after it had disappeared from some
lineage could easily occur without there having been any actual evolutionary reversal at the genetic level, even if the trait in question is directly
coded for by the genotype. An enzyme may reacquire its activity after the
locus that codes for it has mutated to an allele different from the one that
originally coded for it when it was an active enzyme. Activity would be
regained, a cusp would reappear, but the genotype would not be as it originally was.

Evolutionary reversal, then, on any scale larger than a codon or two
at a single locus, is simply a very improbable event. It is no doubt for
this reason that phenotypic evidence of reversal has never been over-

abundant. Perhaps most, if not all, of the possible examples that have been cited really involve traits not directly coded for by the genotype and may, therefore, be observed to change form in time because of some change in a parameter of the external environment that can directly affect the development of the trait in question (see Chapter 8). The principle of irreversibility was first clearly stated by L. Dollo in 1893 and has come to be known as Dollo's law. It is convenient to examine certain other evolutionary "laws" and principles deriving from study of the fossil record at this point.

WILLISTON'S RULE

Dollo's law may be said to have few or no exceptions as we understand it today. Not so general is Williston's rule, named for S. W. Williston, an American paleontologist. This formulation states the principle that, during the evolution of a lineage, serially homologous parts tend to become fewer and increasingly differentiated. Thus, in the transition from primitive reptile to mammal, the tooth row changes from one of many essentially identical teeth to fewer, structurally differentiated ones. Again, the earliest arthropods (the trilobites) show very many, more or less identical appendages, and this is presumably the primitive condition for the phylum as a whole. Modern arthropods, however, show comparatively fewer appendages and a great deal of specialization among them, as, for instance, in the crayfish (see Figure 5-5). During the history of the dicotyledonous angiosperms there has apparently been a decrease in the number of parts of the floral perianth and an increase in their specialization to form petals and sepals. The result of this tendency is seen to be a kind of division of labor among the parts. In the tetrapod lineage, the ichthyostegids have some 150 separate bones in the skull; the mammals have only about 28. In a case like this, reduction in number is perhaps more evident than increased differentiation, although the bones of mammals do tend to be considerably more distinctive, with special processes, and the like, than are the homologous ones in fishes.

It should be evident that this rule is not applicable to organisms lacking segmentation or other serial structures, such as protozoans, sponges, or coelenterates. Furthermore, there are a number of clear exceptions. For example, as mentioned above, the tooth row of piscivorous whales has shown an increase in number and a decrease in internal specialization over that presumably characteristic of their ancestors. The snake lineage has shown an increase in the number of similar vertebrae during the evolution of attenuated body forms. The cacti have experienced an increase in perianth parts and a decrease in their differentiation.

The important difference between Dollo's law and Williston's rule is that the former expresses a fundamental property of change when it in-

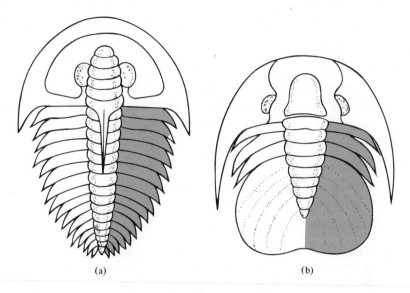

(a)   (b)

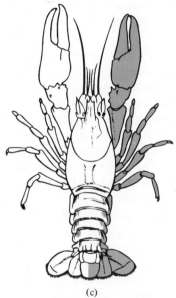

(c)

Figure 5-5 Several arthropods illustrating Williston's rule: (a) a primitive and relatively generalized trilobite; (b) a more advanced and relatively more specialized trilobite (note the result of fusion of many limbs). (c) a crayfish, with a highly differentiated series of limbs. Shaded areas show the repeated series of limbs to be compared. Figures not drawn to scale.

volves some random aspects among numerous dimensions; the latter is only an empirical observation about what certain kinds of organisms have shown during their evolution. In the case of Williston's rule, enough lineages have shown the tendency involved to make it interesting enough to name and think about. The consensus seems to be that this phenomenon has been in all cases related to an increase in overall competence to perform the tasks germane to the structures in question. Thus, the claws of crabs and lobsters perform the function of tearing up meat better than the walking legs can do it. The legs, on the other hand, are better able to effect crawling than either the claws or the swimmerets, which are specialized for providing propulsion. A primitive crustacean with only one kind of appendage, as in the trilobites, could not perform any of these functions as well as the lobster; it could perhaps do all the things that a lobster can do, but its limbs would do them all together and less efficiently. The idea, then, is that a decreasing number of increasingly specialized homologous parts is associated with an increased general competence achieved by a division of labor among the parts. Each part becomes more specialized in function, although the organism as a whole remains ecologically generalized and simply does increasingly better or more efficiently what it has been doing all along.

Although this formulation is plausible, it has all the problems associated with attempting to assign a single general explanation to many diverse occurrences. Even describing them under the same heading may result in obscuring real differences among phenomena. Thus, is a decrease in perianth parts a phenomenon at all similar to a decrease in the number of skull bones? The only connection here is the decrease in number. Natural selection could not favor a decrease in anything for its own sake. The detailed operation of selection in these cases had to have been different, given the difference in the structures and organisms involved. The question then is whether the selective forces in all cases were involved with an increasing competence to perform certain tasks within a given adaptive zone. Natural selection always favors the more competent individuals, provided that "competence" has a positive effect on reproduction (see Chapter 6). The real problem arises in connection with the suggestion that the ecological problems involved were unchanging. Many decapod crustaceans have claws, but many others do not. Claws are special adaptations to specific ecological problems; so are swimmerets, so are canine teeth, and so are molars. Presumably selection favored the evolution of each of these forms in unique ways. It really seems unlikely that crustacean claws are performing functions that were performed by other means in more trilobitelike ancestors. A lobster is probably not a better trilobite than a trilobite; they probably did not occupy similar adaptive zones. The level of explanation provided by the general hypothesis given above simply does not reveal enough about the individual phenomena that it attempts to explain, and it even tends to hide real differences among them.

We might even question the usefulness of formulating the rule. By formulating a rule or law, the implication becomes manifest that there is a single, all-embracing explanation for it. The generalization is not formulated simply as a descriptive device; its very presence invites general explanation. We can examine this better by looking at another erstwhile "law," Cope's rule.

COPE'S RULE

Edward D. Cope was an American paleontologist active in the last part of the nineteenth century. Working with fossil mammals, he was struck by the number of different lineages that independently produced giant forms in the Pliocene and Pleistocene epochs. It was also clear, and other authors have since documented this, that many other kinds of organisms have shown similar tendencies toward increased size during their evolution. Cope formulated this tendency as a primary factor in evolution, as a law. There are, however, a very great many exceptions to this "rule." Even some mammalian lineages do not exemplify it, for example, the insectivores and the insectivorous bats. Also, after a peak of numbers of gigantic forms in the Pliocene or Pleistocene, many mammalian lineages have generally decreased in body size, for example, the ungulates and carnivores. On reflection, we can see that body size is related to adaptive zone. Thus, flying or burrowing organisms would not be expected to exhibit gigantism in their history. A few lineages even of these, however, do show tendencies to increasing body size, but to a much more limited degree than in other kinds of organisms. It is physically impossible to be a very large mole, and the net effects of selection on size in moles can be expected to be toward no change or a decrease in size.

Clearly, in different lineages body size has different meanings. In homeotherms in a cooling environment, as in Pleistocene mammals, increased body size may be evolved as a means to reduce the body surface/body volume ratio, cutting down on heat loss. In predatory animals, the existence of an untapped food source of very large herbivores could contribute to selection for large size. In animals in which males battle one another for the possession of females, selection probably generally favors increased body size. In egg-laying animals that do not take care of their young, selection for increased fecundity could favor increased body size (more eggs). If, for some reason, selection favored increased life span, it would probably also favor increased body size. In view of these considerations, we can ask whether there is any point to formulating increased size during phylogeny as an actual or distinct phenomenon. Does it actually mean something special that a certain lineage of mammals and a certain lineage of mollusks independently show trends toward increased body size?

EVOLUTIONARY TRENDS

We can go further and ask whether it actually means anything special when even a single lineage shows a particular evolutionary trend. Descriptions of trends in one lineage or another abound in the paleontological literature. Thus, there was a trend toward increasing brain size relative to body size in many mammalian lineages, including that of man; there was a trend toward decreasing the size of the forelimbs in two dinosaur lineages; there has been a trend toward decreasing dermal ossification in bony fishes; there have been trends toward the loss of limbs in many tetrapod lineages that took to burrowing. A trend is a long-continued progressive change shown by a lineage, or a tendency shown simultaneously by several sublineages of a lineage, or both. The second kind of trend was discussed in Chapter 4 in terms of parallel evolution.

We may first inquire to what extent trends are real phenomena. In studying the evolution of some lineage, we begin with a detailed picture of the relevant structures of living representatives. We then examine the fossil record and gather together genera that are closest in structure to what we estimate ancestral forms at different levels of time were most probably like. The latter estimate is based on the structure of forms already known or found. Many forms are rejected, because their structure is too unlike that of those already incorporated into the phylogeny being constructed. (Of course, the search is not for actual ancestral species; it is for groups that very likely could have contained such ancestral species.) Eventually, enough material may be assembled so that a phylogeny can be constructed, beginning with a convenient form found in the past. A rough picture of the structural modifications that occurred during the transition from that ancestral form to the modern one(s) can then be formed. Often a trait—body size, tooth shape, position of eye sockets, number of digits—can be seen to become more or less gradually modified in the direction of the condition found in the modern representative(s) of the lineage. Such a more or less continuous change from one form of a structure to another is an evolutionary trend (see Figure 5-6).

The construction of the phylogeny is based on the similarity of form of various structures. Later, we use these similar forms as a basis for observing the directions in which a particular lineage has evolved. The phylogeny is constructed backwards, knowing the future, as it were, and reconstructing a plausible past. Then we use the phylogeny so constructed as if we did not know the outcome and imagine how natural selection, acting at each stage, could have produced the next stage. Here again we know what we are looking for. Natural selection does not "look for" any-

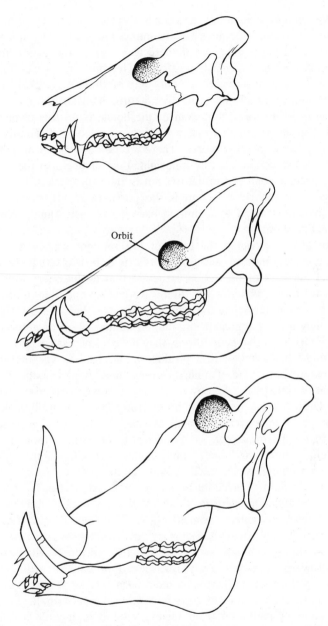

Figure 5-6   A series of skulls showing several trends in the evolution of pigs. Beginning with the Oligocene pig at the top, we can trace a progressive elevation of the orbits, the gradual deepening of the jaw, and a gradual increase in tusk size. Middle figure is a modern pig, the lower, a modern wart hog. Sizes are roughly comparable. (Modified from various sources.)

thing, has no goals in mind, and cannot "perceive" the future. In actual fact, any given ancestor gave rise to many descendants by speciation and divergence. Perhaps only one of these was part of the direct lineage that finally gave rise to a given living form. This one, too, gave rise to many descendants, and again, perhaps only one of these was destined to be on the line of descent of the same living form, and so on. Our interest in a particular living form, say, man or the horse, is, in this framework, unnatural. Our minds in this instance work completely differently from the way that natural selection works. There are no "trends" on which selection acts; trends are constructed by hindsight. Like higher taxa themselves, and lineages, trends exist, not in nature, but in our constructs. The only reality that exists for natural selection is the population of organisms, with its gene pool, constrained to a limited carrying capacity and facing various ecological problems.

But if we restricted the form of our thought processes to conform only to the way natural selection works, it is possible that we could not construct the kind of theoretical structure that evolutionary biology is. As long as we know what we are doing, we can accept that trends have the same kind of reality that lineages do. We can then see clearly that, although they are an aggregate production of the action of natural selection over many consecutive generations, they have no necessary predictive value concerning what selection will produce in the future. The major selective forces may change at any time, resulting in a change in the direction of evolution of some trait, thereby destroying a previous trend or even reversing it. Thus, in one lineage of horses, after some 4 million generations being involved in a trend toward increased body size, the trend was reversed in the beginning of the Pliocene. (Note that, although a true reversal of evolution is highly improbable, reversals of the direction of evolutionary change at the phenotypic level do occur.)

Because natural selection works only in connection with conditions during a particular moment in geological time, how can it be that its action during many consecutive moments in different populations facing somewhat different ecological problems could in fact give rise to a long-continued progressive change in some trait? In connection with a trend such as that toward increased body size, there is a kind of artifact that may enter the picture. If, as seems to have been true in many cases, the first organisms to enter a new adaptive zone are relatively small, then any successful adaptive radiation from them is bound to include a fair number (about half) of large organisms, unless the adaptive zone is one in which large size would not be favored. Because these large organisms have in the past been the ones preferentially discovered in the geological strata, there is a distinct possibility of error. With different kinds of trends there are no doubt other sources of error. Assuming, however, that there are real trends observable in the record, how can we explain their existence?

Any specific organism is adapted to a specific ecological niche and, therefore, has specialized in certain directions. Specialization effectively sets limits to evolutionary potentiality. Thus, it is not likely that elephants will evolve flight or that cacti will evolve aquatic forms—not, that is, for a very long time and unless competitors were removed; that is, unless they were placed in what amounted to laboratory conditions. It is much more likely, given the fact that these organisms are already successfully occupying certain adaptive zones, that they will continue to do so, rather than that they will successfully gain entry into new zones for which other organisms are better preadapted than they. The possible preadaptations of a lineage are limited by its adaptations. Once organisms have become snakes, the path of least resistance is to continue exploring possibilities open to snakes. For example, one specialization involved in burrowing for a vertebrate can be an increase in the numbers of vertebrae. On the other hand, flight—and also jumping, as in frogs—involves instead a shortening of the backbone, possibly accomplished by decreasing the number of vertebrae. A single lineage cannot simultaneously do both. Once the number of vertebrae have increased even slightly during the beginning of fossorial life, the animals have become, because of this increase, less competent to explore the possibilities of flight than similar organisms with fewer vertebrae. On the other hand, they are now even more competent to explore the burrowing way of life and will be more successful at it than organisms with fewer vertebrae. Being more successful at burrowing, individuals in a species will compete with one another in terms of burrowing, and this will lead to increased "burrowing" refinements in other organ systems and to new burrowing adaptations, which will reinforce selection for increased numbers of vertebrae, and so on. Because there is in adaptation this kind of positive feedback, given a long enough period of time, the lineage exploring burrowing could show a trend toward an increased number of vertebrae.

A related factor contributing to the development of trends is the fact that organisms are characteristically not in adaptive equilibrium with their environments. This is for two basic reasons. First, environmental change itself is continuous, and gradual enough in relation to typical generation times so that organisms do not often experience sudden changes in the direction of change. Thus, a general drying tendency on a certain continent takes place over a long period of time or a great many generations, the aridity becoming gradually more severe. In this circumstance, plants that were preadapted to do so will gradually "improve" their xeric adaptations and will show trends in this direction. Second, the fact of intraspecific competition ensures that even if the environment stopped changing, there would still continue a kind of "perfecting" of adaptations to existing conditions, because one individual will invariably be somewhat better adapted than another and will on this basis contribute more offspring to the next generation. It is not likely that this process could carry a trend very far by itself, but then it has not had to, because the environment has not been anywhere unchanging. A trend will, of course, stop when an optimum size or number is reached. It will be an observable trend in the fossil record only if the approach to the optimum, for one reason or another, takes a long time.

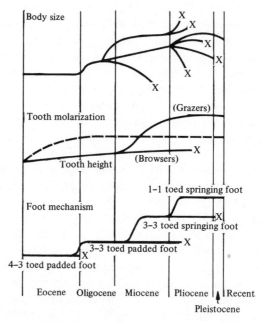

Figure 5-7   Several evolutionary trends traced through time in various horse sublineages, beginning with *Hyracotherium* and ending with *Equus*. X indicates extinction. (Modified from Simpson, 1953.)

Any lineage is likely to show more than one trend during a particular period of time. This can be exemplified by data concerning horse evolution presented by G. G. Simpson (see Figure 5-7). During the evolution of this group there were trends toward increased size, increased molarization and height of cheek teeth, and a decreased number of digits, in the sense that most successful genera were larger, or had higher and more molarized teeth, or fewer digits, or some or all of these traits in combination than their ancestors. A close examination of Figure 5-7 shows, however, that this statement is an oversimplification. Cheek teeth did not continue to increase in height in browsing horses; the trend toward molarization was essentially over before the trend toward larger body size began; the trend toward larger body size was reversed in many lineages of horses; not every lineage of horses showed all these trends at the same time or to the same degree that others did.

ALLOMETRY

Some trends may be correlated with others in various ways. Thus, the number of facets in insect eyes or the number of eggs per clutch in salamanders increases with increasing body size. The relative size of deer antlers increases as body size increases phylogenetically (see Figure 5-8). Relationships of these kinds derive either from physical necessity (insect eyes must be composed of more facets if they are to be bigger) or from evolved developmental relationships between different regions or organs and the entire body (allometric relationships; see Figure 5-9). It is interesting that these allometric relationships frequently remain unchanged over long periods of time and are often as characteristic of a group as any others of its traits. Presumably, if any such relationship began being nonadaptive at extreme values of the independent variable (body size), selection would change the relationship. Thus, in the Irish elk (*Megaloceros*), the antlers of the stag were huge indeed; its body size was at the point where antler size becomes larger than shoulder height (Figure 5-9). Could selection have further increased body size in this lineage in spite of the fact that the antlers would then be relatively even larger? It is clear that if the antlers became too large after a further increase in body size, there would ensue a counterselection against large antlers that would be, perforce, a selection against increasing the body size further, resulting in a compromise, optimum, body size and antler size. This would be the case if, for some reason, selection were unable to change the allometric relationship in question. A species of muntjak shown in Figure 5-9 suggests that the allometric relationship can be modified. It was at one time suggested that perhaps *Megaloceros* became extinct partly because its antlers became too large. The preceding discussion indicates why this could not have been so. It seems possible, however, that allometric relationships may set limits to evolutionary potential in any lineage.

PARALLEL EVOLUTION: II

Trends are frequently discovered and described, as suggested above, during the construction of a phylogeny. In one lineage of perissodactyls we see the gradual emergence of a general "horsiness," which can be analyzed into a number of separate trends, as described above. As we look at a reconstruction of horse evolution, we find that in different sublineages traits characteristic of modern horses were developed to different degrees and at different rates, in parallel fashion; that is, there are a number of parallel trends occurring at different rates. A very interesting case of the parallel working out of a number of trends in related lineages is found in the evolution of the mammalian middle ear.

Reptilian ears, both fossil and recent, involve an eardrum or tym-

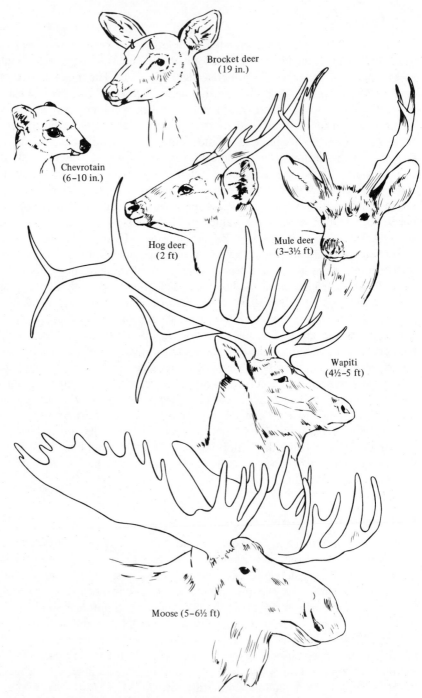

Figure 5-8   Sizes of antlers in different sized living deer. The size given refers to shoulder height of the adult male. Sizes not to scale.

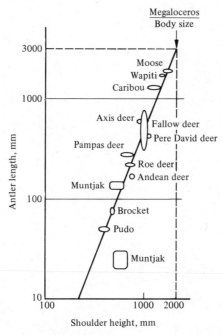

Figure 5-9 Allometric relationship between body size and antler size in various deer. All are living species except *Megaloceros,* for which the body size is indicated at the point where the antlers become larger than the shoulder height. (Data from various sources.)

panum connected to an opening (the fenestra ovalis) in the otic capsule by a slender bone, the stapes (see Figure 5-10). In the mammalian middle ear, the tympanum is contacted; not by the stapes, but another bone, the malleus, which is connected to the stapes by still another bone, the incus. Thus, in mammals there are inserted two additional bones between the stapes and the eardrum. It has been discovered (from embryological evidence; see below) that the malleus and incus are homologous with the reptilian articular and quadrate bones, respectively—bones that function as the reptilian jaw joint. The mammalian jaw joint is formed instead by contact between the dentary and squamosal bones (see Figure 5-10). The change from a quadrate-articular jaw joint to a squamosal-dentary joint can be observed in the fossil record and occurred in a number of different therapsid (mammal-like reptile) lineages independently (see Figure 5-11).

During the Triassic there was a parallel trend among many lines of mammal-like reptiles to increase the size of the dentary bone in the lower jaw in connection with an increase in the mass of jaw musculature. This involved a decrease in the size of the other bones in the lower jaw as well, including the articular that formed part of the jaw joint. This increase in

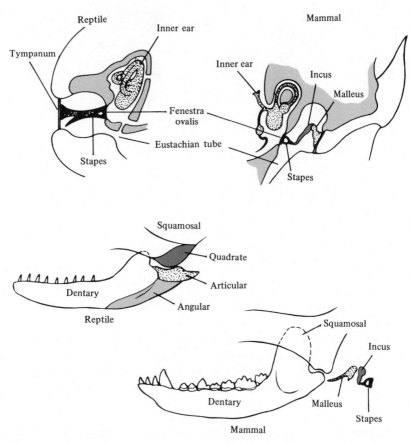

Figure 5-10 Comparison between reptilian and mammalian middle ear and jaw structure. Jaws are viewed from inside. (Simplified from various figures of Goodrich and Romer.)

muscular strength coupled with a weakening of the jaw joint has seemed a peculiar combination and is generally explained in terms of the force of the bite not stressing the joint because of a rearrangement of musculature. At the Triassic-Jurassic border we find three different groups in which this trend has continued to the point where the dentary has become so large that it has become involved with a skull bone, the squamosal, in a second jaw articulation, so that these organisms are intermediate between the reptilian and mammalian conditions of jaw articulation, having both. These three are the morganucodonts, the symmetrodonts, and the icti-dosaurs. Possibly at around the same time, the ancestors of the Jurassic-Cretaceous multituberculates and the Jurassic tricodonts—and, if neither the morganucodonts nor the docodonts are involved in the ancestry of the monotremes, their ancestors as well—were going through a similar stage.

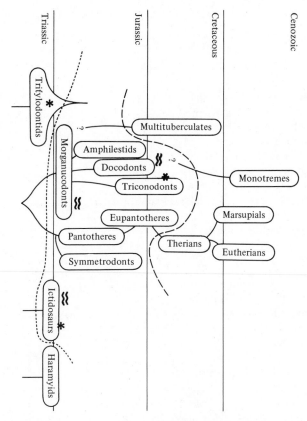

Figure 5-11 Relationships of various mammal-like reptiles and mammals. The dashed boundaries represent the transition to completely mammalian osteology; the dotted boundaries indicate the border between typical mammal-like reptiles and those showing relatively more mammalian traits. A double wave indicates a double jaw articulation. An asterisk indicates reptilian tooth replacement.

(All these groups are considered to be different lineages because of important differences in the cusp patterns of their teeth.) Possibly the descendants of the tritylodontids moved in this direction as well, because these had achieved many other mammalian characteristics—loss of the bony bar behind the orbit, a double occipital condyle, and so on. (It should be mentioned that these, too, and other mammalian characteristics were also achieved by many lineages of therapsids independently.) At a later time in a number of the early mammalian lineages the articular and quadrate drop out of the jaw joint completely. In at least two of these, the therians and the monotremes, these bones seem to have become the malleus and incus of the middle ear.

	Lineages Still Persisting at This Level of Organi-
Grades	*zation*

I. Agnathans

No jaws, no teeth — Hagfishes
Typical vertebrate paired fins absent — Lampreys
Only two semicircular canals
Notochord persisting
Pronephric kidney
No myelin sheaths on peripheral nerves
Lateral line organs present
Well-developed dermal skeleton in early
   forms
Two-chambered heart
Single median nasal aperture

↓   ↓   ↓   ↓   ↓   ↓   ↓

II. Placoderms

True jaws, most with true teeth — Sharks and rays
Paired fins present — Ratfishes
Well-developed dermal skeleton in early
   forms
Lungs present in some
Pectoral girdle present
Mesonephric kidney
Spiral valves in intestines of some
   three semicircular canals

↓   ↓   ↓

III. Bony fishes

Continued trends toward loss and thinning — Lungfishes
   of dermal skeleton — Coelecanths
Continued trend to replace notochord by — Paleoniscoids
   bony centra — Holosteans
Typical vertebrate pectoral and pelvic — Teleosts
   girdles
True ribs
Hyoid arch involved in jaw mechanism

↓   ↓

IV. Amphibians

Middle ear, eustachean tube, stapes, and — Frogs
   tympanum — Salamanders
Tetrapod limbs and five digits — Caecilians
Pelvic girdle fused to vertebral column
Eyelids in adults
Lachrimal glands and ducts

IV. Amphibians (cont'd.)

>True voice in some
>Three-chambered heart
>Ureotelic

↓   ↓

V. Reptiles

>Four-chambered heart   Turtles
>Thick, dry skin   Rhynchocephalians
>Metanephric kidney   Lizards and snakes
>Cleidoic, amniotic egg   Crocodiles
>Necessity for closer thermal regulation
>Articular-quadrate jaw suspension
>Uricotelic except very early embryo

↓

VIa. Birds

>Loss of teeth—beak   Birds
>Feathers and homeothermy
>Wings and hollow, light bones
>Visual specialization in brain
>Increased metabolic activity
>Loss of regenerative ability and continuous
>   growth
>Single occipital condyle
>Uricotelic

↓

VIb. Mammals

>Heterodonty, and loss of continuous re-   Monotremes
>   placement of teeth   Marsupials
>Hair and homeothermy   Eutherians
>Increased metabolic activity
>Two occipital condyles
>Dentary-squamosal jaw suspension;
>   mandible a single bone
>Olfactory specialization in brain; pons
>   present
>Three ear ossicles instead of one
>Enucleated red blood cells
>Loss of regenerative ability and continuous
>   growth
>Live birth (except monotremes)
>Milk glands
>Ureotelic

Figure 5-12   Organizational grades in the vertebrates. Only new additions and changes are listed for each succeeding grade. Arrows indicate the relative number of lineages thought to have made the transition from one grade to the next.

The evidence for the homology of the reptile jaw articulation with two of the mammalian ear ossicles is embryological. Thus, at comparable early stages in the development of an opossum and a lizard, there are two centers of ossification that appear in identical positions in both kinds of embryo. In the former these go on to differentiate into the ear ossicles, in the latter into the bones of the jaw articulation. There is also the circumstantial evidence that as the quadrate and articular disappear from organisms in the fossil record, these organisms take on increasingly the characteristics of mammals, and modern mammals have neither quadrate nor articular but do have the ear ossicles. Exactly how these two bones became intercalated between the eardrum and the stapes is a subject of some controversy. Selection on some parameter of hearing may have been involved, but there are problems with this of explanation. Specifically, it is not clear how an intermediate condition could have done other than interfere with hearing. It is of some interest to note that one morganucodont had a bony process leading from the stapes to the quadrate bone in the jaw joint. It is not clear what the functional meaning of this seemingly strange arrangement was. It is possible that the major selective forces were working in the embryonic stages. Perhaps the two bones were not simply lost or fused with other bones (a common fate of bones during evolution) because they were important members of a chain of embryonic induction centers.

POLYPHYLY AND MONOPHYLY

This example of parallel evolution during the history of mammal-like reptiles, brought about by the more or less simultaneous presence of certain trends in a number of related but distinct lineages, was an early example of what is by now a common problem—the polyphyletic origin of higher taxa. The class Mammalia can be distinguished from other kinds of living vertebrates by a number of characteristics (see Figure 5-12). Only some of these traits can be observed in bones and are, therefore, potentially of use in the fossil record. After much study, it was decided that some bone traits, such as a double occipital condyle, occur too early in the fossil record in organisms that are too much like reptiles in other ways to be considered diagnostic of the class Mammalia. It has been more or less agreed that, as far as skeletal structure is concerned, the cutoff point between reptiles and mammals is best taken to be the change from the quadrate-articular to the dentary-squamosal jaw joint. Arbitrarily restricting the definition of a higher taxon to a single trait (necessary because of the restricted nature of what can be fossilized and because of the gradual change from one form to another) results invariably in the problem that a number of related but distinct lineages may cross the boundary from one taxon to another independently and at different times. The higher taxon becomes, then, not a classical grouping of lineages descended from

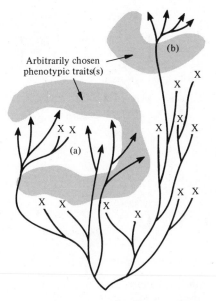

Arbitrarily chosen
phenotypic traits(s)

Figure 5-13   Two taxa, polyphyletic (a), and monophyletic (b). Arrows show lineages that continue to survive.

a single ancestral group (monophyletic) but an assemblage of related lineages descended from a number of different ancestors (polyphyletic) united on the basis of phenotypic characters held in common. Such an assemblage is called a grade (see Figure 5-13). As defined, many of our higher taxa today are turning out to be grades of organization rather than the descendants of single ancestral groups (clades). The problem fundamentally is that the probability of their having been parallel evolution increases as the number of traits used to define the taxon decreases.

GENETICS AND PARALLEL EVOLUTION

It is appropriate at this point to examine the possible genetic background of parallel trends in different closely related lineages. In drosophilid flies, a brown eye pigment (ommatin) is produced by the synthetic pathway shown in Figure 5-14. Several mutations are known in *D. melanogaster* that result in the eyes being bright red instead of the normal dark red. The cinnabar mutation inactivates the enzyme at the step shown in the diagram; the vermillion mutation destroys the activity of an enzyme earlier in the pathway. The brown pigment not being produced, the only pigment in the eye is then a red formed by way of another synthetic pathway. Suppose that we had an environment with several species of these flies, and suppose that red-eyed individuals in all species became selectively favored, for some reason leaving more offspring than the previously normal-eyed

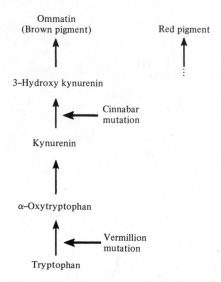

Figure 5-14   Biosynthetic pathway of ommatin in *Drosophila,* showing two of several known mutations causing inactivity of different enzymes in the pathway.

flies. The particular enzyme that is eliminated in any population or species could be different from that eliminated in other populations. The first otherwise nonharmful relevant mutation that occurs in each population is the one that becomes the norm of the population (because it has a head-start on other mutations). These populations evolve in parallel, each achieving the red-eyed condition independently and probably by different genetic means. The role of similar genotypes in this example is obvious. In man, there are a number of different mutations affecting hemoglobin that are apparently selectively favored in regions where malaria is found. Thus, in Africa Hb S is a mutation of the sixth codon of the beta chain, changing the normal glutamic acid to valine. In western Africa, Hb C is a change in the same codon resulting in a change from glutamic acid to lysine. In India, Hb D is a mutation changing the eighty-seventh amino acid of the beta chain from threonine to lysine. In the Far East Hb E is the result of a mutation in which the twenty-sixth amino acid of the beta chain, glutamic acid, is replaced by lysine. These, plus an alpha-chain variant, Hb O Indonesia (α^{116} glu → lys) and a host of different thalassemias (mutations eliminating the beta chain altogether), exist in large-scale polymorphisms with the normal hemoglobin chains in these areas and can be seen as independent answers to the major challenge of the presence of falciparum malaria—again, a case of parallel evolution, this time in different populations of a single species.

On a more complex level, we can consider polygenes (treated again extensively in Chapter 8). If a number of different genetic loci affect the size of some structure or part of one—say, a length—each locus could have an additive effect on size. If each locus can be represented by any of a number of different alleles, each with a different magnitude of effect on the size of the structure in question, then more than a single combination of alleles can result in the same phenotypic size. Suppose that selection were favoring a particular size range of this structure in all the hypothetical species of *Drosophila* above. This size range can be achieved with different combinations of alleles in the different species (see Figure 5-15).

Allele No.	Gene No.					
	I	II	III	IV	V	VI
1	/	/	/	/	/	/
2	//	//	//	//	//	//
3	///	///		///	///	///
4		////		////	////	////
5		₦		₦	₦	₦
6		₦/				₦/

							Genetic make
(a)	2	2	2	4	3	2	= 15
(b)	1	6	1	2	1	4	= 15
(c)	3	3	1	1	1	6	= 15

Figure 5-15 The possible allelic structure in a given population for some polygenic trait. Each gene acts independently in an additive fashion. Each allele has a slightly different magnitude of effect on the trait, as symbolized by I, II, III, and so on. (a), (b) and (c) represent three individuals (considered haploid for convenience) with three different genotypes, all adding up to the same "make" of 15, and therefore all having the same phenotype with respect to this polygenic trait (not necessarily in others in which these genes are pleiotropically involved).

In each of these cases, the phenotype—red eye, resistance to malaria, a size range for a particular structure—is not the direct expression of the physicochemical properties of the primary gene products but results from their interactions with other gene products. Selection favoring a certain phenotypic expression can achieve its "ends" by different genetic means. Therefore, in cases of parallel evolutionary trends in related lineages, we do not have to suppose the grossly improbable event that identical mutations occur and are selected in different lineages; this analysis allows us to accept parallelism as a very probable kind of event from a genetic viewpoint.

MOSAIC EVOLUTION

The phenomenon in which different trends in a lineage proceed at different rates, as in the example of the horse above, has been termed mosaic evolution by Gavin de Beer, of the British Museum of Natural History. Considering bird evolution, and specifically *Archaeopteryx,* he noted how the latter was in some respects more like a reptile—teeth and jaws, tail—and in others more like a bird—feathers, short vertebral column. "Birdness" had evidently been achieved in different systems at different rates. As another example, we may consider the monotremes. They are mammals in that they have fur, milk glands, enucleated red cells, the mammalian jaw articulation, and the three middle ear ossicles. But they are reptilian in having a cloaca, in laying eggs, and in the structure of their shoulder girdle and ribs. In this lineage some aspects of what we think of as "mammalianness" (see Figure 5-12) have evolved further than have others. Looked at from this point of view, mosaic evolution is partly an artefact of our definitions of higher taxa. Thus, there is no necessity that monotremes should keep evolving toward a eutherian reproductive pattern. They may not be doing this at all instead of doing it slowly. We define mammals and then we see how closely and at what rates some lineage approaches our definition, as if this definition were a part of reality instead of a *post hoc* construct. We therefore ask, for example, how far and in what ways did the tritylodontids reach into the mammalian grade of structure? Were the multituberculates more mammalian overall than contemporaneous triconodonts? From the point of view of natural selection these are idle questions. Selection in several lineages had been eliciting similar responses from similar genomes to similar ecological challenges but not necessarily in all organ systems together, nor would it necessarily continue to do so. It would all depend on which were the most successful organisms in each lineage. A trend, even a long-lasting one seen to occur simultaneously in dozens of lineages, cannot be projected into the future; natural selection is totally opportunistic and is not concerned with working out "ideas." We can emphasize this point by looking again at the monotremes. Changing the jaw joint from that characteristic of the reptiles to that found in mammals involves the loss of the muscle that depresses the mandible (because it inserts on the angular, one of the bones lost from the jaw in this transition) and its replacement by another. In the therians the digastric muscle was the newly evolved one. In the monotremes there is no digastric but a completely different new *depressor mandibulae*. Thus, in this respect, monotremes are neither reptilian nor mammalian but something entirely apart.

Nevertheless, it is clear that different traits in a single lineage change form at different rates (regardless of whether or not they are traits of interest in the definitions of higher taxa) over a particular period of geological time. The entire organism does not seem to evolve structurally as a unit. On the genetic level it is becoming clear that different genetic loci can evolve at different rates. For example, different stages in a life-history cycle are characterized in part by the transcription of loci not transcribing at other stages. Selection acts during all life-history stages, and so a locus active only in the embryo may experience more net directional selective pressure (see Chapter 10) over a given period of time than, say, a locus active only in the adult stage, or vice versa. Good examples of mosaic evolution in respect to different life-history stages are the frogs. It is clear that in this order not much significant evolution has occurred affecting adult anatomy in the last 120 million years. There is good indirect evidence, however, that tadpoles have undergone continued evolution into a number of different adaptive zones (see Chapter 8). On the molecular level, the Ingram effect (see Chapter 8) results in one genetic locus being on the average more evolutionarily conservative than another. In birds and other organisms having prominent sexual dimorphism, loci involved in the production of certain hormones, or with the reactivity of certain tissues to hormones, may have been evolving at overall greater rates than the rest of the genome, so that males have become increasingly differentiated.

EVOLUTIONARY RATES

All evolutionary patterns can be analyzed into rate phenomena. Pattern changes are brought about by changes in relative rates of change of different traits. It is fruitful to look specifically at what generalities we can derive from the fossil record concerning evolutionary rates. We have already explored mosaic evolution—the phenomenon of differential rates of change of different traits during a particular span of time in a given lineage. Looking at a specific organ or trait during the entire history of a lineage, it becomes clear that its rate of form change is not constant. This can be seen in the example of the horse in Figure 5-7. Another example is the evolution of brain size in the lineage leading to man: for the first 20 million years or so no significant size change occurred, and then relatively suddenly there ensued a period of some 2 million years of rapid increase in absolute and relative size.

Looking at evolutionary rates in terms of changes in the numbers and frequency of taxa (taxonomic rates) rather than, as we have been, in terms of form changes in organs and traits (morphological rates), we can derive analogous principles. Figure 5-16 shows that the duration of different genera of lungfishes differed significantly over their history. Very early in their history they were producing new genera rather rapidly (that is, they were evolving very rapidly) and so the temporal existence of any particular genus was relatively short. Later the duration of genera increased gradually as the overall evolution of the lineage slowed down.

Thus, overall rates of phenotypic evolution do not proceed at any more regular pace than do the rates of change of form in single traits.

In analogy with mosaic evolution, we also find that different genera in a lineage do not necessarily evolve at similar rates over a particular time span. Thus, over the 30 million years of horse evolution one sublineage underwent enough change so that taxonomists consider there to have been some five or six genera succeeding one another. During this time, another lineage of perissodactyls, the tapirs, produced no new genera; the modern forms are essentially the same genus as those found 30 million years ago.

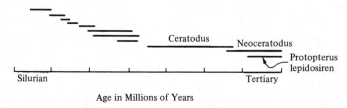

Age in Millions of Years

Figure 5-16 Longevity of genera of lungfishes at various times in the fossil record, showing a decreasing rate of evolutionary change during time in this group. Living genera are named. Some authors consider *Neoceratodus* to be congeneric with *Ceratodus*. (Modified from Westoll, 1949.)

Exceptionally slow rates of evolution (bradytelic evolution), although not common, occur with enough frequency to raise some interesting problems. For example, the brachiopod genus *Lingula* has not changed its shell construction significantly in 400 million years. Because these are denizens of the deep sea floor, it has been suggested that this is somehow related to the fact that the benthonic environments have undergone less change during this period than have other environments. Although this may have been a factor in this case, this explanation cannot be generalized to other "living fossils." The horseshoe crabs (*Limulus*) have survived essentially unchanged for about 200 million years; the Australian lungfish (*Ceratodus*) is also essentially unchanged after a similar span of time; the crocodiles have not changed significantly during the last 100 million years. These three live in some of the most changeable and constantly changing environments there are—sandy seashores and rivers. In these cases perhaps the specific regions involved have not changed with the same speed as others on the earth. There does not, however, seem to be any reason to suppose that this is true of the eastern coasts of China and the United States (horseshoe crabs), of Australia, or even of the tropical regions where crocodiles are found.

Another class of explanations of bradytely involves estimates about the breadth of the ecological niches of the organisms involved. Thus, the Virginia opossum is not significantly different (other than in being larger) from some of its ancestors that lived 70 million years ago. This animal is omnivorous and apparently can tolerate, as a species, a very wide range of climatic conditions. It is currently expanding its range northward in North America and seems tolerant of conditions associated with man's activity. Its survival intact through so many different environments may have to do with its broad ecological adaptability. But degree of ecological specialization is a tricky concept to handle, especially with fossil material. Is *Limulu*s an especially generalized horseshoe crab? Is the ecological niche of *Ceratodus* really very broad? Is it not instead a rather specialized kind of fish? Indeed, relict forms often tend on the whole to be rather more specialized for some peculiar, narrow niche than is the norm for whatever taxon they are part of.

As with the problem of extinction (see Chapter 4), that of bradytelic evolution can best be handled one case at a time. There seem to be no valid general causes of exceptionally slow phenotypic evolution. The best one can do is to suggest that, by chance, a given phenotypic expression has continued in a particular region of the earth to adequately maintain a given adaptive zone; it has not had to change for any number of specific reasons.

EVOLUTIONARY RATES AT THE GENETIC LEVEL

Is *Lingula* today not very different genetically from *Lingula* 400 million years ago? Although we cannot answer this question directly, there are a number of indirect lines of evidence to suggest that, despite their phenotypic similarity, these would probably be quite significantly different genetically. First, in comparisons between living and fossil organisms, only a small portion of the structure can be compared, and these parts may not show the same degree of difference that can be found in other systems. Thus, two closely related species of frogs in North America have different color patterns, different habitat preferences, and different call notes, but their skeletons are indistinguishable. Fossils would tell us little in this case about genetic differences. Also, phenotypic similarity does not necessitate a similar degree of genetic similarity (see above concerning parallel trends, and also Chapter 8). Also, there are a number of closely related species which are different enough genetically to be unable to produce hybrids but which are so similar phenotypically that they were once thought to be members of a single species (sibling species; see Chapter 16).

Let us suppose, however, that the ancient and living forms were identical in every way measurable in phenotypic terms. There is indirect biochemical evidence to suggest that even then they would not be genetically identical or even similar. In comparing the amino acid sequences of a number of proteins from organisms of different degrees of relatedness, it has been found that there is an overall average regular relationship between the estimated time of divergence of two groups in the past and the number of sequence differences. The same has been found measuring the antigenic determinants of the primary gene products using immunological methods. The same has also been found comparing certain regions of the DNAs by the method of DNA-DNA hybridization. In all these cases a similar form of the relationship appears, as shown in Figure 5-17. It seems that a genetic locus has its structure altered gradually as time passes, even though it continues to code for the same kind of protein. There are three reasons for this. First, there are a number of codon synonyms for any amino acid, and the particular code characteristic of a population may change from one to another and not produce a change in the amino acid;

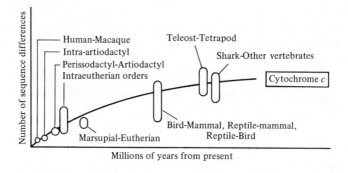

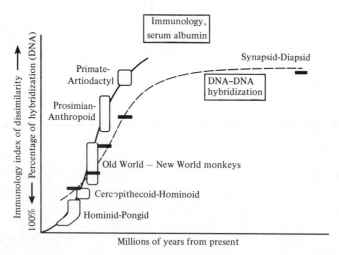

Figure 5-17 The regular relationship between measurable differences in primary gene products of existing organisms and time since the lineages of those organisms diverged in the past. (Cytochrome *c* data from Dayhoff, 1969; serum albumin data from Hafleigh and Williams, 1966 and Sarich and Wilson, 1966. DNA data from Hoyer, et al., 1965.)

a protein could in this way remain unchanged while its code was entirely overhauled. Second, a codon change that results in the substitution of an amino acid by another that is physicochemically similar to the first could occur without giving rise to significant phenotypic effects. Thus, there are mutant forms of hemoglobin chains in man which were discovered accidentally because they have altered electrophoretic properties but which do not produce any known physiological effects different from those of the gene product of the wild-type allele. Different changes of these two kinds could independently occur and accumulate in the genotypes during time in two isolated lineages even if the environments did not change. Environmental change, however, has been continuous on the earth since it was formed. This gives rise to the third reason for the gradual accumulation of genotypic differences in time.

After two groups become isolated from each other by whatever means (see Chapter 14), they will experience different selective pressures, because they are occupying different habitats in different communities. As they evolve in response to these pressures, they come gradually to occupy increasingly different ecological niches and, eventually, different adaptive zones (divergent evolution). Although both of them retain a certain enzyme, say, alkaline phosphatase, the enzyme comes to function in a cellular environment increasingly different from the ancestral one and from that of the sister lineage that also has descended from the latter. In order to retain its function as an alkaline phosphatase, without which the organisms would die, this protein has to adapt to the changing intracellular environment (that is, it has to evolve). Because the intracellular environments of the two daughter lineages are becoming increasingly different in response to various selection pressures that are different for each lineage, the alkaline phosphatases of the two lineages gradually evolve differences in their structure (at places other than the active site). The explanation, then, is that because no two communities or habitats are identical, and because climatic and structural features of the earth are constantly changing, and because there is a continuous succession of ecological catastrophes of all kinds during the history of a community, a genetic locus always or periodically has pressure on it to adapt, and, further, these pressures are usually not identical in different lineages. Thus, in order not to incur a radical phenotypic change—say, loss or impairment of alkaline phosphatase activity—the genotype needs to continue evolving; nothing new is necessarily gained, but adaptedness is retained. And so, the two daugther lineages become increasingly different at most genetic loci in a gradual manner.

MORPHOLOGICAL AND GENETIC RATES

How can this view of the genotype as being continually in flux be resolved with the facts of evolutionary rates as derived from the fossil record, which are that evolutionary change (morphological or taxonomic) is never steady in rate and is in fact markedly irregular. The key to resolving this apparent contradiction is that the phenotype is not a direct expression of the primary structure of the gene products but an indirect expression arising out of the interaction of many gene products and their end products and by-products. Thus, Hemoglobin Zurich is an abnormal hemoglobin in man, composed partly of beta chains in which the normal histidine at the sixty-third amino acid is replaced by arginine. This gives rise to no abnormal phenotypic effect unless an individual carrying this mutation is exposed to sulfanilamide. In this environment the red cells carrying the abnormal hemoglobin lyse, producing a hemolytic anemia. In other words, a phenotypic effect is a result of the interaction of primary gene products with their environments. In order to get a measurable phenotypic effect (or difference) the "right" mutation must occur in the "right" environment at the "right" time and must, furthermore, be selectively favored if it is not to be simply a passing genetic disease and, therefore, not to be registered in the fossil record. Thus, we can visualize the structure of a genetic locus being gradually modified in time and yet giving rise to phenotypic (say, catalytic) differences only intermittently and, in principle, unpredictably (see Figure 5-18).

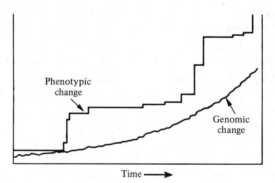

Figure 5-18   A representation of the possible relationship between genomic change and phenotypic change. Phenotypic change only occurs when the right mutation occurs in the right organism at the right time followed by natural selection increasing the frequency of the new allele until it becomes a common one, or the only one in a given population. See text for further explanation.

SUMMARY

In summary, patterns of evolutionary change at both the morphological and taxonomic levels may be similar to both kind and rates in related lineages without invoking similar genotypic changes. Such patterns of change may remain steady in direction for long periods of time, resulting in trends, but this does not mean that these patterns have any objective reality as causative or operative factors in living systems; they are *post hoc* constructs of historical data. As soon as it becomes expedient to do so, the trend will be altered by natural selection. Evolutionary patterns can be analyzed as resulting from rates of change of various parameters of traits. Typically, rates of change of phenotype in the fossil record are irregular and unpredictable, although the underlying genetic changes may be going on at steady overall average rates.

REFERENCES

Colbert, E. H., *Evolution of the Vertebrates.* Wiley, New York, 1955, 479 pp.

Crompton, A. W., "The Cranial Morphology of a New Genus and Species of Ictidosaur," *Proceedings of the Zoological Society of London,* **130:** 183–216, 1958.

————, "The Lower Jaw of *Diarthrognathus* and the Origin of the Mammalian Lower Jaw," *ibid.,* **140:**697–753, 1963.

————, "The Evolution of the Mammalian Jaw," *Evolution,* **17:**431–439, 1963.

De Beer, G. R., *Archaeopteryx lithographica: A Study Based upon the British Museum Specimen.* British Museum of Natural History, London, 1954.

Grant, V., *The Origin of Adaptations.* Columbia, New York, 1963, 606 pp.

Hopson, J. A., "The Origin of the Mammalian Middle Ear," *American Zoologist,* **6:**437–450, 1966.

Huxley, J., *Evolution: The Modern Synthesis.* Wiley, New York, 1942, 645 pp.

Kurten, B., "Evolution in Geological Time," in *Ideas in Modern Biology,* J. A. Moore, ed., Natural History Press, Garden City, N.Y., 1965, pp 327–356.

Rensch, B., *Evolution above the Species Level.* Columbia, New York, 1959, 419 pp.

Simpson, G. G., *The Major Features of Evolution.* Columbia, New York, 1953, 434 pp.

————, "Mesozoic Mammals and the Polyphyletic Origin of Mammals," *Evolution,* **13:**405–414, 1959.

————, *Principles of Animal Taxonomy.* Columbia, New York, 1961, 247 pp.

Stebbins, G. L., "Adaptive Radiation and Trends of Evolution in Higher Plants," *Evolutionary Biology,* **1:**101–142, 1967.

CHAPTER 6

Natural selection

In the conceptual framework of neo-Darwinian, or synthetic, evolutionary theory natural selection is the fundamental and, indeed, the only operating mechanism. Together with chance phenomena (see Chapter 15), it is considered to be the source of organic adaptation and all longer-term evolutionary phenomena, such as those discussed in Chapters 4 and 5. Most of the remainder of this book is explicitly concerned with this process, which will be introduced here.

The concept of natural selection was first presented to the scientific community on July 1, 1858, when Charles Darwin and Alfred Russell Wallace gave a joint paper outlining it at a meeting of the Linnaean Society of London. Both men had independently (Darwin in England, Wallace in the East Indies) come up with the same idea—an occurrence far from uncommon in science. It seems as if ideas become implicit in a situation—a body of known facts and a climate of viewpoint and opinion—and only await formulation in some sensitive and informed mind. Human individuality seems to have little to do with this process except insofar as there are usually not more than a few individuals living at any time with the requisite information, diligence, and generalizing ability.

POPULATION GROWTH

Given unlimited space and resources, a laboratory population of cells—bacteria, yeasts, tissue culture clones—

continues to increase in numbers indefinitely. The increase occurs more or less geometrically—1, 2, 4, 8, 16, 32, and so on—resulting in an exponential increase in numbers of individuals with time, approximately described by the formulation $N_t = N_0 e^{rt}$, where N_0 is the number at the beginning of observation, N_t the number at time t, e a logarithmic constant, and r the intrinsic rate of natural increase of the organism. The latter ranges widely, depending on the organism. Thus, under optimum conditions, for bacteria it can be 50 individuals per individual per day; for beetles, 0.4 to 0.05 individual per individual per day; for rats and mice, about 0.01 individual per individual per day; for man, about 0.00015 individual per individual per day.

If all the offspring of a pair of frogs survived and successfully reproduced (each female producing 2000 eggs every two years after taking three years to mature) in some dozen years there would be no dry land on earth not covered by their descendants. The reproductive potential of frogs is comparatively not especially large, but in a very short time they would be placing severe strains on the resources upon which they depend; if nothing else, they would soon starve. It is clear that the population sizes of organisms do not continue to increase logarithmically. Many experience only slight fluctuations from year to year. Others increase dramatically for short periods and then decrease even more dramatically, running in cycles. Occasionally the populations of some species are presented with the opportunity to colonize new geographical areas, and, during the increase in the species range, there may be large increases in the number of individuals in those populations. The population size in any previously occupied area, however, is usually unaffected.

In laboratory cultures of cells limited in any resource, and possibly in newly colonized areas in nature as well, the exponential phase of population growth is supplanted by one characterized by a decelerating rate of growth, resulting in an S-shaped growth curve overall. This curve (see Figure 6-1) was described by the French mathematician P. F. Verhulst in 1838. Under some conditions population sizes will stabilize at some level or fluctuate around it, experiencing from then on only slight and gradual changes in mean size. This growth curve, for which there are a number of different formulas generated from different assumptions concerning the source and mode of action of the inhibiting forces, is characteristic of biological growth in general. The increase in numbers of cells in embryos during cleavage follows such a curve, as does the increase in size (or growth) of individual organisms. In these cases the inhibition on continued increase involves a number of not clearly understood genetically or epigenetically limited processes. Much more is known about the forces acting on populations to limit their sizes to the carrying capacity of the environment. The carrying capacity of the environment for plants is dictated by the amount of standing room—determined as well by individuals of other

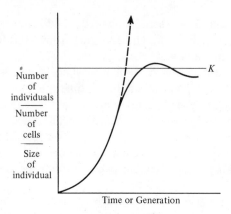

Figure 6-1   The logistic growth curve. Alternative parameters are shown on both ordinate and abscissa to indicate the wide applicability of this sort of curve in the biosphere. K is the carrying capacity of the environment, a limitation used only in the context of population growth, but analogous concepts must be used in other contexts. The dashed line indicates the shape of the curve if limitations such as K were not present.

species—available on the requisite substratum; no new seeds or shoots can continue to develop until an older plant falls. This is the kind of situation prevailing with sessile animals, such as barnacles, as well.

POPULATION CONTROL

Population sizes in animals can be, in the simplest cases, limited directly by the food supply. Here, the number of individuals increases to the point of food shortage. This usually results in decreased reproduction, because that process required energy. However, the situation frequently escalates to outright starvation because of the nature of most animal reproduction. Eggs that began development before the onset of the food crisis hatch out after the crisis begins, and an already very large population becomes too large. Starvation reduces the size of the population, often dramatically, until it is less than the carrying capacity, and then the cycle repeats itself. This situation prevails in some mammals and in some invertebrates as well. Actually, many kinds of animals have their populations controlled in other ways. Thus, in seasonal climates, many kinds of animals—especially insects—have intrinsic rates of natural increase such that they never approach the carrying capacity of the environment in terms of food before deteriorating conditions brought on by the onset of an unfavorable season decimate their numbers. A few individuals survive the bad season (often dormant or semidormant) and start the cycle again with the approach of the favorable season. Other animals have evolved territorial behavior, which, in the cases of some mammals and birds, at least, ensures that the holder of a territory will have sufficient food. In this case the population never approaches the carrying capacity in terms of available food, that capacity being set instead in terms of space as defined by the nervous systems of the animals in question. Related to this is the controversial possibility that some animal populations are limited directly by density. The idea is that, as crowding occurs, it is detected by the animals, resulting in an overstimulation of certain endocrine organs—especially, in mammals, the adrenal glands—so that reproduction is increasingly curtailed as the density (numbers of interactions) increases. This mechanism has been demonstrated in laboratory populations of rodents, but it is not clear whether it operates in nature. What needs to be shown is that such a mechanism begins to operate before the carrying capacity for food has been surpassed. In any case populations proximately limited by density or territorial behavior are, of course, indirectly or ultimately limited by the food supply in that territory size, for example, in any species has evolved in the context of a certain rate of food utilization. An indirect effect of density that may operate to limit population size is the accumulation of waste products. This can be the case in clones of microorganisms living on a decaying body or in a petri dish and might become a factor in some human populations.

INTRASPECIFIC COMPETITION

The existence of a carrying capacity, regardless of how it may be determined in specific cases, forces individuals to compete with others having the same requirements; not everyone can survive. In plants this always involves members of other species as well, but in animals interspecific competition is not always as important as intraspecific competition for reasons that will become apparent below. When the population size is directly controlled by the food supply, individuals compete in terms of who can best and longest survive starvation. When seasonal deterioration controls population numbers, individuals compete in terms of who can best survive cold, or drought, or starvation, or some combination of these during the bad season. When territorial behavior limits population sizes individuals compete for territories, these going to the stronger, swifter, more aggressive, or more energetic individuals. Individuals without territories do not survive long; something eventually kills them. A study of muskrats in Iowa by Paul L. Errington shows that any of a number of proximate agents may be the cause of the death of individuals wandering about without territories, ranging from disease to automobiles and, for only a few, natural predators. A muskrat without a territory is, in a real sense, not a complete muskrat. Although the proximate agents of death may vary widely and are probably determined in large part by chance, the ultimate cause in most cases is the same—inability to gain or hold a territory, or, failure in intraspecific competition. If, as in aging individuals, this failure occurs after a long life during which the individual has successfully reproduced, the meaning of its failure is quite different from that of a young, prereproductive individual, as will become apparent below. Generally speaking, those that survive will be those that reproduce.

VARIABILITY IN POPULATIONS

If all individuals in a population were identical in every way, chance alone would determine which survived and which did not. It was perhaps Darwin's major contribution to point out and document extensively in his book of 1868, *The Variation of Animals and Plants under Domestication,* that populations of living organisms are not composed of identical individuals. Indeed, as has become abundantly clear since then, no two individuals in a population are identical, not even monozygotic twins. Typically, quantitative variation in natural populations can be described by either the normal distribution (for continuously varying traits) or the binomial distribution (for traits that vary discontinuously), the curve of the number of individuals showing particular values for a measurement versus the possible values (the frequency distribution) being bell-shaped (see Figure 6-2). In such a distribution there is typically a single mode, near the mean. If care is taken in sampling so that only males are compared with one another, or only larvae, or only aged individuals, the curve is more likely to be unimodal.

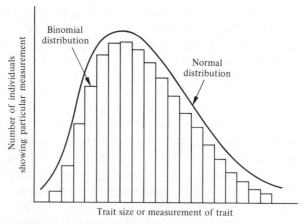

Figure 6-2 Typical distributions of multigenetic traits in sexual populations. The smooth curve is the normal distribution for traits that can be treated as continuous variables; the histogram is the binomial distribution for discontinuously varying traits (such as numbers of vertebrae). Sometimes bimodal curves are encountered, for example, when measurements include both sexes. The curves are shown somewhat skewed in a manner frequently found in natural populations.

There is frequently present another kind of variability—polymorphism. Here, as in sexual differences, there are two or a few distinct, different categories of individuals or traits, with no intermediates. In the human population there are PTC tasters and nontasters, individuals with hitchhiker's thumb and individuals without, and so on. In cricket frog populations individuals may have a green dorsal stripe, a red one, or a gray one. Sweet peas may have white flowers or red ones; snapdragons may have white, pink, or red flowers.

The genetic difference between quantitative traits and polymorphic ones is a matter of whether a change in a single gene can alter the expression of the trait. Flower color can often be described as being under the control of a single gene. In sweet peas, red is dominant and white recessive; in snapdragons, there is no dominance in this trait. With only two alleles in snapdragons, only three colors are possible; but if there were many alleles at this locus, there would be more different shades and colors so that the variation might not be described as a polymorphism. Quantitative traits, on the other hand, are affected by many different genes; they are multigenetic traits, with each locus having a small effect on the final product. (It may be borne in mind that many polymorphic traits are artifacts of description, as in the eye color of man. There are actually many different colors found, and the color depends on the amount of melanin pigment in the iris. This quantitative trait is still for some purposes described in terms of an arbitrarily defined polymorphism, as in state records.)

Obviously a major source of the variability of organisms is genetic variability generated by mutations of all kinds (see Chapter 7) and by sexual recombination. In sexual populations no two individuals are genetically alike, primarily because of the large numbers of independently segregating loci with at least two alleles. Indeed, no two chromosomes would be expected to be identical. However, various and subtle direct effects of the environment (see Chapter 8) have to be taken into account as well when considering the variability of traits in natural populations. As a dramatic recently described example, the monozygotic quadruplets of the nine-banded armadillo have been found to differ significantly at birth in a number of quantitative traits, presumably at least partly because of environmental differences in different parts of the uterus. Nevertheless, even very labile trait have a genetic dimension limiting their possibilities. Four genetically different zygotes developing in the same uterus should be even more different from one another than four monozygotic twins.

NATURAL SELECTION

If all the variability of populations were nongenetic, survival would again be a chance phenomenon, this time determined by chance events during ontogeny. All phenotypic traits of organisms, however, have some genetic component and are heritable to some degree. If an individual has some form of a trait that gives him a better chance of surviving in a certain environment, his offspring will tend to show this form of the trait as well. Differential survival (based in part on the genetically influenced phenotype) leads to differential reproduction, which leads, in turn, to the differential survival (or differential replication) of the alleles responsible for differential survival. In fact, it is customary today to think of natural selection as differential reproduction rather than as differential survival. Individuals do not survive, but copies of their particular alleles may survive in future generations if such individuals are capable of outcompeting (outreproducing) their neighbors. The survival that is important is that of copies of certain patterns encoded in the DNA. The death of individual organisms per se is not itself important, because they all die eventually. What is important is genetic death. Any individual that does not successfully reproduce (either because he dies prematurely, or is sterile, or impotent, or unattractive, or incompetent) is genetically dead. An individual that dies after a long, fertile lifetime—even if his death is due directly to intraspecific combat, say, with a younger individual—may in fact "survive" better than the younger individual. This will all be determined by who eventually leaves the most descendants in future generations and by that measure only.

Survival, then, or success, entails projecting one's particular alleles into the future gene pools of one's population at the expense of alleles from other individuals (if they are different). Note that genes and not genotypes or genomes are passed on to offspring. (In actual fact chromosomes are what are passed on to offspring, but since a very considerable amount of crossing-over and other kinds of chromosomal recombination events (see Chapter 7) take place, parental chromosomes are not typically transmitted intact.) There are some exceptions to this, as in supergenes (see Chapter 9). Since offspring receive from their parents chromosomes and not individual genes, some multigenetic traits from the parents may be passed on to their immediate offspring. In diploid organisms, however, the offspring of sexual reproduction can never have exactly the same total genotype as a parent, and their multigenetic traits are rarely identical. Furthermore, only a few generations are sufficient for a point mutation or a chromosome mutation (see Chapter 7) to alter a chromosome, changing it from what it was in the successful ancestor. But natural selection could maintain intact for very long periods of time particular patterns of nucleotide bases (alleles) at any particular genetic locus if those alleles continued to be associated with genotypes that result in reproductively successful phenotypes.

SELECTION AND EVOLUTION

As one generation succeeds another, there will be at least small changes in the gene frequencies of the gene pool of the deme, or local population. As the number of copies of alleles characteristic of successful individuals accumulate in the gene pool—these having increased their gene frequencies—those of less successful individuals decrease in frequency. This change in gene frequency in a deme is what today we refer to as evolution because it is this that underlies the changes with time in phenotypic traits which are referred to as organic evolution and which are observed in the fossil record. Individuals do not evolve; alleles do not evolve (but see Chapter 7 for the evolution of genes); gene frequencies, and, therefore, the phenotypic characteristics of individuals in populations, and, ultimately, species do evolve.

The mechanism by which gene frequencies change is differential reproduction, or natural selection operating on the phenotypes of individual (parental) organisms. Leaving aside the weeding out of genetic defectives (lethals), selection operates because there is a finite carrying capacity to any environment, allowing only a few individuals to survive; those that survive reproduce; some of those that survive reproduce more than others, resulting in even finer distinctions among genotypes by natural selection. Natural selection works because individuals are phenotypically different and because the phenotype is in part heritable.

DIFFERENT LEVELS OF INTRASPECIFIC COMPETITION: I

Conspecific individuals in the same population compete with one another in two fundamental ways; that is, there are two aspects to intraspecific competition. First, each individual of a species in a population shares the same ecological niche with each other individual in the population and, therefore, is dependent on the same requisites from the environment. This results in a direct, physical competition—every seed eaten by A is not available to B; the capture of A by a predator might postpone B's eventual demise; and so on. This competition may also involve competition for mates, which then grades into the other major way in which individuals compete, that is, for a larger share of descendants in future populations. The latter kind of competition, except when it involves, in males, physical combat for females or else competitive courtship displays (or for females combat for a desirable nesting site) is one in which the individuals do not compete directly or in any physical way. Thus, if for any reason A's offspring die before reproduction, B's offspring (and alleles) have a better chance of surviving in the next generation. These events may well be decided after A and B have both ceased to exist. It is perhaps unfortunate that the term *competition* has come to be used for both of these relationships. Ecological intraspecific competition has an important bearing on reproductive intraspecific competition in that success in the former tends to be associated with success in the latter (just as a successful phenotype is associated with a successful genotype) but the relationship is not always simple. Thus, there might be an ecologically very successful individual who is sterile. But ecological success is always necessary for reproductive success even if it is not sufficient to produce the latter.

This approach allows us to escape from the circular reasoning implied in statements like, "The fittest are those which survive in the most offspring, and those which survive in the most offspring are defined as the fittest." Ecological fitness arises out of the interaction of the unique genotype with a specific environment and tends to lead to reproductive fitness, which works to perpetuate in the gene pool the components of the genotype that gave rise to the ecological fitness.

Although members of different, closely related species can compete with one another on the ecological level, they can never compete in terms of reproductive success. Ecological competition can be almost identical interspecifically and intraspecifically. Reproductive competition can only be intraspecific; it is not possible for heron genes to increase in frequency in crow populations. This reasoning leads us to an interesting difference between sexual and asexual organisms, the latter for the most part being procaryotes. Natural selection is differential reproduction. This definition makes obvious sense only in connection with members of a single deme, a single gene pool, that is, with sexual organisms. From the point of view of the breeding population, there is none in procaryotes or other obligate asexual organisms; there is no gene pool; there is no genetic connection between members of even the same clone, other than the fact that they are likely to be similar genetically because they only recently descended from the same ancestor.

In procaryotes, ecological success also generally leads to reproductive success. Obviously differential reproduction occurs among them, and one individual can be selected to be the ancestor of most of the future generations in a local area on the basis of, say, resistance to some chemical agent in the environment to which most of its fellow clone members are susceptible. But in the absence of a common gene pool, each procaryotic individual is a species by itself (see Chapter 14). With favorable and neutral mutations accumulating in the chromosome as it is passed from one generation to the next, there are changes in the procaryotic genotype with time, but the concept of gene-frequency change is not appropriate, because there is no gene pool. The evolution of procaryotes is directly comparable with that of a single genetic locus or a chromosome (see Chapter 7). On another level, procaryotic evolution is like the evolution of different allopatric populations of a eucaryotic species or of sympatric populations of closely related species with considerable niche overlap in one community. One population may be more successful in expanding its size, while another is less so; one population might supplant another in the sense that its descendants come to occupy the place that the other population's descendants would have occupied if it had left any. This is clearly competition at the ecological level; in fact it is equivalent to what we here term interspecific competition (see also Chapter 3), and it results in a kind of selection. Indeed, it could be seen to fit into the concept of natural selection as Darwin used it.

Today, natural selection is usually taken to mean differential reproduction among members of the same gene pool. Differential survival may refer to the duration of a population or lineage, or of a clone of procaryotes, or of a certain chromosome arrangement or allele. Differential reproduction among members of a gene pool can result in changed gene frequencies in the gene pool; differential survival is concerned only with the duration of a particular configuration of genetic material—an allele, a chromosome, a particular gene pool. Differential reproduction among members of a single gene pool, although it leads to or maintains adaptation and, therefore, determines whether or not a population, species, or lineage will survive or endure, is itself the process that determines only what changes will occur in the frequencies of alleles or chromosomes in gene pools. As such it does not occur among obligately asexual organisms.

EUCARYOTE-PROCARYOTE DIFFERENCES

The achievement of the eucaryotic level of organization (see Chapter 2) can be seen from this point of view, also, to have been a major step in the history of life on the earth. Asexual organisms can only divide and mutate; there is no method for the recombination of genetic material (see Chapter 7) and no opportunity for classic, complete intraspecific competition. (Some forms of genetic recombination have been experimentally produced in laboratory populations of a few procaryotes, but there is no indication that these mechanisms operate in nature.) Adaptation at the procaryotic level is in principle slow, awaiting the accumulation of favorable mutations. This is no doubt why only those procaryotes survive today which have an extremely short generation time, (so that vast numbers of individuals are produced in a short time) and which are, therefore, restricted to the simplest body forms and smallest sizes. Eucaryotes, by having a gene pool, have a means by which genetic material can be tested in the environment in a great many different combinations simultaneously. The most favorable combinations at any time leave the most offspring, and so the character of the gene pool becomes altered in such a way that it remains probable that sufficient well-adapted combinations of gene products will arise from the random recombination of its units. Differential reproduction among members of the same deme leads to the differential survival of alleles in its gene pool such that the survival of the deme itself remains probable. In procaryotes, the differential survival of lineages or clones (involving their differential reproduction) also leads to the differential survival of alleles.

The important difference between procaryotic clones and eucaryotic demes need not take away from the concept of natural selection any of its powerful generality if it is defined as differential reproduction per se. Differential reproduction simply has more dimensions in demes than in clones. This definition can apply as well to the differential reproduction of genetic loci in viral RNA in the experiments of Mills, Peterson and Spiegelman described below and in Chapter 11.

SELECTION DIRECTLY ON THE GENOME

Although it has become most useful to describe evolution as a change in the genetic makeup of members of a clone or in the gene frequencies of gene pools, it is the phenotype that is actually selected in terms of a given specific total environment. Evolution is possible because there is some relationship between phenotype and genotype. This relationship is not usually simple and direct (see Chapter 8), but there is enough correspondence between the two so that we may extrapolate from one to the other. An ecologically and reproductively successful phenotype derives its characteristics in part from its genotype, in which information is encoded that has, in a given environment, produced an ontogeny that has resulted in the successful phenotype. Reproduction of a phenotype results in replication of its genotype.

A special problem arises here in connection with viruses. These systems maintain encoded in their genetic material less information than is necessary for them to complete their development and reproduction alone. They use the protein-synthesizing machinery of other living systems, which they invade, in order to reproduce. No information concerning such machinery is contained in their genomes. They are, however, complete systems in the sense that they have no difficulty in finding the appropriate machinery (or environment) in nature and are in fact specialized in their phenotypes to do so. Not all the phenotypic structure needed to replicate the genotype, however, is present in these systems. An interesting experiment with viral RNA was reported by D. R. Mills, R. L. Peterson, and Sol Spiegelman, of the University of Illinois, showing how this condition could have come about. They placed a viral RNA genome in a test tube containing all the necessary nucleic acid precursors and a specific replicase. The RNA acted as a template and directed the synthesis of more of itself by the replicase. Selection was then instituted for an increased replication rate by shortening the period between the withdrawal of samples of new RNA. After having gradually cut the time allowed for replication from 20 to 5 minutes, they had an RNA that was being replicated some 15 times faster than normal. Examination of the RNA showed why; about 83 percent of the genome had been eliminated from it. Only those genes remained which were concerned with the recognition of the RNA by the replicase. All others, such as those coding for replicase, for coat protein, for lysozyme, and so on, had been lost. Indeed, all the genes coding for the phenotype had been lost when they were no longer needed to bring about replication. All RNA templates that still coded for these now irrelevant phenotypic constituents took longer to replicate than 5 minutes (simply because they were larger than those which had lost these loci) and so their expression was selected against.

In certain special environments, then, selection can operate directly on the genome, but usually it works by way of the phenotype. The latter is necessary to provide an environment in which the genome can replicate. This environment includes the machinery for replication itself and—what most of the phenotype of higher organisms is concerned with—equipment necessary to obtain sufficient raw materials and energy from the external environment.

GENERATION TIMES

An interesting theoretical problem arises when we consider that it is the phenotypes of the previous generation interacting with environmental conditions at a time in the past that determines the nature of the genetic material from which the present generation's genotypes are constructed. It is the length of generation times that allows such a system to work. If the time between generations were so long that significant geological or climatological change could occur during it, the system would not work. Even the longest generation times, those of some woody plants—30 to 50 years or so—are only fractions of tiny segments of geological time. Obviously, systems with too long generation times have been eliminated by evolution, because they could not remain adapted to an environment changing faster than they could evolve. The environment would be presenting them continually with insurmountable ecological catastrophes (see Chapter 3). It is most probable that such systems did not even originate; that is, a generation time less than a certain threshold value was a prerequisite for any evolving system.

DUAL ROLE OF THE ENVIRONMENT

The role of the environment external to living systems in natural selection is an interesting dual one. On the one hand, because of the nature of the epigenetic processes of development and of the nature of primary gene products (for the most part enzymes), the environment of a living system shares in shaping its phenotype (see Chapter 8). The phenotype is in part a direct expression of a particular environment; at low temperatures, for example, the rate of reaction of one kind of enzyme may be slowed down relative to that of another. All gene products have a range of reaction rates associated with a range of environmental situations, and, therefore, all phenotypic traits, whether they are the reaction rates themselves or structural features derived from the interaction of several of these rates at different times, show variation correlated with the variation of external environmental parameters. On the other hand, the external environment has determined as well which particular allelic forms of the genetic material are present at a particular time. It has done this by, to use the excellent analogy of George Wald, editing away copies of alleles that could not interact with past environments to produce viable, well-adapted phenotypes. To use a different analogy, the environment functions as a filter, letting pass only those alleles having the adequate structure, that is, a structure determined to be adequate by the filter.

In a very real sense the environment is supreme in determining the shapes, structures, and functions of living systems. In Chapter 2 an attempt was made to show how the environment determined the details of the emerging structure of the first living systems. What is being pointed out here is that it continues to shape them now. The connection between living systems and their environments is so close that it may well ultimately become more useful not to make a serious distinction between them. Living systems are but one expression of an environment. The fact that they are characterized by individual units (genetically and sometimes phenotypically individual in function; see Chapter 12) has historically led us to emphasize the separateness of the individual units, but this may not always be the most fruitful way to look at these processes.

SELECTION FOR FECUNDITY. COUNTERSELECTION

The picture that we have so far derived of natural selection can be caricatured by a picture of the survivors in any generation (the adaptively adequate, or fit) firing gametes into the next generation. The most successful will be those whose gametes become involved in producing the largest proportion of the next generation's gamete-firing survivors. At the simplest level, once ecologically relevant fitness has been achieved, the way to maximize one's share of the next generation should be to increase the number of gametes that one produces (this argument does not involve procaryotes). If one produces a greater share of the gametes in the next generation, then, if the chances of one's gametes surviving are no worse than those of other individuals, the chances are that they will go to produce a larger proportion of the next generation's breeding individuals than will those of other individuals. Natural selection, then, should maximize the reproductive potential of all organisms (r selection; see Chapter 9). And yet, not all organisms produce tremendous numbers of offspring in the way of many marine invertebrates or some fishes, such as the cod. Indeed, many organisms produce relatively few offspring each year. Many mammals, for example, bear only one to three young every year or every other year. It would be a mistake, however, to imagine that in such mammals selection was not working to maximize reproductive potential.

In cases in which organisms produce demonstrably large numbers of eggs or seeds per unit time, the propagules are always simply dispersed into the environment and left to fend for themselves. They are characteristically small and weak. They invariably suffer tremendous mortality, and the probability of an individual's survival increases with age. In order to produce many offspring, the amount of energy—yolk, parental care—devoted to each propagule must be less unless the size of the body of the parent can be increased significantly. Selection obviously does not always favor increased body size—in moles, for example, or insectivorous birds or some kinds of parasites or, for other reasons, in whales or elephants). Body size is a trait impinged on by numerous selection pressures; although larger individuals may have greater fecundity, they may be hampered in some ecological sense so that there is instituted a counterselection against increased body size. Fecundity, like body size and, indeed, like all phenotypic traits, is the compromise product of many selection pressures and is more or less in harmony with the rest of the ecological and evolutionary "strategy" of the organism.

An interesting example in this area is afforded by clutch size in birds, as explored by David Lack, of Oxford University. Hen birds typically are physiologically able to produce many more eggs than are found in normal clutches. This can be proved by removing eggs from a nest. When this is done, the hen replaces the losses, bringing the clutch size back to near what it was before the theft. They can repeat this over and over again. Why, with this power of laying so many eggs, do birds usually have rather small clutch sizes (from one to ten or so)? Or, what selection pressures operated to produce the homeostatic mechanisms that act to limit clutch size to some number less than the maximum physiologically possible? The answer, as found by Lack, illustrates the very important principle of selection for an optimum, rather than a maximum.

There is a binomial curve of clutch size in any population of birds, with most hens producing the mean number for the population or species. A few individuals produce less than the mean number typical for the species, and these tend to leave fewer offspring in the next generation. Selection thereby tends to eliminate alleles that favor the smallest clutch sizes by selecting against individuals that carry many of them for this multigenetic trait. Every generation, however, new individuals are generated by recombination who have a low total "make" of alleles at genes affecting clutch size (see Figure 6-3). These individuals are again disfavored by selection, but the alleles in question will probably not be eliminated from the gene pool (for reasons discussed in Chapter 9 and 10), because it is only extreme combinations of them that produce too low clutch sizes. On

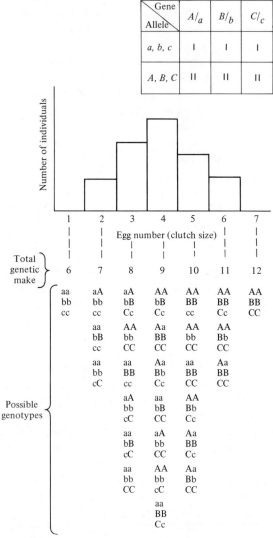

Figure 6-3 A hypothetical system of three polygenes coding for egg clutch size. Each gene has two possible alleles capable of contributing different amounts of "make" to the clutch size. Some genotypes are severely selected against (for example, A/A, B/B, C/C), but are regenerated every generation by genetic recombination.

the other hand, it was found that individuals that laid more than the mean number of eggs also did not contribute as many breeding individuals to the next generation's population as did those closer to the mean. The important point here is that in birds it is not sufficient simply to drop eggs into the environment. They must construct nests; they must incubate the eggs—if too many, they will not cover them sufficiently; they must feed the nestlings—if too many, the parents will be unable to satisfy the hunger of any of them. Lack especially concentrated on this last point and shows that nestlings in large clutches in some species have a higher than average mortality, while in other species, they fledge about as well as others but soon succumb to weather or predators, because they are weak in comparison with nestlings from more moderate-sized clutches. Thus, clutch size is closely tailored to other aspects of the life and physiology of birds and to the availability of food organisms for the nestlings. Concerning the latter, it was found that clutch sizes generally were somewhat larger in North Temperate regions than in tropical ones for similar kinds of birds. This was found to correlate with the density of insects during the nesting season, which was greater during the northern summer than at any time in the tropics. In temperate conditions the parents need not expend so much energy per insect in catching them and can, therefore, raise more offspring. Selection then accumulates alleles at those genes which tend to favor intermediate clutch sizes. Again, gene recombination always generates some individuals that tend to produce larger than normal clutches.

Optimal numbers of offspring are also selected in other groups. Only when it is possible to drop propagules into the environment and forget them is it possible to increase the number of offspring to the maximum physiologically and anatomically possible. Whenever some energy is expended in providing for the young (parental protection, nest building, milk secretion, yolk production, and the like) there will be fewer young by a factor at least equivalent to the energy expended in these ways. In other words, a counterselection is set up against increasing fecundity, which, other things being equal, is the general drift of selection due to reproductive competition. Other things are rarely equal in the complexities of nature, however. Individuals in a population compete with one another at the ecological level, and any selective pressures set up by reproductive competition alone are countered if they interfere with adaptation at the ecological level. Another obvious situation of this kind occurs in sexual selection (see Chapter 12). An organism is a balanced outcome of many selective pressures and is from one point of view a phenotypic harmony of them and from another it is an adequate compromise between them. Fecundity or sexual attractiveness is increased, not to the maximum, but to the maximum possible under the physiological, anatomical, and ecological realities of the moment.

A good example of the interrelatedness of various features of organisms can be found when considering fecundity in lizards. The fence lizard (*Sceloporus olivaceous*) deposits 10 to 23 eggs per clutch as many as three times a year. It has a very low survivorship, however (see Figure 6-4), and only poor ability to autotomize (drop) its tail. Fecundity is high and mortality, by predation, is high. The desert night lizard (*Xantusia vigilis*), on the other hand, has only two offspring once a year. Its survivorship is relatively high (high enough so that we could say that the phenomenon

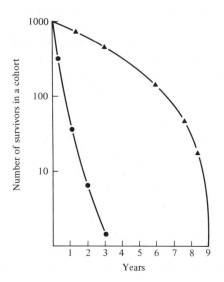

Figure 6-4 Survivorship curves for two lizards. ● indicates *Sceloporus olivaceous;* ▲ indicates *Xantusia vigilis.* (Modified from Zweifel and Lowe, 1966.)

of senescence is present—see Figure 6-4), and it readily autotomizes its tail and regenerates a new one rather quickly. It can be calculated that about 50 percent of predator attacks are survived by means of tail autotomy in this lizard (the tail drops off and wiggles about, attracting the predator, or if the latter has seized the lizard by the tail, it is left holding that organ only). Thus, the species with long life expectancy and good defense against predators has a low but adequate fecundity. The reproductive potential of the species with little predator defense and a short life expectancy is high. Selection obviously is not maximizing fecundity in both cases to the same extent. The species with fewer offspring bears them live; the eggs originally are quite large, and further nourishment may be received from the mother in the oviducts. Thus, in this case more energy is devoted to each ovum, thereby necessitating that there be fewer of them. The strategy works partly because at birth the young lizards are larger and more capable of dealing with the world than are the smaller products of the smaller eggs of *Sceloporus.*

Among salamanders, fecundity is closely tied to body size. A small species simply has a smaller ovary, and so the strategy of producing many smallish eggs is not optimum. On the whole, smaller salamanders tend to produce fewer, larger ova and to protect them by parental defense of the nest site. Selection tends in smaller salamanders to increase ovum size; in larger ones it tends instead to increase clutch size.

RUDIMENTARY POPULATION GENETICS

At this point we back off and approach natural selection from another point of view, the mathematical, or from that of population genetics. To do this at all, and certainly to do it at this level, it is necessary to grossly oversimplify by restricting our view to, for example, one gene at a time and to two alleles at that locus. Needless to say, the principles involved can be applied, at much more expense in space and energy, to any number of genes and alleles.

HARDY-WEINBERG FORMULATION

The central theorem of population genetics is the Hardy-Weinberg formulation, which describes (and predicts) the situation that obtains at equilibrium, that is, when there is no (postzygotic) selection, no mutation, and no gene flow. (This formulation, or law, affords another example of two individuals independently arriving at the same idea at the same time. The mathematician G. H. Hardy, in England, and the geneticist W. Weinberg, in Germany, both independently published this mathematical description of equilibrium gene pools in 1908.) This description is concerned with sexual populations, or demes, only. The breeding structure must be random (panmixis must obtain), and the population size must be infinitely large or, as an approximation, large. The latter stipulation is necessary to rule out gene-frequency changes due simply to chance fluctuations of gene frequencies (Chapter 13). Such chance effects are possible, although small, even in quite large populations if genes or alleles are truly neutral with respect to natural selection and in this respect too the formulation may be unrealistic. Of course, equilibrium conditions probably never obtain in nature either for more than a fraction of geological time at any given locus. Probably most sexual populations are panmictic over a reasonable number of generations, although there most certainly do occur transiently in such populations various preferential mating phenomena (assortive mating). But the purpose of a "law" such as this is to describe an ideal situation, against which real populations can be measured.

Given equilibrium conditions, then, let p be the frequency of allele A and q be the frequency of allele a at some genetic locus. Then $p + q = 1$; $p = 1 - q$. By probability theory for independent events the probability of an individual's receiving A from both his parents $= p \times p$, or p^2. The same is true for a, and so the probability of an individual's being homozygote a/a is q^2. The probability of being a heterozygote and getting A from one's father and a from one's mother is pq, while the probability of getting a from one's father and A from one's mother is qp, or, the probability of being any kind of heterozygote is $2pq$. Therefore the proportions of the different homo- and heterozygotes in a deme can be described by the formulation

AA	Aa	aa	genotype
p^2	$2pq$	q^2	frequency

which is the Hardy-Weinberg law. Given the gene frequencies (actually allele frequencies), we can, using this formulation, predict the genotype frequencies. Or, given the genotype frequencies, we can derive the gene frequencies.

Now, at equilibrium conditions, this law states that the gene frequencies will not change. This can be shown by considering the situation in the next generation. $p(A)$ would be determined by the number of gametes carrying A, and this would depend on the number of carriers of A in the population. Thus, $p(A) = p^2 + pq$ (one half of $2pq$) $= p^2 + p(1 - p) = p^2 + p - p^2 = p$, as before, and $q = 1 - p$. Thus, in the absence of any force tending to change gene frequency—mutation, selection, gene flow—it does not change from generation to generation. Of course, random fluctuations can occur in any but infinitely large populations.

ADAPTIVE VALUE

We want now to see how selection affects the equilibrium frequency. For this, a way needs to be found to quantify selection, and this way is the concept of the selection coefficient s and its converse, adaptive value or Darwinian fitness, w. Because the fitnesses of competing individuals are relative values (if the best is eliminated, say, by chance, the next best becomes the best), the adaptive value of the best can be set at 1 (or 100 percent), with the less than best fitnesses being less than 100 percent by an amount determined by the selection against them, or by $1 - s$. Thus, $w = 1 - s$. For the best genotype $s = 0$, and $w = 1$. Applying this to the Hardy-Weinberg formula, we multiply the genotype frequency by its adaptive value:

AA	Aa	aa
$p^2(1 - s_1)$	$2pq(1 - s_2)$	$q^2(1 - s_3)$

This is so because the degree of reproductive success would be decreased not at all for the most fit ($w = 1$) and to a greater and greater degree as s increases in value. In the case in which A is a dominant, and $1 - s$ for AA and for Aa equals unity

AA	Aa	aa
$p^2(1)$	$2pq(1)$	$q^2(1 - s)$

gives the genotype frequencies in the next generation. In the case of no dominance, with A superior to a in a simple way,

AA	Aa	aa
$p^2(1)$	$2pq\left(1 - \dfrac{s}{2}\right)$	$q^2(1 - s)$

gives the genotype frequencies in the next generation. Note that the selection coefficient is defined in the negative sense of selection against. Strictly speaking, selection does not produce anything; it weeds out genotypes produced in other ways; it is a negative force; traits cannot be selected *for* except indirectly. Frequently, in speaking or writing, we refer to something having been selected *for;* we continue to do so for the sake of convenience only.

To carry the argument further, actual numbers must be applied to these formulations. To do this with the least inconvenience, we must choose the simplest situation from the point of view of doing calculations. *A* is dominant to *a,* and *AA* and *Aa* are favored ($w = 1$ for both of them). For *aa*, $s = 1$; that is, no *aa* individual ever reproduces (such an individual has a genetic disease, $w = 0$). Suppose that $q(a)$ is 55 percent when the environment changes in such a way as to convert *aa* to a condition of genetic disease (no real change would be likely to be so drastic). What would be the genotype frequencies and gene frequencies in the next generation? Beginning with

AA	Aa	aa
0.20	0.50	0.30

derived from $q(a) = 0.55$, $p(A) = 0.45$, we note that *A* gametes come from both *AA* and *Aa* individuals in the proportion 0.20 from *AA* and 0.25 from *Aa* (only half of the gametes in this class being *A*) totaling 0.45, while only half of the *Aa* class gametes provide all the *a* gametes in proportion 0.25. Then $p(A)$ in the next generation equals 0.45 out of a total of 0.70 (instead of 1.00) or 65 percent. The gene frequency of *a* is then 35 percent, having gone to that from 55 percent in one generation. If we plug all this into a computer and allow the process to continue for many generations, we derive Figure 6-5:

Generation	$q(a)$	Generation	$q(a)$
Parental	0.55	*f*10	0.085
*f*1	0.35	*f*15	0.059
*f*2	0.26	*f*20	0.046
*f*3	0.21	*f*30	0.026
*f*4	0.17	*f*40	0.024
*f*5	0.15	*f*50	0.019
*f*6	0.13	*f*100	0.010
*f*7	0.11	*f*500	0.002
*f*8	0.10	*f*1000	0.001
*f*9	0.09		

Figure 6-5 The change in the gene frequency of a recessive allele *a* with maximal selection against it ($s = 1$).

This is plotted on a graph in Figure 6-6.

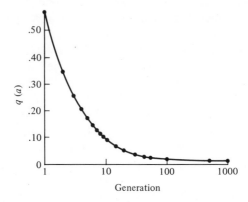

Figure 6-6   The change in the gene frequency of a recessive allele *a* with maximal selection against it ($s = 1$).

The interesting point to notice about this is that even with a constant maximum selection coefficient operating against the double recessive genotype, the effectiveness of selection against the recessive allele decreases steadily, approaching an asymptote. The asymptote represents the mutation rate at which *a* is produced from other alleles (from *A* in this case). The approach to this would be even slower if the selection coefficient were less than one or if heterosis (Chapter 9) were a factor. The approach would be faster if *a* were dominant or if there were no dominance. Obviously, alleles are not easily selected out of a population, although their frequency can be decreased to 1 percent in rather short order (50 to a 100 generations). Using less extreme selection coefficients, Haldane showed by similar calculations that on the average it should take some 300 generations for one allele to replace another in a population (go from 1 to 99 percent frequency). This calculation is discussed again in connection with the concept of genetic load in Chapters 9 and 10.

PRE- AND POSTZYGOTIC SELECTION

It is sometimes possible to collect data about single gene products from natural populations, and such data provide an opportunity, by using the Hady-Weinberg formulation, of testing whether selection is operating on these genes in these populations. Thus, if the number of alleles is known, their gene frequencies in one generation can be derived from this formulation if frequency data are obtainable for at least one genotype, and then they can be compared with the gene frequencies in the next generation. Two kinds of differences are possible as the effects of selection, depending on what life-history stage the effective selection is working at. If the main selective forces are operating at the gamete stage (prezygotic selection), the next generation will have the possible genotypes in a ratio predicted by the Hardy-Weinberg law, but the gene frequencies will differ from those of the parental population. If, on the other hand, the main selective forces are working after zygote formation (postzygotic selection), the genotype frequencies will not approximate those predicted by the Hardy-Weinberg law. Of course, because other forces can change gene frequencies, we need to be able to rule out the effects of chance fluctuations, of gene flow, and of mutation. Because new mutations do not appear in appreciable numbers of individuals until after they have been favored by selection for some generations, the latter need not be a serious problem. In very large populations chance effects may not be large enough to influence the data. The effects of gene flow, and also the possibilities for assortive mating, need to be assessed by the type of population structure characteristic of the organisms being studied. Thus, comparisons from one generation to the next should give fairly reliable data about selection. As the number of generations between comparisons increases, the effects of frequency-altering forces other than selection increase.

DIFFERENT LEVELS OF INTRASPECIFIC COMPETITION: II

Prezygotic selection (see Chapter 11 for an example) results from competition between gametes and could be one aspect of the indirect competition between individuals in a deme, as opposed to direct competition, when females mate with more than one male. A major component of indirect competition is in terms of the adequacy of the internal homeostasis of an individual as determined by genes acting on various physiological processes, such as rate of heartbeat under particular conditions, efficiency of mitochondrial oxidative phosphorylation, the characteristics of the cell membrane, the ability to produce a hormone in optimal amounts, or the ability of an embryonic inducer to complete its function. We may for present purposes oppose such traits to a second category of those involved directly in obtaining food (beak size, speed, tooth structure), or mates (antler size, color of feathers, courtship behavior) or in escaping from predators (alertness, stamina, protective coloration). It is in terms of this second category that individuals compete with one another, both ecologically and reproductively, in a direct way. Selection working on traits in the first category in effect determines which individuals will survive long enough to compete with one another more directly.

Are the traits in these categories different in the way selection acts on them? Genetic diseases ($s = 1$) can be found in genes affecting both kinds of traits. Thus, the evidence indicates that any variants of cytochrome c in mammals would result in genetic disease and that there is consequently only one wild-type allele for this enzyme. An example in the other category is chondrodystrophic dwarfism (normal trunk, dwarf limbs) in man. Individuals carrying this dominant trait may be physiologically fit, but, at least under some cultural conditions, would not compete well with normal individuals either reproductively or ecologically. While physiological fitness is a *sine qua non* for Darwinian fitness, it is not by itself sufficient.

SELECTION AND DIFFERENT KINDS OF TRAITS

If some genes are characteristically represented in populations by only a single wild-type allele, their variability is low (restricted to a few selectively neutral new mutations). Other genes may have greater variability. Thus, in populations of *Drosophila* flies the number of different allotypes of enzymes involved in important biochemical pathways was found to be less than the number of alternative forms of enzymes not so involved— proteolytic enzymes, for example. But the variability of genes is only indirectly related to the variability of multigenetic traits (see also Chapter 8). The variability of physiological traits ranges widely (coefficient of variation—standard deviation of the mean $\times$ 100/mean—ranges from 7.1 to 304 in a randomly chosen set of physiological traits in animals, as opposed to a range of from 4 to 10 for anatomical traits in mammals; see Figure 6-7). Part of this wide range of variability can be explained in terms of some of the measurements being only indirectly related to some final, important physiological function. Thus, blood pressure may be very closely monitored by natural selection, showing fairly low variability from one individual to the next under identical conditions, while heart rate, or capillary diameter, or the force of ventricular contraction could be very variable in the same population. Each of the latter contributes to the final important product (blood pressure), which is arrived at by different combinations of values for the subordinate components in different animals (by analogy the value 6 can be arrived at by 12/2, 36/6, 24/4, and so on).

Another factor contributing to the variability of physiological traits is life-history stage. Thus, in the annelid worm *Nereis,* the ability to regenerate lost segments decreases as the animals approach sexual condition. The coefficient of variability for regeneration increases as the ability to regenerate decreases, or, as regeneration becomes less important in the lives of individuals, it becomes more variable in the population. Another example is the diameter of the seminiferous tubules in the lizard *Anolis.* As the breeding season approaches, not only do the tubules increase in diameter, but the diameter becomes less variable among individuals. Selection obviously can work on a process or structure only when the latter is functioning. In the winter there is effectively no selection on genes involved with the seminiferous tubules, and the effects of their gene products

V

0   *V range 0–10*

Anatomical or morphological traits in vertebrates

Swimming stamina, trout   Seminiferous tubule size in breeding
Stroke amplitude of fly's wing   male lizards

Heart rate of a bird

Body temperature of tropical
lizards

V range 10–20

Antler length, adult deer
Tail regeneration in young *Nereis*
Aortic blood pressure in swimming
   fishes
Systolic blood pressure in a bird

Skin sloughing period in lizards

V range 20–30

Heart rate of swimming fishes
Oxygen utilization by snails

V range 30–40

Spotting pattern in skin of toad   Seminiferous tubule size in male
                                                         lizards out of breeding season
                                                      Testicular growth of artificially
                                                         stimulated birds out of season

Length of life of red cells in sheep
Number of eosinophyls in blood of mice

V range 40–60

Tail regeneration in sexually mature *Nereis*

60   Heart rate of horseshoe crab *Limulus*

Figure 6-7   The coefficients of variability (**V** = 100 times the standard deviation divided by the mean) for various sorts of traits in animals.

must only not interfere with the rest of the physiology of the organism. Selection has produced gene products that produce a certain size of seminiferous tubule under conditions prevailing (internally and externally) in July. We return to a discussion of this material in Chapter 13.

Anatomical traits, such as jaw length or relative length of forelimb, and certain physiological traits, such as stamina, swimming speed, or blood pressure, apparently do not vary much between individuals. In the case of anatomical traits this is related to the fact that once developed, these traits are relatively fixed, and the finished product of the interaction of gene products affecting them is being continually exposed to the selective environment. In the case of physiological or behavioral traits, those with low variability in populations seem to be those which are either constantly occurring (blood pressure) or which are frequently evoked by stimuli from the selective environment (stamina). Physiological and behavioral traits showing wide variability seem to be those which do not recur or recur infrequently and those which form only one component in some homeostatic system, all of whose parts can vary. On the average, then, homeostatic traits, and possibly some of the genes affecting them, might be expected to be somewhat more variable within populations than traits involved with obtaining perquisites from the external environment.

SELECTION AND POPULATION INCREASE

Intraspecific competition in the direct sense derives from the fact of a limited carrying capacity. Whenever this limitation is alleviated, the selection coefficients of alleles affecting traits involved in obtaining perquisites from the external environment should decrease, and the population size should expand. Indeed, selection under these circumstances emphasizes increases in fertility (r selection; see Chapter 9). One way in which this can happen is when species are introduced into a new community that contains available niche space without native competitors or with native species occupying only part of the niche. In the latter case the new species will expand its population rapidly if the native cannot effectively compete.

Occasionally a species experiences a dramatic increase in number in its own community as a result of some ecological catastrophe. Examples are fairly plentiful in connection with the widespread alteration of natural environments by man. Thus, the previously rare coral-feeding crown-of-thorns starfish is·undergoing tremendous increases in population sizes all over the Indian and Pacific Oceans. The key factor seems to be the local destruction by man of coral communities in connection with building harbors. Normally all of the huge numbers of offspring of any female starfish perish as larvae as they settle out on the substratum in preparation for metamorphosis into starfish. At this point they are fed upon by filter-feeding invertebrates, including the corals, and only a very few happen by chance to settle upon a spot not occupied by a filter feeder. Having survived this process, the starfish gradually gains almost total immunity to further predation. Various organisms feed on starfishes, but they rarely devour an entire adult animal. If a few appendages are left, or a few lost, the rest of the starfish is regenerated from the pieces. Regeneration probably precludes reproduction in the same year, but eventually the starfish reproduces again. Thus, the picture that we have of this situation is of a population of long-lived adults (the life expectancies of most echinoderms are not known) producing a large number of offspring, the great majority of which perish before metamorphosis, and a few young individuals who probably gain life expectancy as they grow. Then the picture is modified by man so that a significant number of larvae metamorphose. Because the food supply is abundant (the population size was controlled almost completely by predation) the number of adults begins to increase. This results in the gradual destruction of the adult food supply (the coral reefs) so that the population may come to be controlled by food supply. This, however, is unlikely in this case, because corals apparently do not have a power of growth and reproduction commensurate with the ability of these starfish to eat them. If this is so, the starfish could become extinct over much of its range.

The situation before the interference of man is interesting from the point of view of intraspecific competition. In what sense do any of the larval forms, or the adults, compete wtih one another? In adults, reproductive competition is clear; females produce a maximal number of eggs because chance alone determines which will survive metamorphosis (the more eggs, the more chances for survival). Ecological competition, however, does not seem to be a factor in these populations. There is no shortage of food or space, or of mates for adults. This is true also of insect populations that are controlled by climatic deterioration. During the good season the populations are so far below the carrying capacity that no ecological competition occurs. It is clear, then, from these examples that ecological competition is not a major factor in every deme; it is probably a negligible factor in the clones of procaryotes as well. Populations of these kinds, with numbers controlled only by external agencies independent of density, are characterized by reproductive competition alone, and this often results in maximal reproductive potentials being evolved, rather than the optimal ones discussed above. Organisms with such populations can be thought of as having become specialized to reproduce maximally (see Chapter 9, r selection, for further discussion of these matters). From such considerations as these, one can postulate that the degree of variation of traits germane to obtaining perquisites from the environment in populations like those of the starfishes would be relatively high, because competition for these perquisites is mineral; that is, reproductive success does not hinge on ecological success, which is more or less guaranteed in individuals that do not carry genetic diseases affecting internal homeostasis. This hypothesis needs to be tested.

CHARACTER DISPLACEMENT

It is important to realize that the selective environment is a complex concept, involving features of the physical environment, the members of one's population or deme, and organisms of other species. The responses of organisms to changes in the first and last are essentially similar in that these determine the arena in which intraspecific competition occurs at the ecological level. The role of the physical environment in this respect is fairly obvious and is touched on in many places in this book. The role of individuals of other species is less obvious and can be exemplified by the phenomenon of character displacement in relation to ecological exclusion (see Chapter 3).

Consider a species X, having a population in a certain community. A species Y, having a very large niche overlap with X, begins to become established in that community too. Perhaps Y was expanding its range slowly or arrived relatively suddenly, having been brought onto the scene by man, for example.

Focusing on an obvious aspect of the ecological niche, suppose that the food organisms of these species are identical in kind—insects, seeds, and so on—each species being an indiscriminate feeder within a category. In keeping with the usual situation in living systems, we will suppose that there is a normal curve of food-particle size taken by individuals of each

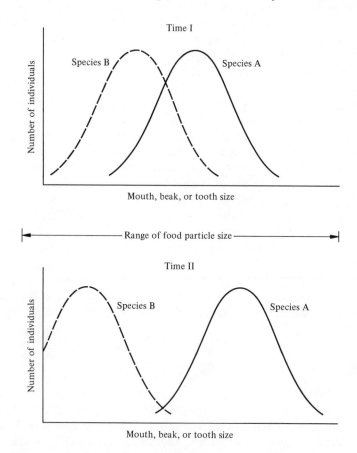

Figure 6-8   Character displacement. See text for explanation.

species (every species is specialized to some extent), and, in the situation being described, that there is significant overlap in the normal curves of X and Y (see Figure 6-8). This figure suggests the degree of ecological competition that will occur between members of these two species in respect to food in our community. Note that some food particles are outside the usual reach of each species. Any individual in either of the species is more restricted in the size of food particles it can accommodate than is the population as a whole. In each population there will be some individuals that can feed on larger particles than can others; this may be because they are themselves somewhat larger or have a larger mouth (beak or tooth) or because of some less easily defined factor. Now, if there is some expenditure of energy in trapping these food particles and if the number of particles of each size is about the same, individuals on the ends of the normal curves in each species that lie outside the range of the other species will have to expend less energy overall in feeding than do conspecific individuals that must compete, not only with members of their own species, but also with those of the other species that feed on the same food particles.

The effect of this situation is to make food particles of extreme sizes relatively more common (or more easily accessible). Individuals in each species that feed on the most common or most easily accessible food supply will have an advantage over individuals that feed on a less common food supply and, because they have more energy to spend for reproduction, will tend to leave more offspring in the next generation. These new individuals will, of course, carry the alleles of their parents, including those at genes involved in the determination of food-particle size. The effect of this process (natural selection) over a number of generations will be to move the normal curves of food-particle size of these two species apart (see Figure 6-8). The species will become adapted to each other's presence in the community; they will become, in fact, coadapted.

They will also become more highly specialized in this respect than they were before the displacement; this would be true if, as is usual, the range of food-particle size is not infinite. We can assume that, before meeting, in the absence of severe competition, each species had adapted to the mean size of food particles in their environments. This is one explanation of the fact that they had a large niche overlap in this area (by parallel evolution, if they are closely related). Having met, they were forced to restrict their feeding to prey of a more restricted size range than was the case before.

Consider this process to be operating at many areas of niche overlap and niche contact for all species in a community, and you begin to get a notion of the complexity of coadaptation in a community. Consider, for example, the possible responses by the food particles to the events that we have just considered. It is for this reason—the fact that other living species are evolutionarily reactive—that it is worthwhile distinguishing them from the physical environment. For example, we could as easily have obtained the same results with species X by having some change occur in the environment so that larger food particles became relatively more scarce. That would not be character displacement and it would not lead to considerations of community coadaptation.

Coadaptation via interspecific competition

There are those who maintain that among plants processes leading to coadaptation are unimportant and that these organisms adapt simply to the physical environment—soil, temperature, seasons, altitude, latitude and so on. They maintain that if a seed falls in a spot where these features are optimum, the plant will grow. They contend that interaction with other plants is restricted to competition for light. Every individual plant must grow as fast as it can to avoid being overshadowed by some growing faster. Some are adapted to growing in dimmer light than others, but they consider this to be simply a feature of the physical environment, although the lack of light is determined primarily by overshadowing plants. Because plants do not have so varied needs as animals, it may well be that they interact less with one another. But this should be a difference in degree only. Consider, for example, the fact that woodland spring bulbs are adapted to flower and grow rapidly early in the spring before the leaves of trees appear. It is possible that there has been no reciprocal adaptation by the trees to the bulbs, although one can cite the various subtle chemical effects (allelopathy) of one sort of plant on another either through the air or in the soil. Consider also the various micorhizal relationships between fungi and plants varying from pine trees to orchids. Such symbioses undoubtedly involved some coadaptation. Consider the adaptations of jungle epiphytes to living on trees and the reciprocal adaptations of trees in evolving easily sloughed bark so that when the weight of epiphytes reaches a certain level, the bark falls away. Animal-plant relationships, of course, lead to various adaptations by both—high-crowned teeth for chewing tough leaves, thorns and bitter taste, and so on. Although these are clearly coadaptations (some very close ones occurring between plants and insects, see Chapter 3), they do not result from character displacement, which is better restricted to reciprocal changes in similar features of similar organisms (or organisms with similar needs) in response to interspecific competition—which has to be at the ecological level, of course.

In Chapter 3 the topic of interspecific competition was broached, and we may pursue it briefly here. Species X and Y above, after meeting in one community, began to compete on an ecological level. This competition led to a change in the outcome of reproductive competition between conspecific individuals, which is what actually constitutes evolution. Thus, interspecific competition can lead to evolutionary results only by way of intraspecific competition on the reproductive level. An individual of species X competes ecologically with other individuals of species X and, in less numerous ways, with members of species Y, but it competes reproductively only with members of species X (except its mate).

SUMMARY

To sum up, natural selection is differential reproduction. In sexual demes it has dimensions beyond that in asexual clones, because the former possess a gene pool. Reproductive success (Darwinian fitness) depends upon an individual's being physiologically fit (not having a genetic disease) but is not guaranteed by such fitness and is further determined by competition for food, mates, predator escape, and level of fertility. Optimal values of fecundity are often better than maximal because of the necessity for ecological competition. Such competition involving members of different species is an important factor in community coadaptation and in divergent evolution, because it can determine the outcome of reproductive competition in one of the species at least.

REFERENCES

Baker, H. G., and G. L. Stebbins, eds., *The Genetics of Colonizing Species.* Academic, New York, 1965, 588 pp.

Dobzhansky, T., *Genetics and the Origin of Species,* 3d ed. Columbia, New York, 1951, 364 pp.

Errington, P. L., "An Analysis of Mink Predation upon Muskrats in North-Central United States," *Research Bulletin of the Iowa Agricultural Experimental Station,* **1943:**797–924.

Fisher, R. A., *The Genetical Theory of Natural Selection.* Dover, New York, 1929, 291 pp.

Ford, E. B., *Ecological Genetics.* Wiley, New York, 1964, 335 pp.

Grant, V., *The Origin of Adaptations.* Columbia, New York, 1963, 606 pp.

Hamilton, T. H., *Process and Pattern in Evolution.* Macmillan, New York, 1967, 118 pp.

Lack, D., *The Natural Regulation of Animal Numbers.* Oxford, Fair Lawn, N.J., 1954, 343 pp.

Li, C. C., *Population Genetics.* The University of Chicago Press, Chicago, 1955, 366 pp.

Mayr, E., *Animal Species and Evolution.* Harvard, Cambridge, Mass., 1963, 797 pp.

Mettler, L. E., and T. G. Gregg, *Population Genetics and Evolution.* Prentice-Hall, Englewood Cliffs, N.J., 1969, 212 pp.

Mills, P. R., R. L. Peterson, and S. Spiegelman, "An Extranuclear Darwinian Experiment with a Self-duplicating Nucleic Acid Molecule, *Proceedings of the United States National Academy of Sciences,* **58:**217–224, 1967.

Sheppard, P. M., *Natural Selection and Heredity*. Harper & Row, New York, 1959, 209 pp.

Smith, J. Maynard, *The Theory of Evolution*. Penguin, Baltimore, 1966, 335 pp.

Stebbins, G. L., Jr., *Variation and Evolution in Plants*. Columbia, New York, 1950, 643 pp.

Wallace, B., *Topics in Population Genetics*. Norton, New York, 1968, 481 pp.

Wright, S., *Evolution and the Genetics of Populations*, vol. 1. The University of Chicago Press, Chicago, 1968, 469 pp.

Zweifel, R. G., and C. H. Lowe, "The Ecology of a Population of *Xantusia vigilis,* the Desert Night Lizard," *American Museum Novitates,* no. 2247, 57 pp.

CHAPTER 7

Sources of variability

Without phenotypic variability among individuals, no evolution, or continued adaptation to the environment, could take place, and existing living systems would become extinct. New biogeneses might occur, but they too would cease functioning in a changing world in the absence of variability among the individual units making them up. Actually, there is no problem obtaining the required variability; indeed, an important problem of living systems is to minimize the effects of various forces tending to disorganize (and, therefore, render less uniform in a population) structures at both the genetic and phenotypic levels. A certain degree of uniformity of phenotype in a given population is imposed by the nature of adaptation. In today's organic world, adaptation involves some degree of specialization. In a particular species, all traits—behavior, anatomy, physiology—are coadapted as part of a smoothly functioning system different from other systems adapted to, or specialized for, other ecological niches. An individual too different in some way from the majority of its population would most likely (1) be less well adapted for the species ecological specialization (or less specialized) than other members of the species or than "normal" members of any species, and so be unable to compete anywhere successfully or (2) be internally unbalanced in the sense of having lost internal, or physiological, coadaptedness. In sexual organisms, uniformity of phenotype is required for the further reason that individuals must interact with one another, and this is possible only on the basis of shared or complementary traits. A similar argument can be made in respect to territorial and social organisms. In a continually changing environment, however, slight phenotypic variability must be maintained, largely because no phenotype remains optimum for long. This means that some of the continually generated variability must be used by biological systems in the way that Darwin suggested. In this chapter the nature and sources of this variability, and problems associated with it, are explored.

PRIMARY GENE PRODUCTS

The primary gene products are specific proteins, most of them enzymes. The one gene–one enzyme hypothesis that grew out of work on the biochemical genetics of molds in the 1940s—especially the work of G. W. Beadle and E. L. Tatum of the (then) Rockefeller Institute, in New York—can be accepted with slight modification today. Not all proteins are enzymes; collagen, for example, is a primary gene product, as are albumen and fibrinogen. Some hormones are primary products of translation, such as growth hormone and proinsulin. In the latter case the protein is split into fragments by proteolytic enzymes, and some of the fragments rejoin to form active insulin. Similarly, some primary gene products are proenzymes—trypsinogen, pepsinogen—which are not activated into enzymes until a fragment has been removed. Furthermore, a great many enzymes are composed of subunits which are the primary gene products and which by themselves are inactive. Thus, a more general formulation of the above idea is one gene–one polypeptide, with a large part of the different polypeptide products of gene translation forming some part of an enzyme.

Further refinement is necessitated by the realization that not all genetic loci are involved in coding for mRNA, which is destined for translation into protein. The transcription of some loci produces tRNA and of others rRNA. A further complication arises when we incorporate the idea of various regulator genes, whose transcription produces RNA's that function as, or along with, effector molecules acting to block or release the transcription of loci (structural genes) that are involved in producing mRNA. This idea further involves in some systems—for example, procaryotes—operator genes as the site of action of effector molecules, each operator attached to one or more structural genes. Thus, today we may say one structural gene–one polypeptide.

SINGLE GENE TRAITS: PHENOTYPE-GENOTYPE RELATIONSHIPS

The primary structures of the primary products of gene translation are only indirectly related to the phenotype as the latter is "perceived" by natural selection. Structures such as a tooth or a behavior pattern or an excretory product or a color change are multigenetic traits, because they are the results of the interaction of a large number of different primary gene products acting in concert and sequentially. In general it is probably rare for natural selection to work for a long time and with high intensity on a single genetic locus, although we can arrange to set up artificial selection experiments to do just this (see Chapter 10). Consider, for example, sickle-cell anemia in man. Here the beta chains of hemoglobin are different from the usual (wild-type) beta chain by a single amino acid (see Chapter 11). The result of having such hemoglobin is a severe anemia that has an enormous effect on the Darwinian fitness of the individuals involved. One could describe this as an example in which selection is operating on a single gene with high intensity. But consider the etiology of the anemia. The abnormal hemoglobin is not as soluble as normal hemoglobin and precipitates in the deoxygenated state. This deforms the red cell, which then bursts, releasing its hemoglobin, which is then destroyed in the blood serum. As a result of this, not enough oxygen reaches the tissues per unit time, and a clinical anemia results. But it is possible to implicate both the rate of respiration and the properties of the red cell membrane in this disease, because a less intense metabolism or a tougher red cell membrane would have precluded any physiological effect of the mutation, and then natural selection could not have detected its presence. The red cells of some deer populations, for example, are regularly sickled without bursting. From this point of view, single gene traits can be detected by natural selection, only because they are present in a specific genetic background. In another background (or in another external environment) they would have minimal physiological effects and low selection coefficients.

To pursue this further, consider that there are known mutant forms (allotypes, or allotypic variants) of human hemoglobin beta chains that are not associated with any clinical symptoms whatever in the heterozygote —hemoglobin G, hemoglobin Dhofar (found in a professional soldier), and the like. Hemoglobin Zurich is also without clinical associations in the heterozygote unless the individual carrying it is given sulfanilamide. In this new internal environment, the abnormal hemoglobin incurs a high selection coefficient, because its presence now causes anemia. Then we have hemoglobin S, discussed above, associated with morbid physiological effects in the normal human internal milieu, but in which it is possible to imagine some genetic changes at other loci—for example, loci affecting the red cell membrane—so that the presence of Hb S would no longer result in a genetic disease. Thus, for example, an individual with both Hb S and hemoglobin Memphis (an alpha chain mutation) was found to have a less severe anemia than individuals with only Hb S. The alpha chain mutation has acted as a suppressor mutation. Next we can consider methemoglobinemia, Hb M, represented by a number of allotypes caused by mutations of amino acids directly involved with the heme active site of hemoglobin so that the iron in the heme becomes oxidized and cannot combine with oxygen. Much more drastic changes in either genetic background (internal milieu) or external environment would be needed to render this mutation unperceived by natural selection. An example of a system in which this mutation would have essentially a zero selection coefficient associated with it might be certain antarctic fishes living supercooled under the ice. These have no hemoglobin at all in their blood, presumably because their metabolic use of oxygen is at so low a level that it can be supplied in sufficient quantity simply dissolved in the plasma.

From the present point of view, then, a single gene trait exists when there is a mutation at a genetic locus such that the new allotype is functionally so different from the wild-type product in the given millieux (both internal and external) that its presence alone is enough to throw the organism out of the previous homeostatic equilibrium into a new one—a different color, for example—or completely out of equilibrium, as in a genetic disease. Interestingly, with more refined methods of examining proteins, it has become possible for biologists to identify single gene alterations, such as Hb G, that seem to have little or no physiological effects. Such allotypes may have hardly any selective disadvantage (or advantage) associated with them, and, in the absence of refined biochemical methods, would quite probably not have been detected at all. It used to be necessary for a mutated gene to have a clearly perceptible physiological effect to be recognized as such by biologists, and, although this is no longer true, it is well to realize that a genetic alteration with little physiological effect may not be detected by natural selection (see Chapters 10 and 13). This would depend entirely on whether the individual of which it is a part reproduced better or worse because of its presence.

WILD-TYPE ALLELES AND POLYMORPHISMS: ALLELIC STRUCTURE OF POPULATIONS

The phenotype, then, is a complex manifestation of the interaction of the gene products of the entire genome (see Chapter 8 for further amplification). Nevertheless, in the interest of simplification we often deal with genes, both intellectually and experimentally, as if they were separate entities, and it is useful to do so here in treating of gene mutation. Looking at any specific genetic locus, there are typically two or possibly three different patterns that it may show in a population. There may be only a single form of the locus found in the population, one "wild-type" allele. Such is the case in most human populations with the hemoglobin beta chain locus, for example, where there may be only a few, rare, heterozygotes carrying non wild-type alleles at this locus. Thus, in man the number of abnormal hemoglobin chains (both alpha and beta) occurring in Europe has been estimated as about 0.4 percent and in Korea as about 0.06 percent. The rare alleles are rare presumably because natural selection maintains the wild allotype in preference to any of their gene products. There is the possibility that one of them may be destined to replace the wild type in time, because the individual carrying it is going to reproduce successfully at a great rate because of its presence, and because his offspring may continue to do so. Nevertheless, we may say that in the recent history of the population the allele that is the current wild type has been favored by selection above all other mutations that have appeared. It should be pointed out, however, that in large populations any new mutation, even if it results in genotypes considerably more fit than the original wild type, most likely will be lost by chance (see Chapter 15).

On the other hand, there may not be a single wild-type allele characteristic of the population but instead two or three commonly found alleles. The locus is then polymorphic. Thus, in both the alpha and beta hemoglobin chains there are characteristically two alleles found in populations of orangutans. There are at least three ways to interpret this kind of pattern. First, one allele may be replacing the other, and the condition that we find is characteristic of an intermediate stage of replacement (see Chapter 10). Or selection may be maintaining both alleles in the population by favoring the heterozygote (see Chapters 9 and 11) or by alternately favoring one allele in one season or microhabitat and the other in another, with heterozygotes forming in the intermediate seasons or habitats (see Chapter 9 and below). In contrast with rare mutants all such alleles can be referred to as wild types.

A pattern of variability involving a whole different order of magnitude is found at some genetic loci. There may be as many as nine alleles of a specific kind of esterase in *Drosophila* populations and even more of them in populations of certain butterflies. This is not something peculiar to this kind of enzyme, because in fish populations there are characteristically only one or two alleles of specific esterases in any population—only three in the huge populations of skipjack tuna occupying the Pacific Ocean, for example. For genetic loci with four or more alleles present in significant frequencies there is no longer any point in referring to wild-type alleles.

THE NUMBER OF POTENTIAL ALLELES

Considering the nature of primary gene products, it is perhaps surprising that all loci are not characterized by a large number of allelic variants. Thus, a polypeptide composed of 110 amino acids, if it could become modified at every amino acid residue by replacing it with any of the 20 other common amino acids, could potentially exist in 20^{110} different allotypic forms. Because the active sites of enzymes and other biologically active polypeptides typically cannot become modified without causing severe genetic diseases, we can say that only 100 of the residues are potentially modifiable, giving 20^{100} different possible allotypes. Moreover, typically amino acids cannot be replaced by just any other ones, and replacements tend to be conservative in terms of molecular size, charge, degree of hydrophobicity, and so on, so that, on the average, perhaps any residue can be replaced by any of only three others. This brings the number of potential allotypes down to 3^{100}. Recent work with hemoglobins has shown that replacements can occur much more readily on the surface of a polypeptide without resulting in serious genetic disease and that almost any change of residue that is "inside" the three-dimensional structure of the molecule is strongly selected against. If we assume that half of the amino acid residues of the polypeptide are on the surface and half inside, the figure for potential allotypes is down to 3^{50}, or 3×10^{26}. If 90 percent of these are selected against more or less severely in the existing external and internal milieux—for example, for not giving adequate kinetic rates at existing pH values or temperatures—we would be down to the astronomical number of 3×10^{25} possible allotypes for the population.

The first thing that we may note about this is that there are no populations having this many individuals and that, because of the limitations on the number of individuals that can exist in any population, the potential variability of a gene product cannot begin to be expressed. Next, we may note that, because living systems have been present for only some 3×10^9 years on the earth, given an organism with a generation time of only 1 hour and a constant population size of 1 million individuals and with complete replacements of all allotypes by others each generation, only 2×10^{19} different allotypes would have occurred so far since the origin of life. This example involves mutation rates of 100 percent; that is, every gamete in every generation has a new mutation at this locus. A figure closer to reality would be one mutation at a locus for every 100,000 gametes in each generation (see below). This would generate a maximum of only 10 new allotypes per hour instead of 1 million, and there would so far have been generated only 2×10^{14} out of all the possible allotypes at the locus. Thus, mutation rate is very important in determining how much of the potential variability is actually expressed.

EVOLUTION OF THE GENE

We can now consider exactly what is involved in the replacement of one allele by another in a population, or what is involved in the evolution of a genetic locus. It has been argued that only populations evolve and that it is impermissible to consider that genetic loci evolve. Instead, it is said, they only change their allele frequencies, which change results in the evolution of the population. From this point of view it is not possible to consider what is involved in the fact that the wild-type cytochrome c gene of man is 50 percent different from that of black bread mold, or what in addition is meant when it is noted that the cytochromes c of man and some other mammal differ by only 10 percent of their amino acid sequence. If we wish to treat these data in an evolutionary sense, and today we must, it becomes necessary to think in terms of evolving genes.

Consider that the hemoglobin beta chain locus of man differs from that of a species of gorilla by one amino acid residue: the 104th residue from the amino end in man is arginine, in the gorilla it is lysine. Another way of putting this is that the overwhelming majority of individuals in each species have the species-characteristic allele at this locus or that the frequency of the human wild-type beta chain is very high in human populations and that of the gorilla is very high in gorilla populations. No human has been found with gorilla beta chains, although these differ by only a single residue. Presumably whenever this mutation occurs in man, it results in a pathological condition, because out of the 30 or so residues with known mutations on the beta chain—even if sick, the individuals have lived long enough to be detected—none occurs in this region; the nearest one, at position 102 (Hb Kansas), causes serious morbidity, because this residue is involved in the structure of the heme active site and also in the contact between the two alpha-beta half molecules. The ancestral populations must have had predominantly either human beta chains, gorilla chains, both, or neither. In the first two cases one population (either human or gorilla) can be visualized as continuing to have selection maintain its beta chain as the predominant allotype in the populations; the other population, after becoming separated from the first (see Chapter 14), experienced selective pressures such that this allotype was no longer optimal—perhaps the result of other evolutionary changes was to alter the internal milieu of the red cells, for example. New mutations continually occur at this locus. Perhaps several of these were slightly better than the original one in the new internal environment. This means that individuals carrying them reproduced better than those carrying the ancestral allotype. Several of these may then have increased in frequency, forming a polymorphism, the population going through a stage in which there were several wild-type alleles common at this locus. With continuing change in the environment, one of the latter might come to be slightly favored over the others and increase in frequency at their expense. This could, of course, be the original ancestral one as well as any of the others, but we shall assume that it is one of the new ones. After a time both descendant populations or species come to be characterized by different beta chains (see Figure 7-1). If the human chain was the ancestral one, then the human beta chain locus has not evolved and the gorilla locus has. Because the brown lemur, the spider monkey, the rhesus monkey, and the chimpanzee all have the same amino acid as man at this residue, this, in fact, is what seems to have occurred in this case.

There is something more here than simply gorilla and man having

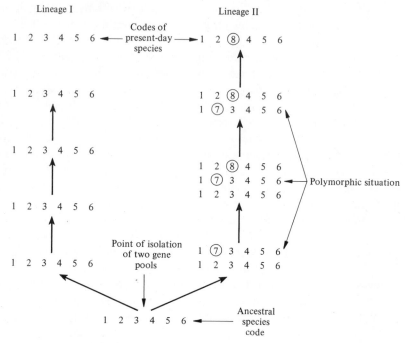

Figure 7-1 The evolution of a genetic locus (represented by six codons) in two lines descended from one ancestral population. Circled codons code for amino acid substitutions.

different wild-type alleles at the same locus. The allele common in the gorilla is not an allele found either commonly or rarely in man; it simply does not exist in human populations as far as anyone knows. This is much more obviously true considering the common human and spider monkey beta chains with six amino acid sequence differences. The predominant human form of this locus could not possibly be one of the alleles, rare or otherwise, in spider monkey populations; the locus in question has evolved since the time of the common ancestral anthropoids. A genetic locus can evolve in this way because fundamentally it is a complex entity. There are so many different possible forms of a genetic locus that many of them come to be realized in time. Different ones, both because of selection and by chance, come to be realized in different populations. With enough time after two lineages have diverged, a large number of sequence differences accumulate and, by both chance and selection, they will be different ones in the different lineages. After two lineages come to be characterized by allotypes that are different by five or more amino acid residues, mutation can no longer in one or two steps convert one of the alleles into the other. At this point the locus has obviously evolved away from the ancestral condition. In this case, after enough quantitative differences have accumulated, we may say that qualitative distinctions become manifest.

It should not be concluded that alleles at a single locus in one population can never be different by more than one codon. In cattle, for example, there are two common allotypes of the hemoglobin beta chain, and these differ from each other in three sequence positions. There is no need to invoke any special history for this situation beyond what was outlined above. Suppose that we had an ancestral population with two common alleles at a locus and that they differed in only a single codon. Now one of these undergoes a mutation so that there are now three allotypes, two of which differ by two amino acids. Subsequent environmental change may now favor these two with a subsequent loss or reduction in frequency of the third. To accumulate still another residue difference, the process can be repeated (see Figure 7-2). In a case such as this, however, there must have been a long-standing polymorphism at this locus. Where only single amino acid differences are involved there might have been a long-standing polymorphism but it cannot be demonstrated. Where two or more sequence

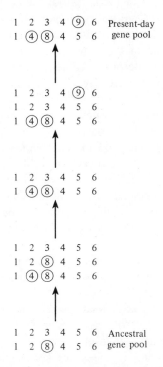

Figure 7-2 The evolution of a polymorphic genetic locus (represented by six codons) in a lineage showing how alleles can differ by more than a single codon. Circled codons code for amino acid substitutions.

differences characterize the allotypes of a polymorphism, this locus must have been polymorphic for a fairly long time.

Polymorphisms may indeed traverse species barriers. For example, the hemoglobin beta chains of sheep and goats are characterized by two common allotypes A and B. These differ from each other by about seven residues, in each case; the A allotypes of the two species differ from each other by only four residues. It seems that here the beta chain polymorphism goes back beyond the time of the common ancestor of sheep and goats or that the polymorphism is older than the species. This, of course, is not more surprising than that the cloven hoof is older than these artiodactyl species.

GENE MUTATION

Looking now more specifically at gene mutation, we may first note that this process is the fundamental source of variability in all living systems and that it is the only source among procaryotes, with possibly a few exceptions. Gene mutation was detected and appreciated as such before we had any knowledge of the nature of the hereditary material. Genes were postulated as discrete entities by Gregor Mendel in 1866 and by most biologists after the rediscovery of his work in 1900. Shortly thereafter Hugo De Vries conceived of the notion—on the basis of what we know today to have been misinterpreted observations—that these hereditary units could spontaneously change. By 1925, H. J. Muller, at Indiana University, was already accurately measuring mutation rates and using x-irradiation to stimulate increased rates—a technique discovered earlier by T. H. Morgan at Columbia University.

This early work was based on indirect evidence of gene mutation detectable at the level of the phenotype, and the operational definition of gene mutation was "any qualitative change in phenotype that could be inherited in patterns predicted from Mendelian principles." Thus, if bright red eyes appeared on a fly in a population characterized by dull red eyes and if, upon mating this fly with a normal-appearing one, half or none of the offspring had bright red eyes or, further, if none had bright red eyes but, upon crossing two of them, one third of their (F_2) offspring had bright red eyes, one could say that a mutation had occurred at the eye-color locus or gene.

There was a difficulty with this operational procedure in that if dull red eyes result from the production of a brown pigment and if the production of this pigment involved a pathway made up of several enzymatic steps (as indeed is the case; see Figure 5-14), a mutation affecting any of the steps so as to eliminate its functioning would result in bright red eyes, the latter pigment being produced by another biochemical pathway. It was possible to circumvent this difficulty in two ways. First, if two homozygous bright-red-eyed flies were mated and very few of their offspring had dull red eyes, because of crossing over during gametogenesis in one or both of the parents, it would be clear that the bright red eyes had been formed by two different mutations, and separate stocks could be isolated and maintained (see Figure 7-3). Second, if the steps in the pathway of brown-pigment synthesis were known, and the intermediate products available, one could supply the missing intermediate compounds to larval flies and obtain imagos with normal dull red eyes. For any given stock one simply supplied all the intermediates one by one until one worked, and this would divulge which step had mutated so as to

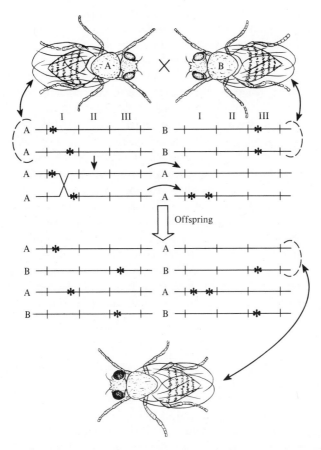

Figure 7-3 Two ways in which homozygous flies when crossed can produce some offspring that will have active the pathway determined by genes I, II, and III, which was inoperative in the parents. The genes are shown next to each other for convenience only. The chromosomes are labelled as coming from either fly A or B. The asterisk indicates a mutation that impairs the activity of the gene product.

have its function eliminated. In this way, then, one could identify several genes that had effects on eye color and work out which one had mutated in a particular strain (see Figure 5-14).

There were still some serious limitations on this view of mutations. A mutation came to be associated with the loss of activity of some enzyme, because this is how one detected them; that is, there were wild-type genes coding for active enzymes, and there was a mutation—in principle always the same mutation—that eliminated the activity of an enzyme, giving only two states possible for every gene. Furthermore, because most mutations tended to be recessive (see below), it was difficult to detect them in diploid organisms, and this led quickly to microbial genetics (because many microorganisms are haploids or have haploid life-history stages), thereby restricting fine-scale genetics to such organisms.

The simultaneous increase in our knowledge of the hereditary material and of the structure of proteins, leading ultimately to an elucidation of the genetic code, and the invention of relatively easy methods for examining protein structure—primarily electrophoresis and immunology—as well as the more laborious methods of peptide fingerprinting and amino acid sequencing, has led us to a very much more detailed view of mutation and has given us the means to detect a great many more of them at any locus. The number of possible different alleles at a locus, each of them a result of a mutation, was discussed above and shown to be astronomical. Thus, an enzyme in the pathway leading to brown eye pigment in flies can mutate in an almost inconceivably large number of different ways. Perhaps the overwhelming majority of these mutations would impare the activity of the enzyme, but even so there would be a vast number of possible mutated forms that had results short of complete elimination of activity. Because it is possible for a codon to mutate to another one coding for the same amino acid (see Figure 2-13), it is even possible to have a mutational event that is in principle completely undetectable by either man or natural selection.

Substitutions

If the mutational event results in the substitution of an amino acid by one of different charge (39 percent of all possible single-step substitutions result in this) or if substitution occurs at a residue such that the configuration of the protein molecule changes, the new allotype may be electrophoretically different from the wild type and can, therefore, be detected even if it has no physiological repercussions in the system. This means that we are no longer dependent on an operational view of mutation; we can in principle detect almost every mutation that occurs at any locus in a system, although practically this would be impossible.

The phenotypic results of substitutions at a particular codon and amino acid residue vary in the magnitude of their effects. A mutation to a synonymous codon results in no change in the primary gene product and cannot be detected by natural selection. This occurs in 36 percent of all possible single-step mutational events. The evolutionary potential, however, of different code synonyms is not identical. For example, arginine is coded for by CGA and also by CGC (mRNA codons). CGC can mutate in one step to serine, AGC, but CGA cannot arrive at a code for serine in one step. It is hard to estimate the importance, if any, of this kind of potential difference between synonymous codons.

A substitution can result in the replacement of an amino acid by one of similar properties and, therefore, be a conservative substitution. The properties of amino acids that seem to be important in this respect include polarity—an amino acid can be polar or nonpolar (hydrophobic); amino acids can be basic or acidic or dipolar; they can be chemically very reactive, such as histidine, or very unreactive, such as valine; they can be very bulky—valine, leucine—or occupy little space—alanine, serine. Conservative substitutions (other than at active sites or crucial configurational residues) can lead to allotypes having very little selective difference from the wild type. If there is a polymorphic situation, with more than one wild-type allele in a population, some of them may differ in this way. Some of the known abnormal human hemoglobins not associated with clinical conditions are presumably in this category. Some replacements found in hemoglobins and not associated with overt morbidity, however, are physicochemically not conservative. But they all occur on the surface of the molecule as opposed to its interior. Because of this, the importance of the physicochemical classification of amino acids to natural selection is uncertain; it appears that the spatial position of a substitution is as important as the actual change involved.

In any case, another category of substitutions is that in which a very different kind of amino acid replaces another, the so-called radical substitutions. As just pointed out, the position of the substitution in the molecule is probably as important as its specific structure. Any substitution at an active site or at some important configurational site can be radical. Even if a radical substitution results in detectable negative effects on the physiology of an organism, we cannot always assume, therefore, that selection will eliminate the apparently less good allotype. For example, there is some evidence that, of those abnormal human hemoglobins which are common in certain places because selection favors them in the context of chronic blood parasitism in the populations, those resulting in the severest clinical problems in homozygotes are the ones that are most effective in protecting heterozygotes from the blood parasites (see Chapter 11). It seems that, as with medicine, treatments that are the most effective are also the most dangerous.

A form of substitution that always has a radical effect on the physiology and Darwinian fitness of an organism is the so-called nonsense muta-

tion. The mRNA codons UAA, UAG, and UGA are called chain termina-
tors in that when one of these appears in its turn on the ribosomes, the
polypeptide being synthesized is broken. 19 out of the 45 code words (42
percent) are capable of mutating to a nonsense triplet in one step; but
some half of the amino acids have codons such that nonsense mutations
cannot occur at positions that they occupy. When this kind of mutation
occurs, eliminating some enzyme activity, it is often possible, if the chain
terminator did not occur early in the chain, to detect the partially synthesized gene product immunologically. In man, some of the thallasemias may
be caused by mutations of this kind. In these diseases, there may be no
hemoglobin beta chains, for example, or no alpha chains.

DELETIONS AND INSERTIONS

Another type of gene mutation that can occur is a break resulting
in a deletion or insertion, thereby decreasing or increasing the size of the
gene. As a general rule any random deletion or insertion of bases is most
likely to result in a frameshift (see Figure 7-4), thereby converting the
information in the gene from this point on toward the end coding for the
carboxyl end of the polypeptide chain into nonsense. If, however, the deletion or insertion involves just one codon or some whole-number multiple
of one codon, the structure of the preexisting polypeptide is unaffected
other than by inserting or deleting a few codons (or amino acids from
the polypeptide). It seems that the probability of an insertion or deletion
of just the right number of bases might be a very improbable event, but
there is evidence that it has occurred during evolution. (see Figure 7-5).

First, the likelihood of the survival of a deletion is slightly increased
at the ends of the gene. Looking at cytochromes c from different organisms, it is clear that the vertebrate enzyme is shorter than that of invertebrates and that the loss (or addition) was at the amino end (see
Figure 7-5). This means that in a population ancestral to all living vertebrates, a deletion may have occurred which was subsequently favored by
selection and which, therefore, came to be the common form of this locus
in that population. Comparing all eucaryotic cytochromes c with the cyto-

```
      ... GTC|AAC|CTA|TAG|GAT|GAT|TCT|CAT|AAA ...
          Deletion
              ↓
(a) ... GTC|ACC|TAT|AGG|ATG|ATT|CTC|ATA|ATT ...
          Insertion
                ↓
(b) ... GTC|AAC|CT[G|GAT|ATA|GGA|TGA|TTC|TCA|TAA|ATT ...
```

Figure 7-4   Representation shows the drastic effects of most random insertions
and deletions. In (a) a single base has been deleted; in (b) four bases have
been inserted; in both cases (and in most others) this results in a frameshift,
converting the information in the gene to nonsense.

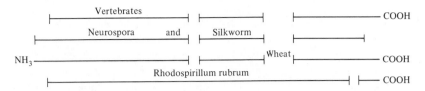

Figure 7-5 Comparison of sizes and sequences of various cytochromes *c* to show where deletions or insertions must have occurred in one or another lineage. The ancestral sequence is not determinable. The breaks are necessary to obtain maximum homology when aligning the sequences next to each other.

chrome *c*2 of the bacterium *Rhodospirillum rubrum* shows that one needs to invoke two internal deletions in the former in order to homologize it maximally with the latter (see Figure 7-5), or one must invoke insertions into the bacterial enzyme. In comparing the sequences of the various hemoglobin subunits from different species, it is again clear that one must invoke numerous deletions or insertions as having occurred and having been preserved by selection.

Perhaps the process of insertion or deletion is what must be invoked to explain certain peculiar evolutionary events at the level of biochemical pathways. Thus, primates, a bat, a bird, and the guinea pig are all unable to synthesize ascorbic acid and, therefore, require it as a vitamin (vitamin C) in their diets. In some of these forms, at least, the enzyme gulonolactone oxidase is missing from the biosynthetic pathway. If this were caused by a typical nonsense substitution, it seems probable that in at least one primate, for example, the enzyme activity could return by chance mutation of the chain-terminating codon to a code for an amino acid or by a suppressor mutation. But if the loss were caused by a deletion of significant parts of a locus or by an insertion of even a small piece resulting in a frameshift, one might not expect significant back mutation. In this case a gene might come to be effectively lost from a lineage.

MUTATION RATES

In his early work on mutation rates, Muller found that there was a spontaneous mutation rate to lethal alleles on the X chromosome of *Drosophila* flies of $\frac{1}{1000}$ gametes. Subsequent work has shown that the rate of mutations producing simply some detectable effect on the phenotype for all chromosomes in *Drosophila* is on the order of $\frac{1}{20}$ gametes, or about 5 percent, and that the mutation rate to lethals for all chromosomes together is about $\frac{1}{400}$ gametes. Therefore, one can expect every 200 zygotes to include a maximum of one doomed to fail by reason of the spontaneous mutation rate alone. Interestingly, about half of all human zygotes abort very early in development, but much of this seems to be due to "nongenetic" faults, such as overripeness of the ovum prior to fertilization.

Mutation rates are commonly described on an average per gene basis. Looked at this way, *Drosophila* is found to have 1 detectable mutation per 10,000 loci, or a rate of 0.01 percent per locus. Haldane calculated the spontaneous mutation rate in man on the basis of statistics concerning dominant genetic diseases, such as retinoblastoma, chondrodystrophy, Huntington's chorea, and others, and found it to range from 0.01 to 0.001 percent per locus. Comparable rates have been shown in corn (1/20,000 to 1/100,000 loci). The feeling now is that eucaryotic mutation rates typically range from 10^{-4} (0.01 percent) to 10^{-6} (0.0001 percent) per locus. Procaryotes show much lower mutation rates, those of the bacterium *Escherichia coli* ranging from 10^{-7} to 10^{-10} per locus, for example. Molds also apparently have much lower mutation rates than higher eucaryotes. It is probable that this difference is an artifact associated with the difference between single-celled and multicellular organisms, the generation time of the latter being much longer in terms of cell divisions. Thus, if 20–30 cell divisions occur in the germ line of an organism, as in man, prior to breeding, there is this much more chance of having a mutation occur, and they will accumulate in the individual's germ line. Single-celled organisms, on the other hand, divide only once. When mutation rates are normalized on a per division basis, those of procaryotes and eucaryotes come out about the same. Thus, it is an essential aspect of multicellularity for organisms having it to generate more variability per generation than do single-celled organisms; the latter have the strategy of producing more generations during a given period of time, with less variability per generation (see also the end of Chapter 2). Presumably the mutation rate characteristic of both on a per cell basis was some optimum arrived at (or imposed) during the early stages of organic evolution.

There are two ways in which it is possible to estimate the numbers of genes present in a species. One approach involves the mapping of genes by taking advantage of crossover events (see below). This approach in eucaryotes has led to estimates ranging from 5,000 to 10,000 loci per organism. The other approach involves actually measuring the amounts and dimensions of the genetic material. This has led to an estimate of some 100,000 genetic loci per eucaryotic organism. Combining these data with those concerning mutation rates allows us to note that every individual probably carries one (in principle) detectable mutation; that is, of all the ways in which two individuals are different from each other, one of them at least is present because of a gene mutation. We have already noted that estimates for *Drosophila* came out to $\frac{1}{20}$ individuals carrying a detectable mutation. This estimate was based only on genes whose allelic differences were clear-cut and qualitative. Other estimates, based on genes having only small quantitative effects, run as high as 35 percent of polygenes (see

Chapter 8) having undergone a detectable mutation per gamete. Again, every individual is a mutant.

If one considers, not detectable mutations, but those which occur having no phenotypic effects (to synonymous codons) or only slight effects (conservative substitutions at noncritical residues in the primary gene product), that is, mutations that could not be detected by classical genetic methods, it is clear that the mutation rate of more than one per gamete has to be an underestimate. Thus, 64 percent of mutations only are to non-synonymous codons, raising the figure to more than three mutations per two individuals. Because perhaps half of mutational events occur at the surface of a primary gene product and, therefore, may have very small selection coefficients associated with them, the figure easily goes up to a minimum of three mutations per individual for every detectable one. In this way it is possible to estimate that the mutation rates arrived at by classical indirect methods are at least threefold lower than the actual rates.

BACK MUTATION

A word may be said about reverse, or back, mutation. Because each mutation event is a random phenomenon, mutation rates calculated in terms of any mutation at a locus have to be much greater by many orders of magnitude than the rate at which a specific codon mutates in a specific way. Mutation rates per nucleotide base pair in eukaryotes have been estimated at 10^{-9} to 10^{-10}. This rate must be further decreased by multiplying it by the inverse of the number of different ways a single codon can mutate ($\frac{1}{9}$). It is not surprising, then, that back mutations are rarer than mutations away from the wild type. This fact alone biases evolving organic systems in the direction of divergence after gene pools become separated. Phenotypic revertants to wild type, however, occur at rates greater than those of reverse mutation. The phenomenon involved here is the so-called suppressor mutation. What exactly is involved is not entirely clear (probably more than a single mechanism), but a second mutation at a different genetic locus can restore the activity, lost by mutation, of the primary gene product of a locus, thereby "suppressing" the first mutation.

DOMINANCE

Many and perhaps most new mutations in diploid organisms are recessive, or, what is saying the same thing, wild-type alleles tend to be dominant. In haploids, of course, all genes are expressed equally. Recessive mutations cannot be detected until they are present in homozygous form, and, because of this, many alleles that in haploid systems would have been eliminated immediately can become incorporated into the gene pool of diploids, there to form part of the relatively increased potential variability of these kinds of populations.

Dominance is apparently not an absolute attribute of an allele, however. Thus, for example, it has been possible by elaborate genetic experiments to transfer a small section of a chromosome carrying a dominant allele of a given gene from one species of cotton to another. In the new gene pool the allele that was dominant in the species from which it came behaved as a recessive. Another example shows that selection can alter dominance. Thus, in a normally white species of moth an allele at a locus was found such that when present in homozygous condition, it resulted in yellow wing color. Two stocks were then set up with a small percentage of yellow recessive genes in each. Every generation fairly mild artificial selection was instituted favoring yellow wing color in one case and white in the other (see Chapter 10 concerning how such experiments are run). After a period of time yellow became dominant in the population in which selection had favored it; it remained recessive in the other. An observation from nature may be noted here as well. Thus, in England a certain moth has undergone intensive genetic study (see Chapter 10). Individuals from many different populations were crossed, and it was discovered that some dominant alleles for the same trait became recessive when placed in different genetic backgrounds.

To explain dominance and also to account for the fact that it seems to be a property that can evolve, several theories have been advanced that can be seen now to be merely aspects of a single theory, which we may call the coadapted genome theory. R. A. Fisher, in England, and H. J. Muller, in the United States, independently proposed the idea that an allele becomes dominant gradually with time as natural selection builds into the genome "modifiers" or modifier genes that favor the activity of its gene product. The allele that eventually becomes the wild-type dominant is seen as having an initial high adaptive value; that is, other possible alleles are selected against fairly severely. In time the genome gradually becomes reorganized so as to prevent the expression of other (harmful) alleles (Fisher) or so as to magnify the expression of the favored dominant (Muller); that is, all along the way individuals are favored who have alleles at other genetic loci so that they create conditions (internal milieux) that favor the expression of the desirable allele and/or discourage the expression of harmful or less favorable alternatives.

Both Haldane and Sewell Wright, of the University of Wisconsin, proposed another way of looking at this, focusing on the primary gene product. They proposed that the wild-type gene product is at least twice as active as other possible gene products of the same locus. Then, in a heterozygote, the phenotype is created more by the gene product of that wild type than by those of other alleles that may be present. Considering that enzymes are each adapted to optima regarding pH, temperature, substrate, feedback end-product concentrations, and the like (see Figure 7-6), we can see that this result could be attained by a process such as proposed

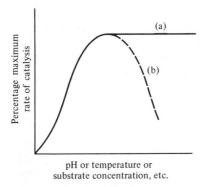

pH or temperature or
substrate concentration, etc.

Figure 7-6 Typical curves for the catalytic rate of enzymes in different environments. In some cases (a) the rate approaches a maximum and remains there, in others (b) it reaches a maximum and then drops off again after optimal conditions are surpassed.

by Fisher and Muller. Another very important possibility is that wild-type enzymes may be more stable in existing internal milieux, other allotypes lasting only a short while before becoming degraded.

Dominance is not a commonly detected phenomenon at the molecular level, being noted as a possibility, with other possible explanations, in only four or five cases out of many hundreds so far. The usual situation at the level of primary gene products is codominance: if two different allotypes are present in a system, they are detected in equal amounts by methods such as electrophoresis. Of course, this is not so in cases in which mutations have occurred, eliminating the gene product (nonsense mutation or frameshift) or its activity (mutation at the active site, for example). An instructive example is sickle-cell anemia in man. Here a single codon substitution results in a gene product with different charge so that hemoglobin (Hb S) made with these altered beta chains has an electrophoretic mobility different from the wild-type hemoglobin (Hb A). Homozygotes for this allotype have a severe anemia (see Chapter 11) that does not appear in the heterozygotes; that is, sickle-cell anemia is a recessive trait. On the other hand, the electrophoretic picture shows equal amounts of Hb A and Hb S in the red cells of heterozygotes. In other words, the phenotypic trait being scored need not be, and probably usually is not, a direct result of the physicochemical properties of a single primary gene product.

Concerning other mechanisms that may be involved in dominance, there is the possibility that less favorable alleles become associated with operator genes (in systems in which these exist) that are refractory in their response to effector substances, leaving only the wild type(s) with easily induced operators. There is also the possibility that different codons for the same amino acid are not exactly equivalent in that each is associated with a different transfer RNA. Perhaps some tRNAs are produced in less quantity or can operate less efficiently on the ribosomes so that an allele that necessitates their use would be translated into polypeptide at a lower rate than other alleles. Perhaps those individuals were most successful in leaving offspring that had sustained mutations at various loci such that alleles with high selection coefficients became associated with such tRNAs at one or more of the codons of which they were constituted.

GENETIC LOAD

The fundamental source of all biological variability is gene mutation. A new recessive mutation in a diploid population, however, does not contribute immediately to this variability except, especially in a small population, when any kind of heterozygote at a locus is favored over any kind of homozygote. It is not clear just how common this situation is. In the classical viewpoint, the new recessive mutation is not detected by natural selection until by chance it has increased in frequency to the point where homozygotes carrying it begin to be produced. (In a large randomly breeding population the fraction of homozygous individuals is the same as the square of the gene frequency; see Chapter 6.) At this frequency selection begins to eliminate the allele; the homozygotes suffer genetic death. The equilibrium allele frequency depends on the mutation rate, the intensity of selection against the homozygote, and the degree of recessivity. This can perhaps best be appreciated considering a dominant lethal ($s = 1$). As soon as such a mutation appears, the individual carrying it suffers genetic death; therefore, the frequency of this allele exactly equals the mutation rate (see Chapter 10 for further discussion using other values of s). Any change in the intensity of selection or any decrease in dominance increases the frequency at which the allele is maintained in the population. A fully recessive gene is not detected until homozygotes carrying it appear. If $s = 1$ against these homozygotes, the allele is maintained at a frequency just below that needed to produce homozygotes; any less stringent selection allows the allele to be maintained in the population at higher frequencies.

Apparently, then, in diploid organisms there are a large number of lethal, semilethal, and subvital alleles present in the gene pools. In 1950, Muller pointed this out in the context of human populations and referred to "our load of mutations," cautioning against increasing this load by means of radiation pollution. In human populations many genetic deaths are actual deaths or deformities of individual persons, and the viewpoint of Muller was that in this context genetic death is an evil.

Genetic death, however, is absolutely essential in any living system in order to maintain adaptedness. The human problem, aside from attachments to individuals, derives from the fact that at any instant every population present on the earth is adequately adapted to its environment, which is how we perceive our own populations, and so a mutant individual has almost by definition a lower Darwinian fitness than one with a wild-type genotype. But living systems evolved in constantly changing environments, and only those systems survived which constantly generated new variability. Even if the environment ceased changing (for human beings this could be achieved to some significant degree perhaps by means of technology), genetic recombination (see Chapter 6 and below) would continually generate less fit genotypes even if the percentages of outright lethals became reduced to their lowest possible levels. We are stuck with our genetic system.

The picture that we have, then, of diploid gene pools is of one or a few selectively equivalent or almost equivalent wild-type alleles at each genetic locus and also a genetic load composed of a host of recessive or partially recessive subvital alleles, fewer semilethal alleles, still fewer lethals, and very few dominant subvital, semilethal, and lethal alleles. We may then inquire as to how many alleles can be present at a genetic locus in a population at a single time. Electrophoretic methods have demonstrated from 2 to 12 or so at a locus, with the majority of cases being at the low end of the range. This is an underestimate, because not all different allotypes are electrophoretically different. Motoo Kimura, now of the Japanese National Institute of Genetics, and James F. Crow, of the University of Wisconsin—and also, independently, Sewall Wright—have shown that the number of different alleles that can be maintained per locus in a population depends on the population size, the mutation rates, and the intensity of selection favoring heterozygotes (see Chapter 9). An increase in any of these parameters allows more alleles per locus to be maintained in the population. For example, with s against the homozygotes at 0.01, it would take about 600,000 breeding individuals to maintain 10 alleles at each locus if the mutation rate u was 10^{-7} per locus. At a mutation rate of 10^{-5}, it would take only about 200,000 breeding individuals, and at $u = 10^{-4}$, only about 80,000 breeding individuals would be necessary. These considerations are taken up again in Chapters 9 and 10; it is sufficient here to note that mutation rate has an effect on the number of alleles that may simultaneously occur at a locus in a single population.

CHROMOSOME MUTATIONS

Although gene mutation is the fundamental source of all new variability, it is not the only source of genotype variability. In diploid sexual organisms important processes resulting in increased variability occur at the level of the chromosome, involving breakage and subsequent repair of the DNA strands by enzymes known as repairases. This kind of phenomenon can have an effect even on a single locus. In molds and plants it has been possible to detect intracistronal crossovers by producing heterozygotes for two different nonsense mutations at some locus and carefully scanning the offspring of matings of these with homozygotes for one of these mutants and observing the reappearance of individuals with wild-type activity at frequencies significantly higher than the known mutation rates at this locus (see Figure 7-3). Crossing over, however, increases with the distance between the markers being observed and, on these grounds, is apparently a stochastic phenomenon in the same sense that gene mutation is.

Crossover events apparently take place during the synapsis of homologous chromosomes during meiosis and are capable of generating tremendous variability among the gametes. Thus, if an organism has four chromosomes, there are 2^4, or 16, different genotypes possible among the gametes in the absence of crossing over. Every crossover results in two new chromosomes so that, with a fairly high frequency of crossovers—12 percent has been measured in corn, for example—it becomes possible to match variability to the large number of gametes produced in most diploid species. It would be advantageous to an organism if all its gametes were different genotypically. Because only two of them are likely to survive, this strategy provides for the greatest likelihood that the two "chosen" will be well adapted (as opposed to simply adequately adapted) to the existing environment—or, that they will be as relatively fit as possible. If the organism produced 1000 gametes, gene mutation would be likely to significantly alter the genotypes of 5 percent (as in *Drosophila*), or 50 of them. Our four-chromosome organism then could produce 66 different kinds of genotypes in its 1000 gametes. If crossing over is about 12 percent (as in corn), the number of different genotypes in these 1000 gametes approaches a minimum of some 300, or about 30 percent.

A probably very important category of chromosome mutations is referred to as unequal crossovers or nonhomologous crossovers. If there are two somewhat similar genetic loci lying next to each other on a single chromosome, mispairing can take place at meiosis so that a locus pairs with the similar, nonhomologous locus on the other chromosome rather than with its homologue (see Figure 7-7). Crossing over may take place between these nonhomologous loci. An interesting example is found among the abnormal human hemoglobins. There is a class of abnormal beta chains that were found upon sequencing to be composed of the amino end of the normal human delta chain and the carboxyl end of the normal human beta chain. These are called the Lepore-type hemoglobins, and at least seven different ones have been discovered in different parts of the world. Apparently the delta locus lies right beside the beta locus, and because these differ in only 10 codons from each other (out of 146), they are so similar that mispairing readily occurs.

This example might be evolutionarily unproductive, but there is another case in human genetics in which such an event has apparently led to a selectively favorable allele. Haptoglobin functions in a chain of

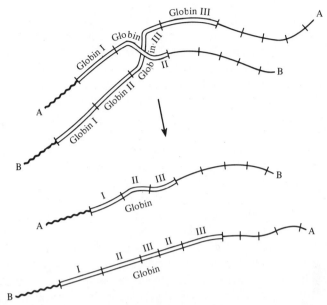

Figure 7-7   Unequal or nonhomologous crossing-over in two different chromosomes A and B, resulting in one deficient (and probably lethal) chromosome, and one with an extra globin. This process can result in a kind of gene duplication.

events to recover heme pigments from degraded hemoglobin in the blood. It is composed of two chains, alpha and beta. The alpha chain is polymorphic in many populations, and there are three common alleles: Hp 1^f, Hp 1^s, and Hp 2. The gene product of Hp 2 is almost twice the size of the products of the other two alleles and, on sequencing, was discovered to be composed of most of the Hp 1^f product at the amino end and most of the HP 1^s product at the carboxyl end. Both homozygotes and heterozygotes for this allele have increased ability to recover heme pigments from the blood (this being a rare case of functional dominance at the molecular level), and the Hp 2 allele is favored by selection in a number of populations in Europe and India (not in Africa). Because Hp 1^s and Hp 1^f differ from each other by only a single codon, this is effectively a case of gene duplication undoubtedly caused by an unequal crossover.

GENE DUPLICATION

Gene duplication has long been known in classical genetics—in *Drosophila,* for example, and in corn. In *Drosophila* it is possible to see various genetic loci along the chromosomes by examining the unusual polytene salivary chromosomes. Certain mutations in *D. melanogaster* were eventually linked to duplications of short sections of various of the four chromosomes. It was discovered that one effect of irradiation, besides stimulating the usual kinds of gene mutations, was to produce numerous such duplications, often translocating them to other chromosomes. One such locus, the "bar" gene, showed that, once the duplication was present in homozygous form, it would undergo an unequal crossover (see Figure 7-7) about once in 1600 times, resulting in one normal fly (one copy of the locus) and an "ultrabar" individual (triplicate bar locus). Thus, once a duplication is present (and we do not know how this comes about), it is easy to see how further multiple duplications can occur, because then there is a clear mechanism for mispairing, as in the Lepore hemoglobins.

Examination of the amino acid sequences of many kinds of proteins has shown that some mechanism of gene duplication has been, and is, operating to generate the raw material for the evolution of new enzymes. Thus, hemoglobin alpha, beta, gamma, and delta chains and myoglobin from man are identical in a number of ways and, comparing just the hemoglobin subunits, have at least about 50 percent of their homologous amino acid residues identical. Some 60 percent of the homologous amino acid residues of bovine trypsinogen and chymotrypsinogen are identical. Many different dehydrogenases from different sources have an almost identical 10-residue sequence around the active site. There is evidence that many different cysteine-active proteases from plants have related sequences, while the serine-active proteases of procaryotes and animals show equivalent similarities. In addition, a number of proteins have been shown by computer analysis to have sequences composed of multiples of some fundamental sequence (see Figure 2-11). As an example, the heavy chains of immunoglobulins seem to be composed of what amounts to two fused light chains (as in the Hp 2 of haptoglobin), while the latter themselves seem to be composed of two fundamental units similarly fused (see Figure 7-8). These data taken together make a strong case for the need to invoke some process of gene duplication as an important phenomenon in the history of life. This process, whatever it is, is still going on, as evidenced by some laboratory strains of the house mouse in which all the individuals have two hemoglobin alpha chain loci, the gene products of which differ from each other by only a single amino acid residue out of 141, indicating a very recent duplication. There is now evidence that human fetuses show two gamma chains that also differ by only a single residue.

The idea, then, is that after a duplication has occurred, the two loci begin to evolve independently and thereby diverge, most of the duplicates probably losing the ability to code for operational enzymes by mutation. In time they are expected to become more and more different, and the functions of some of them may change or be regained differently from what they originally were, for example, a malate dehydrogenase might be converted to an isocitrate dehydrogenase. Thus, there is a clear functional difference between myoglobin and hemoglobin, and yet they seem to have descended from a single ancestral gene (the lamprey has only a

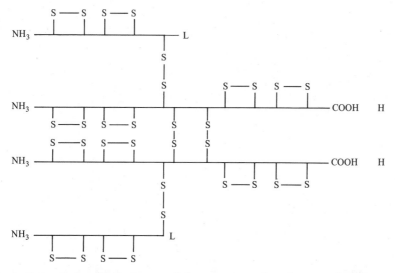

Figure 7-8   A simplified picture of the structure of an immunoglobulin molecule. The light chain (a single gene product) could have served as an ancestor of the heavy chain (another single gene product) by a process of gene duplication and fusion. H is the heavy chain L is the light chain. S–S represents disulfide bridges extending from cysteine residues in the primary structure.

single globin, as different from myoglobin as it is from hemoglobin). The advantage of a mechanism of this kind is that a new gene product can be invented without destroying the old ones. The duplicated genes may most of them never lead to anything, and this too would not matter, because the essential original gene product is still present and is still being maintained in functional condition by natural selection. The types of duplication seen in *Drosophila* and the fact that the human hemoglobin beta chain locus is found right next to the gene that it most closely resembles (the delta chain locus) suggest that the duplication process produces a series of tandem genes next to each other on the DNA of a single chromosome. In time some of them might become translocated to different chromosomes. Thus, there is some evidence that the genes from human beta and alpha hemoglobin chains are on different chromosomes. Their is direct evidence that some genetic loci—for example, the ribosomal RNA loci—are reduplicated many hundreds of thousands of times in eucaryotes. The acquisition of a new function for a duplicated gene must probably await its becoming linked to a new regulator gene as well as simply becoming structurally different.

Gene duplication is apparently the source of molecular isotypes. Many proteins, including many enzymes, exist as multimers—some are dimers, some tetramers, and so on. More than one gene is often involved in producing the enzymatic function. Thus, vertebrates have *l*-specific lactate dehydrogenase activity in their tissues. But two genes are involved

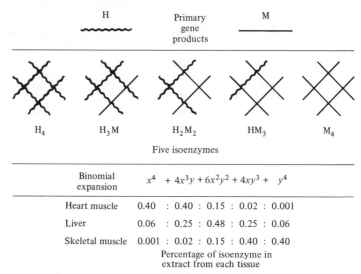

	H		Primary gene products		M

H_4	H_3M	H_2M_2	HM_3	M_4

Five isoenzymes

Binomial expansion	$x^4 + 4x^3y + 6x^2y^2 + 4xy^3 + y^4$

Heart muscle	0.40 : 0.40 : 0.15 : 0.02 : 0.001
Liver	0.06 : 0.25 : 0.48 : 0.25 : 0.06
Skeletal muscle	0.001 : 0.02 : 0.15 : 0.40 : 0.40

Percentage of isoenzyme in
extract from each tissue

Figure 7-9 Isoenzymes of lactate dehydrogenase in vertebrates. H and M are primary gene products that associate at random in the cell. The proportion of each type of primary subunit in the cells determines the amount of each type of isoenzyme that will be present in the tissue (examples are shown).

in coding for this activity in most cells. The gene products of these two interact at random to form tetramers of five different kinds (see Figure 7-9), termed isoenzymes. By controlling the relative amounts of monomer subunits found in different tissues (differential transcription or differential stability of the monomer product), different tissues may contain different spectra of isoenzymes. Because it is known that the kinetic properties of these isoenzymes are different, this system allows lactate dehydrogenase activity to become adapted to different intracellular milieux. The amino acid composition and molecular weights of the two kinds of lactate dehydrogenase in a single organism suggest that the two genes involved code for very similar polypeptides analogous to the beta and alpha hemoglobin chains.

MORE CHROMOSOME MUTATIONS

Another kind of nonhomologous crossover event is that which may take place between nonhomologous chromosomes. This could, perhaps, occur after a gene or short segment has become translocated to a new chromosome. Translocation itself is a still further kind of chromosomal mutation with genetic and evolutionary implications. In general, this could be a mechanism for the buildup of various kinds of gene regulation systems. As a possibility, consider a system such as that found in procaryotes involving a regulator gene and/or an effector substance that acts to initiate the transcription of a structural gene by way of an operator gene. By means of translocations, one type of operator could become associated with several different sorts of structural genes (if it duplicated first), thereby building up a battery of different genes all of which might be triggered by some single event, such as the appearance of an inducing substance at an appropriate stage in ontogeny. Or a classical sort of operon could be fabricated by linking structural genes from different chromosomes (perhaps involved in one biochemical pathway) to a single operator.

An interesting form of chromosomal mutation about which more is said in Chapter 9 is the inversion. Here some part of the linear sequence of genes is reversed from the way that it usually occurs. The mechanism for this must involve multiple breaks and "erroneous" repair of the genetic material. An important result of the presence of such mutations in the gene pools of organisms harboring them is a significant decrease in the variability generated by meiotic crossing over. The products of crossing over between an inverted and a normal sequence are inviable, the process resulting in acentric fragments and dicentric chromosomes, as well as in deletions, depending on the kind of inversion involved (see Figure 7-10). For this reason, crossing-over can only result in viable genomes when it takes place between the homologous chromosomes of individuals that are homozygous for the same inversion, and different inversions will be maintained more or less intact with respect to each other. The more different inversions of the same chromosomes that are present in the population, then, the less chance will there be of a viable crossing over occurring and the less variability will be generated during meiosis. Inversions are, therefore, dampers on the generation of variability.

SEXUAL RECOMBINATION

So far the discussion has centered on the generation of new sort of genetic material—mutated genes, mutated chromosomes. In procaryotes and other asexual organisms this is the only kind of variability possible at the level of the genome. Sexual recombination results in the rapid reshuffling of existing chromosomes, and, as such, is a mechanism for amplifying whatever genetic variability exists in a population. Using the example given above, of an organism with four chromosomes and 1000 ova, it was possible with gene mutation and crossing over to generate some 300 different haploid gametes. Thus, only some 30 percent of gametes represented new genotypes. But as soon as sexual recombination occurs *all* the zygotes will be gentically different from one another and from their parents. This is based on the premise that the other parent will be in some (even slight) way different at each chromosome and on the observation that gamete pairing is fundamentally a random phenomenon. In this way genes are tested by natural selection in as many different genetic backgrounds

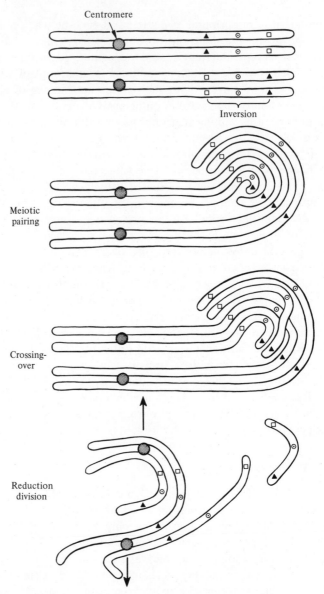

Figure 7-10 The results of crossing-over involving an inversion. The cells involved will die; therefore progeny will never show this cross-over, that is no living progeny underwent such an event.

as possible in as short a time as possible (one generation). The rapid generation of genotypic variability is the fundamental meaning of sexuality. In a large enough population, this process alone would continue for a very long time to generate new genotypes even if mutation stopped.

PARTHENOGENESIS

Some eucaryotes have secondarily lost some or all of the advantages of sexuality, becoming either self-fertilizing or parthenogenetic. Self-fertilizing organisms (many plants) obtain variability from meiotic crossing over in addition to mutation, but the effects of this too are lost in some nonmictic forms of parthenogenesis. There are a number of forms of parthenogenesis, but all involve homozygosity—in some cases due to typical haploidy, in others (effective haploidy) caused by endomitosis before meiosis (as in triploids—Chapter 11, or allopolyploids—Chapter 14). In some organisms—plant lice, water fleas—several parthenogenetic generations during a favorable season are followed by a sexual generation in which genetic variability is renewed. Sexuality is somehow triggered by a deterioration of conditions for life, and the resulting eggs are capable of overwintering to give rise to next year's new parthenogenetic population. Parthenogenesis in these cases has been interpreted as a specialization allowing the greatest possible reproductive effort for each female during the good season in the delays necessitated by courtship and the like, are not involved and in that each female gives rise only to daughters. Females that gave rise also to males would soon lose the race to contribute the greatest number of offspring to the next generation because males are not capable of producing new individuals. The basic function of males in a biological system is to provide genetic variability. In systems in which, because the conditions of life are relatively so easy, genetic variability can be discarded temporarily, the need for males is obviated. Once again, variability is necessary only because conditions anywhere on the earth will ultimately change.

GENE FLOW

Another source of genetic variability that can be tapped by sexual organisms is gene flow. This can in some cases involve different species (hybridization). The movement of alleles from one deme to another by means of occasional reproductive contacts (emigration-immigration) constitutes gene flow between the demes. Members of the same deme share the same gene pool, and gene flow is not in question in matings between them. Members of different demes could, however, be genetically different in a number of ways, owing either to random divergence (see Chapter 13) or to natural selection molding the demes in somewhat different adaptive ways. Alleles present at some locus in one deme may not be present at all in another, isolated from the first. The effect of gene flow here is something like that of mutation, but the mechanism arises out of sexuality.

There are a number of problems and questions concerning gene flow, however. It has recently become clear that barriers to gene flow between different species are not so absolute as was once thought. Many of the hybrids, however, lead nowhere evolutionarily, usually being phenotypically intermediate and so fit for neither parental niche (see Chapter 6). Even where hybrids occur in fair numbers over long periods of time, they may simply be a by-product of contact between two species and may represent nothing more than part of the portion of individuals in every generation of both species that are excluded from life by the limitation of a finite carrying capacity (that is, they are part of the genetic load of the population). As such they are not even a "wastage of gametes," as was once maintained. Very gradually individuals that tend to breed with members of the other species should become fewer and fewer in the area of contact, because they tend to leave poorly adapted offspring or none. New areas of contact between the species may, however, form from time to time. The question is how many alleles actually do make the transition from one species to another. Species that are genetically so close as to be able to form even subvital hybrids inevitably have many identical alleles at numerous loci anyway, as has been proved in *Drosophila*. Thus, although the potentiality for gene flow between species exists, the actual amount of it in real demes is unknown.

The actuality of gene flow between different demes of a single species is also open to question. In some species, say, some migratory birds or the Atlantic eels, the entire species breeds in a very restricted region regardless of where the individuals spend their nonreproductive time. Every river on the eastern seaboard of the United States has (or had) Atlantic eels in it. They are isolated in different drainage systems, but when it comes time to reproduce, they all migrate to a place in the ocean near Bermuda. Presumably gene flow is an irrelevant concept in a case like this—all American eels are panmictic. On the other hand, there are a number of species, for example, those which live in caves, that are broken into a number of completely isolated demes that can never come into contact. Gene flow between them is impossible without the advent of large-scale catastrophic events. The interesting question, however, is just how much contact actually occurs between demes of the more "usual" kinds of organisms.

There is now evidence, for example, that many butterflies, potentially very mobile, have populations restricted to single valleys. Individuals are capable of flying to the next valley but do not do so, apparently being inhibited by various microclimatic features, such as insufficient relative humidity in the region between populations. Again, many wind-pollenated plants have their pollen transported for only surprisingly short distances so that if there is a discontinuity between individuals of some hundreds of yards, this could be sufficient to isolate adjacent populations from each other (provided that the seeds are subsequently also not very motile). Again, there is now mounting evidence that frogs and other amphibians in temperate regions home to a specific breeding locality—a pond, and the like—every year. Thus, although they may spread out from that pond for foraging during the summer and for hibernation during the winter and at these times encounter individuals from other demes (and compete with them in an ecological sense), they always return to a restricted region for breeding. Couple this with the fact that amphibians have very low vagility anyway, because they are slow and require high relative humidity in order to travel, and one can readily imagine that genetic contact between adjacent populations may not be very great from year to year.

On the other hand, frogs do move about a bit. Thus, immature leopard frog individuals frequently migrate away from the breeding pond where existing territories are saturated. They move only on rainy or misty nights and can move up to some 70 meters per night. At this rate, and considering the average number of nonwinter rainy nights to approximate that found in the north-central United States, it can be calculated that leopard frog populations could traverse the area between northern Mexico and Michigan, (as during an expansion of their range) in as little as 5000 years, or 1600 generations. If adjacent breeding sites are about 300 meters apart, as in an area in northern Michigan, an individual could reach another deme's pond in two nights. This should be enought to keep gene flow going on at a lively rate. In this case, however, it may be best to define the breeding populations as including all ponds in an area that can be reached in a week or so by migrating young individuals, and then consider how much gene flow might occur between this group of ponds and some other adjacent group separated from the first by a large river, say, or some other barrier difficult but not impossible for a frog to cross. There is in fact a very real difficulty in defining the actual limits of any deme or breeding population. Until this can be done in a particular case, it is impossible to discuss gene flow.

Traditionally, genotypic or phenotypic similarity between adjacent populations of a species has been accounted for partly on the grounds of gene flow between them. Thus, if one can demonstrate that an allele is found in high frequency in all populations of some frog or mouse in western Texas and that this allele decreases in frequency as one samples animals progressively eastward until another allele becomes predominant, one interpretation would be that, although selection was maintaining one allele in the west and another in the east, gene flow was carrying these alleles outward, away from the regions where selection originally favored them. This reasoning is based in part on the obvious fact that a new mutation must first occur and accumulate in some population before selection can favor it. Individuals carrying it to nearby demes is the most probable way that it will spread, because the frequency of occurrence of a specific mutation is very low (around 10^{-10} or so). Once in the new deme, selection again favors it, because this deme is located near the first and is subject to similar selective pressures. As the allele travels away from its region of favorability, it gradually loses its selective advantage and does not occur in as high a frequency as in that region.

An alternative viewpoint could be that the population carrying an allele at high frequency experiences a large increase in numbers resulting in emigration during a period favorable to a range extension. After range expansion has continued for a while, conditions change so as to be less favorable to the species, and populations are maintained only in pockets within the previous range with very little gene flow between them. Those in one region may all independently experience selection favoring a given allele; those in other regions experience either less intense favorable selection or unfavorable selection and come to show a different allele predominating at this locus. Because most environmental parameters, (with the major exception of soil types) change from region to region gradually, the result of selection could very well be a gradually changing allele frequency across a geographic transect. Histories like both of those given above have probably occurred in one or another specific case. It is unlikely that a single historical pattern is appropriate to all kinds of populations.

If there is little contact between two demes at some given time, this does not mean that contact will not become reestablished before genetic divergence has proceeded so far that interbreeding is no longer possible.

SUMMARY

Gene mutation is the fundamental source of all biological variability. In sexual organisms chromosome mutations, taking place mainly during meiosis, result in the further production of variability by producing new kinds of chromosomes, which are the units in which variability is transmitted. Sexual organisms have the further process of sexual recombination, in which the existing chromosomes are recombined in vast numbers of different ways, allowing a large amplification of variability in a single generation. New alleles may also arrive in a deme from other demes (or, occasionally, even from different species) by gene flow. The importance of this process depends on the population structure of the species.

REFERENCES

Brues, A. M., "Genetic load and its Varieties," *Science,* **164:**1130–1136, 1969.
Bryson, V., and H. J. Vogel, eds., *Evolving Genes and Proteins.* Academic, New York, 1965, 629 pp.

Dobzhansky, T., *Genetics and the Origin of Species,* 3d ed., Columbia, New York, 1951, 364 pp.
Ehrlich, P. R., and P. H. Raven, "Differentiation of Populations," *Science,* **165:**1228–1231, 1969.
Fisher, R. A., *The Genetical Theory of Natural Selection.* Dover, New York, 1958, 291 pp.
Grant, V., *The Origin of Adaptations.* Columbia, New York, 1963, 606 pp.
Ingram, V. M., *The Hemoglobins in Genetics and Evolution.* Columbia, New York, 1963, 165 pp.
Jukes, T. H., *Molecules and Evolution.* Columbia, New York, 1966, 285 pp.
Kimura, M., "Evolutionary Rate at the Molecular Level," *Nature,* **217:**624–626, 1968.
———— and J. F. Crow, "The Number of Alleles that Can Be Maintained in a Finite Population," *Genetics,* **49:**725–738, 1964.
Li, C. C., *Population Genetics.* The University of Chicago Press, Chicago, 1955, 366 pp.
Livingstone, F. B., *Abnormal Hemoglobins in Human Populations.* Aldine, Chicago, 1967, 470 pp.
Nei, M., "Gene Duplication and Nucleotide Substitution in Evolution," *Nature,* **221:**40–42, 1969.
Perutz, M. F., and H. Lehmann, "Molecular Pathology of Human Hemoglobins," *Nature,* **219:**902–909, 1968.
Stebbins, G. L., *Variation and Evolution in Plants.* Columbia, New York, 1950, 643 pp.
Strickberger, M. W., *Genetics.* Macmillan, New York, 1968, 868 pp.
"Structure, Function and Evolution in Proteins," *Brookhaven Symposia in Biology,* no. 21, 1969, 428 pp.
Watson, J. D., *Molecular Biology of the Gene.* W. Benjamin, New York, 1970, 662 pp.
Wright, S., "Polyallelic Random Drift in Relation to Evolution," *Proceedings of the United States Academy of Sciences,* **55:**1074–1081, 1966.
————, *Evolution and the Genetics of Populations,* vol. 1, *Genetic and Biometric Foundations.* The University of Chicago Press, Chicago, 1968, 469 pp. (Vol. 2, 1969, 511 pp.)
Zuckerkandl, E., and L. Pauling, *Molecular Disease, Evolution and Genic Heterogeneity* in M. Kasha and B. Pullman, eds., *Horizons in Biochemistry.* Academic, New York, 1962.

Expression of variability

This chapter treats the relationships between the genotype and the phenotype. The genotype can be seen as a developmental program selected for its ability to encode hereditary information for an adequate phenotype over a relatively small range of environmental conditions. The phenotype is the actualization of the genotypic message under a specific set of environmental conditions; it is the result of an interaction between the genotype and the environment and is represented by any function or structure of an organism that could possibly come under the influence of natural selection. Any structural or functional feature of an organism that can have no bearing on its relative ability to reproduce need not be (but often is) considered as a part of its phenotype. Whether there can be structures or functions not germane to reproductive success is a controversial point (see Chapter 13). Possibly many of the more recently observed chemical features of organisms, such as the electrophoretic mobility of one of their enzymes, fall into this category of attributes of living systems. It is difficult to see how natural selection could be operating on a characteristic of an enzyme that could not possibly appear while the enzyme is in the living system. This is not to say that even slight differences in electrostatic charge of proteins could not have selective significance, only that the change in electrophoretic mobility also caused by such a change cannot have such significance. It is important to keep this distinction in mind since what is measurable (for example, electrophoretic mobility) may not be biologically meaningful, even though it may be linked with some phenotypic effect that is meaningful.

SICKLE-CELL ANEMIA IN MAN

An instructive example giving us a feeling for the complexity of even the conceptually simplest kinds of phenotypic events is sickle-cell anemia in man. This disease is triggered by the presence of a hemoglobin (Hb S) containing beta chains that differ by a single amino acid from the kind usually present in people in most parts of the world (the result of a single substitution mutation). The "abnormal" chain contains valine instead of glutamic acid at the sixth position from the amino end of the polypeptide. This could have been brought about by a changed messenger RNA mutated to the triplet GUA or GUG instead of the normal GAA or GAG. This in turn is the result of a single mutational event replacing thymine with adenine in the second position of the sixth triplet of the beta chain structural gene. Hemoglobin S (Hb S) tends to precipitate out of solution when it is deoxygenated, forming elongate crystals. Since these molecules have altered solubility characteristics, an important factor at this level is the concentration of hemoglobulin in the erythrocytes, which is about 34 percent in man. This is a very concentrated protein solution and is probably related to the amounts of oxygen needed by the tissues per unit time. A concentration exists at which Hb S is completely soluble. If the needs of the organism for oxygen had not been so great, the concentration of hemoglobin in the red cells might not have evolved to be as high as it is and then this particular mutational event would not have resulted in a malfunction. Thus, all the genes affecting respiratory rates are involved in the malfunction.

The formation of hemoglobin crystals produces a characteristic deformation of the red cell to a shape resembling a crescent. In humans this somehow weakens the red cell membrane to such an extent that it lyses. In other mammals sickle cells appear in some populations but do not lead to erythrocyte lysis (some deer populations, for instance). Therefore, some characteristic of the human red cell membrane is also involved here, and so the genes responsible for that characteristic are also involved in the malfunction. The loss of red cells and their hemoglobin, which is quickly broken down after lysis, leads directly to anemia, which again is a function of the amount of oxygen needed by the tissues per unit time. Hence it can be seen that the disease state triggered by a single base substitution really involves a fairly large segment of the total genome.

Going further, we can note that this abnormal allele of the beta chain gene is present in high frequency in some populations, that is, that natural selection in some places is favoring, or at least not eliminating, this allele.

We pursue this further in Chapter 11, but note here that all the genes involved in the susceptibility to attack from mosquitoes and parasitization by the malaria sporozoan are involved in this complex evolutionary reaction triggered by the simple thymine-to-adenine mutation in one gene.

Hemoglobin S is different from normal hemoglobin (Hb A) electrophoretically, but that is not a phenotypic difference. The phenotypic difference, insofar as it resides with the beta chain locus, is the altered solubility of Hb S.

CHANGING PHENOTYPE DURING ONTOGENY

The phenotype at any given stage in a life history cycle can be taken to be a function of the action of the gene products of genes being actively transcribed at that stage, plus gene products of genes that were transcribed in the past into long-lived mRNA, plus gene products that were produced by translation of short-lived mRNA in the past but which are themselves long-lived, plus the end products of the action of long-since-broken-down gene products of loci that were active in the past, plus the immediate external environment (temperature, and so on). The actions of these gene products and their interaction with end products and with some specific environment produce the phenotype at a given moment. This view emphasizes the importance of developmental phenomena (or the ontogenetic history of the individual whose phenotype is being considered) in the concept of phenotype.

Contemplation of the obvious differences between a tadpole and a frog, or a caterpillar and a butterfly, demonstrates the importance of life history stage for the phenotype. Developmental differences between stages are not always of such large magnitude but are probably of the same kind. Different loci are evidently expressing themselves during different stages. One aspect of development is the switching on and off of genetic loci by various kinds of effector molecules. Another aspect is the gradual, more or less permanent turning off of loci; for example, in all cells but erythroblasts the genes coding for hemoglobin subunits are turned off, apparently permanently. As cells become functionally specialized, they seem to undergo a permanent decrease in the number of actively transcribing loci, but may retain some restricted ability to switch activity from one locus to another. Some genetic loci may be actively transcribing for only a few hours or a few days of the life span; others may be more or less continually active; still others may be active intermittently. Loci that are active during only part of the life of an organism can undergo evolutionary change (that is, be impinged upon by natural selection) in the same way as any other locus. This means that some feature of a gastrula, for instance, can evolve independently of some feature of an adult, or that tadpoles can evolve independently of frogs. This last system is an interesting example.

Adult frogs of different species are, with the exception of a few kinds, more similar to one another than are their tadpoles. It seems that the recent evolution of external features in this class of vertebrates has taken place largely in the tadpole stage and has to do in part with their adaptations to an extraordinary range of microhabitats. Some live in ponds, some in streams, some in thin films of water flowing over rocks in waterfalls, some in the tanks of bromeliads, some in terrestrial or arborial nests, and so on. Some are adapted to scrape algae from surfaces, some are pelagic filter feeders, some are carnivores. Tadpoles have expanded into a number of adaptive zones almost rivaling those of the fresh-water teleosts, whereas frogs remain restricted to only a few adaptive zones. One could say that the overall rate of evolution of external features in frogs has been much slower than in tadpoles. Although organisms having larval stages show these effects more strikingly than others, there is no reason to restrict the concept to them. For example, some genes are active only in the cotyledons of plants and not in any other tissues, which means that they are active only during seedling stages. Being able to selectively activate different genes at different times has had an important impact on evolutionary flexibility. Any genes operating at a given time must be coadapted. A gene selected to function in the environment of an embryo would probably not be optimally fit to function in the adult environment. Indeed, without the ability to switch off one gene and turn on another, development (and therefore complex multicellular organisms) could not exist at all. Tadpoles have at least two kinds of hemoglobin subunits; frogs have two others, quite different. The eye pigment of tadpoles is porphyropsin (opsin plus retinine 2); while that of the frog is rhodopsin (opsin plus retinine 1). During metamorphosis some enzyme must be activated capable of reducing one of the double bonds in retinine 2, converting it to retinine 1. Tadpoles excrete mostly ammonia. During metamorphosis the gene coding for the enzyme arginase is activated, ultimately activating the ornithine cycle which produces urea, which then becomes the major nitrogenous excretory product in the frog. Of these examples, the hemoglobin one is perhaps the most instructive in the sense that it involves replacing a protein with another very like it but different.

The need for replacing one protein by another very like it at a different life history stage may be one of the explanations for the prevalence of multiple molecular forms of enzymes in living systems. In the human embryo hemoglobin is constructed of two alpha chains and two epsilon chains. At a point early in development, the epsilon locus is turned off and the gamma locus turned on, and the fetal hemoglobin consists of two alpha chains and two gamma chains. Shortly after birth the gamma locus stops transcribing and the beta locus is activated, so that adult hemoglobin is composed of two alpha chains and two beta chains (Fig. 8-1). One of the adaptive meanings in this complex example can be shown by noting

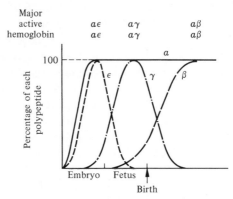

Figure 8-1   Diagram of the changes with age of the amounts of various hemoglobin subunits during human development. (Modified from Ingram, 1963.)

that fetal hemoglobin is more easily saturated with oxygen, thus making possible the efficient transfer of that molecule across the placenta to the embryonic bloodstream. Note that the alpha chain locus is always operative, pointing out that not *all* proteins need to be changed from one stage to the next. Nevertheless many are, and in different tissues of adults one form of an enzyme may replace another. For example, in heart muscle and other highly oxygenated tissues one form of lactate dehydrogenase is found, whereas in skeletal muscle and other relatively anoxic tissues another form is found. These two are the products of different genetic loci and are isoenzymes. In frog embryos only the form of the enzyme found in adult heart muscle is present. Shortly before hatching, the other locus is activated in some tissues. In the adult some tissues have only one form, some the other, and some have both, and so differentiation involves in part the activation of different isoenzymes in different tissues (see Figure 7-9). The two different kinds of lactate dehydrogenase have been demonstrated to be catalytically different, and there are strong suggestions that the functioning of each kind is appropriate to the levels of oxygen in the cells containing it. Because isoenzymes are the products of different loci, they evolve separately and can provide raw material for specializations at different life history stages as well as in different cell types. Undoubtedly the source of genetic raw material for the production of isoenzymes has been the process of gene duplication discussed in previous chapters.

Thus, there is a stage-dependent differential expression of portions of the genome during development that allows characters expressed at different stages to evolve separately in the process of adapting to different environments. The genomes of higher organisms have information encoded for several different "kinds" of organisms together with a sequential program invoking one after the other.

MULTIGENETIC TRAITS AND POLYGENES

Some phenotypic traits of organisms can be adequately described as single-gene traits in that alteration of a single gene has profound or measurable effects on the organism. Hemoglobin S would be such a trait and so would other phenotypically significant allelomorphisms at any locus. With adequate data these traits are observed to behave in simple Mendelian fashion when organisms are experimentally bred, or adequate data about them from natural populations are found to be consistent with Mendelian predictions. These instances can usually be described as showing discontinuous variation. An example would be the ability to taste phenylthiourea or lack of this ability in humans. The kinds of phenotypic traits most usually used in making taxonomic distinctions, however, vary in continuous fashion, involving sizes, lengths, color shades, ratios, and the like. These traits invariably are affected by the gene products of many different loci; they are multigenetic traits. Each locus involved may have only a tiny effect on the final structure of the trait being observed, although some may have larger effects than others. The genes coding for these kinds of traits are often referred to as polygenes and the traits as polygenic traits. The number of genes involved in any given complex trait is seldom known and probably varies from a few to nearly the entire genome. Behavioral characteristics may be examples in which much of the genome is involved. Thus, for a call note to be given at exactly thus-and-so pitch would involve the sizes of larynx, pharynx, mouth, the lung capacity, and other anatomical features plus, of course, the behavioral impulses of the central nervous system, as well as various reflex systems and sensory systems capable of triggering the call response at appropriate moments in the flow of time. That kind of trait is not usually considered as being coded for by a set of polygenes. The polygene concept is usually restricted to genes coding for a single numerical dimension of some anatomical character. Behavioral traits are best referred to as multigenetic.

PLEIOTROPY

It is not likely that the gene product of any locus affects only a single trait. Being an enzyme in a mesh or web of many other enzymes, its rate of utilization of substrate(s) and its rate of production of end product(s) probably have implications for many other kinds of traits. Such a gene would show pleiotropy. It is likely that *all* genes show some degree of pleiotropy, ranging from undetectable to considerable. In general, loci that act early in development have larger and more widespread pleiotropic effects than do loci that begin transcribing at later stages. This results from the epigenetic nature of development; for example, should an unfavorable mutation of a gene having an important effect on the neural crest cells of a gastrula appear, there could later result anomalies in pigment cells, various autonomic ganglia, the adrenal medulla, sensory neurons, myelin sheaths, and various bones and cartilages of the visceral skeleton mainly in the head. The neural crest cells are precursors of all these organs and cells and more. Mutations are known in rats that affect cartilage development at early stages. One of these leads to death in several different ways, probably depending on the genetic background, mainly involving inability to breathe either because the lumen of the trachea is occluded, or because the ribs are too thick to be readily expanded, or because the nasal cartilages obstruct the passage of air (Figure 8-2). A mutation of a gene affecting development only in its later stages would probably not affect traits as different as the trachea and the adrenal medulla.

These considerations provide us with an explanation for why embryos of disparate organisms resemble each other more than do the respective adults. There are simply less restrictions on the evolution of genes which express themselves during later stages of development. Thus, the notochord of vertebrate embryos performs certain functions in early development which are absolutely essential to continued epigenesis. The entire organization of the embryo depends on its characteristics (that is, genes affecting

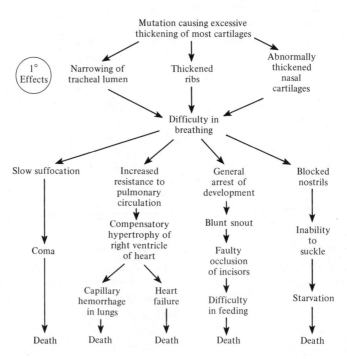

Figure 8-2 Various causes of death, showing pleiotropy, caused by the "emphysemic gene" mutation in the laboratory rat. (Data from Gruneberg, 1938.)

it are highly pleiotropic (and so most mutations affecting its structure and/or functions tend to be eliminated. The selection coefficients of new mutant forms of genes affecting this organ must be very nearly unity in each kind of vertebrate, and the organ itself as a result is evolutionarily very conservative. Thus, some loci are evolutionarily more malleable and tend to be more variable in populations than others. Molecular considerations have led to an analogous concept. Alpha chains of hemoglobin in mammals form part of the embryonic, fetal, and adult hemoglobins. The gene coding for this polypeptide is highly pleiotropic in comparison with that coding for the beta chain which functions only in the adult. One would predict that alpha chains, since they must interact with more gene products in more different environments, should be more conservative evolutionarily than any of the other subunits. This has been found to be so after extensive amino acid sequence analysis of many mammalian hemoglobins. Furthermore, there are about half as many known alpha chain variants in human populations than there are beta chain variants. The former probably produced nonviable individuals more often than did the latter. Also, four different beta chain variants are involved in extensive polymorphisms in various places in the malaria belt, but only a single alpha chain has been capable of forming such a polymorphism. This difference in the evolutionary potentials of loci with gene products that must interact with many other gene products in comparison with loci whose gene products need interact with only a few other gene products has been termed by Ingram effect after V. M. Ingram, a biochemist at Harvard University, who first called attention to this possibility in connection with hemoglobin subunits.

Thus, although genetic loci operating at different life history stages can evolve independently of each other, there are probably more restrictions on the evolution of loci operating very early in development because of their far-reaching pleiotropic effects. Because of this, the very earliest stages of different organisms tend to be more similar to one another than later stages. Widespread pleiotropic effects should perhaps also be expected in loci that impinge upon the functioning of endocrine glands whose hormones frequently perform a multiplicity of tasks in different portions of the body. At present, amino acid sequence studies do not reveal any suggestion that peptide hormones have been evolving more slowly on the average than have other polypeptides, but it may be significant in this respect that hormones from quite distantly related organisms are capable of eliciting "normal" responses. Also, the amount of data about peptide hormone amino acid sequences is miniscule.

FIXED TRAITS

Some traits of organisms may be said to be fixed in the sense that, once they appear, they show no further changes during the life of an individual organism. The numbers and sorts of adult teeth present in a mammal would be such traits as would be the numbers of vertebrae in fishes. Other examples are the color patterns on the wings of moths and butterflies, the numbers and positions of wing veins in flies, the number and positions of sepals and petals in plants, and the shape of an adult bird's beak. These traits are not necessarily coded for by a single locus or only a few loci. They are, however, the results of interactions between gene products and the environment over a relatively short span of time. During this span of time, these traits show some degree of responsiveness to modification by altered environmental conditions and so are not "fixed" in the sense of being completely determined by the genotype (Figure 8-3). For instance, it was found that sibling whitefish embryos reared in different temperatures had different means and ranges of vertebral numbers. Many of the genotypes of the species in question were able to interact with temperatures that were somewhat abnormal for the species to produce vertebral numbers somewhat abnormal for the species. Once the vertebral number is determined, however, continued changes in temperature will not further modify the trait. Similar experiments with *Drosophila* showed that there is a critical period during which alteration of environmental factors, such as temperature, results in altered patterns of wing veins. The critical period is some portion of the time during which the anlage of the structures bearing the traits are being formed or during which the chemical background for the anlagen is being developed. Thus, fixed traits are those that develop only once during the life of an organism. During the period of their development, no matter how short, the genotype is "open to suggestions" from the environment. In a population as a whole, such traits are normally or binomially distributed, each individual differing slightly from each other individual according to its genotype (presuming the same environment for all). Changes in external environmental conditions may result in shifts in the range and mean of population data for some such trait

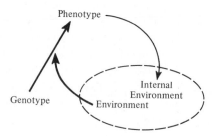

Figure 8-3   Generalized concept of the relationships between genotype and phenotype. The internal environment is caused in part by the actions of many or all genes other than the one being examined—by what is known as the genetic background.

without there having been any evolution at all. Should the environment return (after not too long a period) to previous conditions, later generations would again show the previous mean and range for the trait in question. Any individual genotype has a range of response possible to it, and the particular response finally elicited depends on the particular external environmental conditions. Needless to say, these considerations do not apply to the primary structures of enzymes, which are completely determined by the genotype; they apply to the processes and structures produced by the interactions of primary gene products under given conditions.

RANGE OF RESPONSE

At the molecular level the range of response possible with a single genotype can be understood in terms of the known characteristics of the primary gene products, the enzymes. Probably all enzymes have pH and temperature optima, with rates of activity decreasing on either side of the optimum. In addition, a number of enzymes show optimal substrate concentrations as well; that is, the phenomenon of substrate inhibition whereby increased activity is obtained by increasing the substrate concentration only up to a certain point, beyond which activity decreases again. Therefore, in different environments an enzyme will show different rates of activity (see Fig. 7-6). Alterations of primary structure may, of course, also alter the activity of an enzyme. Thus, there are known mutations which affect the wing veins of *Drosophila* in the same way as do environmental shocks at the critical period of development. Presumably the mutations are affecting the rate of activity of some enzyme, which rates can also be altered by environmental change. The term phenocopy was coined to denote a phenotypic change brought about by altering the environment such that the change mimicked the phenotypic effect of a known mutation. Presumably different allotypic forms of an enzyme are differentially sensitive to activity alteration by changes in various factors of the environment. Again we see how the phenotype is the product of the interaction of genotype with environment.

The idea of a range of response of single genotypes to different environments should not be raised without mentioning plants. These organisms are extraordinarily malleable under environmental influences. A typical observation would be that sibling seeds from a locality in the mountains planted at different altitudes will develop phenotypes quite different from each other but like the phenotypes of conspecific plants in each of the new (for them) altitudes. Aquatic plants are often referred to in this context because some (the water buttercup, the arrowleaf, and others) develop one type of leaf under water and a quite strikingly different kind of leaf in air, often on the same plant (Figure 8-4). Indeed plant phenotypes are so "unstable" in this sense that plant taxonomists have had difficulty in finding traits on which to base their systematic judgments.

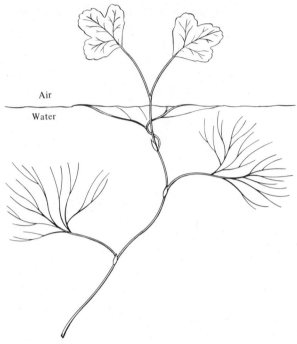

Figure 8-4 Phenotypic variation in leaf structure of the water buttercup.

CANALIZATION

Although some traits in some organisms are highly unstable, there are others in some organisms that are very stable indeed. Examples would be the presence of four limbs and only four limbs in a typical tetrapod, or of five (or, rarely, less by loss) digits on the limbs of the same, or of two eyes and only two eyes. Variations on traits of this kind are dubbed monsters partly because they are so rare. The reason for their rarity probably lies in part with the high adaptive value of the normal form of the trait and the high selection coefficients associated with any other-than-"normal" allele of any gene affecting the trait, meaning that any individual monster usually does not survive long enough to be visible, and certainly does not breed.

Another kind of explanation for the stability of some traits has been suggested by C. H. Warrington of the University of Edinburgh and by J. M. Rendel. This viewpoint sees fixed, multigenetic traits as existing in a sense independently of the particular alleles present at the loci in question. The main point is that the contribution of each separate locus is so small that alteration of the structure of its gene product does not usually result in a significant alteration of the trait. Traits built on a substructure of such a great many different loci that a significant change in one of them will not usually alter the trait are referred to as strongly canalized. Changes in external environmental influences will also have little effect on a strongly canalized trait, probably because any alteration in the environment does not affect every gene influencing the trait, either because different genes produce different kinds of enzymes or because different enzymes act at different times during the development of the trait, while significant changes in environment often do not last beyond a short period of time. In this sense a strongly canalized trait is buffered. Should a genetic or environmental effect disrupt the development of a canalized trait because the effect is unusually large magnitude, then selection will place the individual in question at a reproductive disadvantage *if* it lives; thus if the disrupting influence were a genetic one, it would promptly disappear from the gene pool of the population. If the disrupting influence were an environmental one, either it will not occur again or those individuals who breed in proximity to the offensive environment will not leave many offspring. In this way the population will gradually "move away from" the potential influence of that environmental stimulus. Alternatively, only those individuals will survive in proximity to the environmental influence who have genotypes exceptionally capable of withstanding its influence and of producing "normal" phenotypes in spite of its presence.

GENETIC ASSIMILATION

The mode of evolution of a canalized trait has been experimentally investigated and has been called by Waddington genetic assimilation. Starting with wild stock *Drosophila* flies, Waddington produced a phenocopy of a mutation (crossveinless) in which one of the veins in the wing was missing. The phenocopy was produced by subjecting pupating larvae at a certain (critical) stage in their development to an environmental shock, namely, a short period of time at an abnormally high temperature. In the first generation the treatment produced a small percentage of individuals missing the wingvein. These individuals were selected to produce the next generation (that is, all normal flies were discarded; the selection coefficient for them was 1.0). This procedure was carried on for a number of generations. The control for the experiment was to rear a certain number of offspring each generation without subjecting them to the heat shock. Thus, the experiment consisted of selecting for the parental function only those individuals whose crossvein development was relatively easily disrupted. By the fourteenth generation a few individuals appeared that were crossveinless *without* receiving the environmental shock. By the sixteenth generation these represented 1 to 2 percent of the population of untreated (or control) offspring. Intense inbreeding of some of these individuals resulted in populations with an average of 18 percent untreated crossveinless flies. Thus, after selection, a trait which could be elicited in only an individual here and there in normal populations by rather severe environmental shock was transmuted into a frequently occurring, spontaneous trait in the experimental populations. Roger Milkman, of the University of Iowa, has presented a plausible description of what was happening here and we will utilize his approach in the following discussion.

Suppose that the wingveins are affected by a number of different loci. Each of these loci can be represented by a number of different alleles (see Figure 5-15). Each allele for any locus will have a different magnitude of effect on the character in question, or would produce a different amount of genetic "make" for wingvein; allele 1 might tend to favor no wingvein, while allele 3 favored a partially developed one, and allele 5 favored a fully developed one (Figure 8-5). If an individual had what amounted to allele number 5 at each of the loci in question, it could become all but impossible to eliminate the wingvein by environmental shocks of magnitudes that would not also kill the organism. On the other hand, if an individual had the equivalent of allele number 1 at every locus, no wingvein would develop under any environmental conditions. A natural population is seen as having a mixture of alleles 1 through 5 at each locus in each individual, giving a spectrum of individuals, most of which are fairly well buffered under normal environmental conditions to produce a crossvein; that is, there would be very few individuals with very low total genetic make for wingvein. Because pleiotropy exists, the selective "reason" why a particular allele is present at a given locus may have relatively little to do with wingveins, but presumably the presence of the vein is favored (since it appears in all wild-type flies), and so the percentage of alleles 1 through 4 or so would tend to be low in the population as a whole. Because of the immense numbers of alleles potential at a given locus, it is presumably possible to have alleles equally favored for some major function and differing only in their (pleiotropic) effects on wingveins, and so selected on that basis only. Through recombination, some individuals will receive from their parents a combination of alleles adding up to an unusually low total, and these would lack a wingvein if they met an abnormal environmental condition at the critical period for wingvein development. These were the individuals detected by Waddington in his first generation. By choosing these to be the parents of the later generations he actually selected a mix of alleles quite unrepresentative of the wild population—a mix with an unusually high percentage of low-make alleles. By continuing the selection he gradually increased this percentage until it was high enough to produce by recombination some genotypes totaling so many low-make alleles that they produced the crossveinless condition spontaneously. This can be taken as a general formulation applying to polygenic systems. Usually selection would favor some average total make in such a system, thus determining the magnitudes of the effects of the alleles that are present (for example, by favoring the alleles with the smallest effects

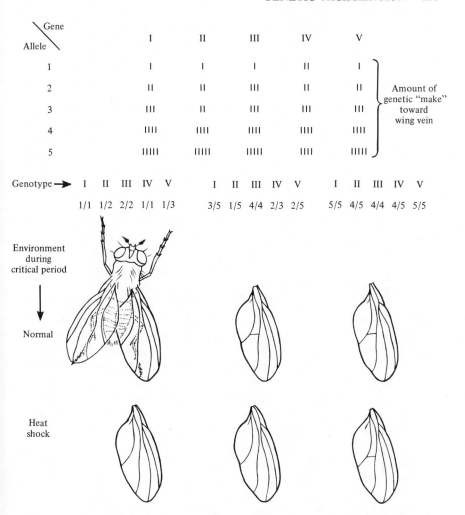

Gene / Allele	I	II	III	IV	V	
1	I	I	I	II	I	
2	II	II	III	II	II	Amount of genetic "make" toward wing vein
3	III	II	III	III	III	
4	IIII	IIII	IIII	IIII	IIII	
5	IIIII	IIIII	IIIII	IIII	IIIII	

Genotype ➤

I	II	III	IV	V		I	II	III	IV	V		I	II	III	IV	V
1/1	1/2	2/2	1/1	1/3		3/5	1/5	4/4	2/3	2/5		5/5	4/5	4/4	4/5	5/5

Environment during critical period

↓

Normal

Heat shock

Figure 8-5 Hypothetical genetic explanation for crossvein inheritance in *Drosophila*. See text for explanation.

most), while recombination would continually generate total makes on either side of the favored mean. In this kind of system any grouping of alleles totaling up to the desired mean effect would be selectively favored without stipulating which is present at a given locus. That, of course, would be determined by other pleiotropic functions of the locus.

In Waddington's experiment, then, selection broke down a canalized trait (presence of a certain crossvein) and built up a new, not so strongly canalized one (absence of the wingvein). Waddington called this genetic assimilation (a term which can do quite well for this phenomenon) to emphasize that what had originally been a purely phenotypic effect (phenocopy) became assimilated into the genome of the population.

BALDWIN EFFECT

Waddington later supplied some possible evolutionary examples of genetic assimilation of which we can cite one here. Most tetrapods have skin that is capable of becoming callused after repeated exposure to friction. Normally the skin does not form calluses except after prolonged rubbing, but ostriches are hatched with callused rumps. This is clearly an adaptation to sitting on hard dry soils and rocks, on which the hatchlings can sit without damage to their rears. Other birds, however, are not hatched with callused rumps; this trait seems to have been evolved only by ostriches. It seems as if a potential trait which normally does not appear in the absence of a certain level of environmental stimulus has become a fixed genetic trait in ostriches. One can form a model of the history of this trait using the *Drosophila* experiment cited previously.

The form of this argument is as follows: An environmental change occurs such that, say, one-half of the individuals in a population have genotypes that under the new environmental conditions cannot regulate to produce phenotypes with high adaptive values (say, their skins are always too thin), and the other half can produce phenotypes that are adequate (say, thick to very thick skins). There will be an initial fairly rapid change in gene frequencies as alleles that tend to produce the now undesirable phenotypes are reduced in frequency or removed from the population. This will be an evolutionary event. But the acceptable phenotypes are not new ones; they are phenotypes which appeared in the old populations at one side of the normal curve for this trait under some conditions. They were mostly unusual phenotypes, but are now common ones. Any new mutations, in order to be selectively favored under the new environmental conditions, must be able to function well with the range of phenotypic expression now present in the population. As such mutations accumulate at any and all genetic loci, they tend to increasingly canalize that trait in its new (formerly unusual) range of structure because individuals that are produced with a tendency toward the old range of the trait will contain some of these new alleles, which, because they are coadapted with those older alleles that could produce the newly favored range of phenotypes, cannot function well with alleles that tend to reconstruct phenotypes located in the lost part of the old range for the character. In this way, what was originally a possible phenotype for some individuals under extreme environmental conditions becomes the usual phenotype for most individuals under the new (formerly extreme) environmental conditions. Now, should environmental conditions return to something like their original range, the ranges of phenotypes in the population would not return to their old values because the alleles that would be needed to do this no longer exist, having been replaced by new ones which operate better in the context of the phenotypes produced by *some* of the old alleles interacting with the new environmental conditions.

Notice that this formulation adds a new dimension to the discussion. Initially we saw the shift in the range of phenotypes present in a population to be caused only by the loss of some of the alleles present and recombination of those that were left to form a new range of phenotypic expression. Since the absolute numbers of alleles is less than initially, there will be less phenotypic variation as well as a new mean (see Chapter 10). If this were the only process involved (as it probably is in laboratory experiments), a return to the old environmental conditions would result in a shift in mean back toward the old mean (not necessarily reaching it since some alleles have been lost). This is because the only evolutionary event to have occurred is loss of some of the initially available alleles, and even those present do not in most combinations result in the unusual but now desirable trait when not stimulated by unusual environmental shocks. If, however, a long period of time is allowed to pass during which the environmental shock is present—and therefore most of the genotypes present do produce the new (formerly unusual) phenotype—mutations will occur and be tested by natural selection. They will occur in the context of the new phenotype and will tend to make it independent of the environmental shock (which is still present). Note that some combinations of old alleles would also make the trait in question be independent of the environmental shock, but recombination could produce only a limited number of such individuals because of apparent limitations on homozygosity (see Chapter 10). The new mutations, however, if they are favored, will spread in the populations, and as they accumulate will tend increasingly to *fix* the trait in a genetic sense. They will tend to create a new coadapted genome (see discussion of dominance in Chapter 7).

Phenotypic variation allowing an initial adaptation to changed conditions before the advent of new mutations, followed by new mutations which genetically fix the phenotypic change, has been termed the Baldwin effect after the nineteenth century American biologist J. M. Baldwin. He proposed an essentially similar scheme in opposition to natural selection. We can see, however, that it is actually only a special case of natural selection in that it is selection working with ranges of phenotypic response rather than with completely determined traits, with polygenic traits rather than with single-gene traits.

The ability of a single genotype to produce more than one phenotype, depending on external conditions, has been termed phenotypic plasticity. Conceptually it is the opposite of the notion of canalization, where many different genotypes produce essentially the same phenotype. Both effects probably serve to inhibit large-scale evolutionary response to temporary environmental fluctuations. Thus, the ability to regulate developmental responses such that the same phenotype is produced even in the presence of some new alleles or of a somewhat altered environment will allow the population to retain its adaptiveness to some general environmental complex. For instance, in a short period of unusually cool weather, some new alleles could be selected as a response to cold without altering the basic adaptations of the organisms. On the other hand, phenotypic plasticity could produce among many of its phenotypic effects some that were adaptive to the altered environment that caused the response.

The problem with the plasticity idea in this context is that the phenotypic responses that occur will have no necessary relation to the needs of the organisms in the altered environment. The assumption can be made that *some* of them may be. This is a major problem with the Baldwin effect. There will be no guarantee that the phenotypic responses given will be appropriate. One simply assumes that given a large enough population, or simply by chance, some of the responses will be appropriate to the altered conditions. But it is well to remember that loss of a wingvein will probably not improve the performance of flies in hotter climates. We are assuming in the Baldwin effect that the unusual phenotype elicited in some individuals by an environmental change is a selectively favorable effect, or else individuals showing it would not reproduce and there would be no Baldwin effect to observe or discuss. In any case, this sort of effect, should it occur, can also impede large-scale evolutionary responses to temporary environmental changes that do not last long enough for new alleles to be selected so that the unusual phenotype becomes fixed. When conditions return to their previous state, the population phenotypes will also return close to their previous states. In this case there may have been a price to pay, however, in the loss of some of the alleles originally present in the population.

PARALLEL EVOLUTION: III

Since most traits of organisms are of the polygenic or multigenetic type, it is clear that significant alteration of many different genetic loci could result in their modification. Furthermore, alteration of a number of *different* genetic loci could alter a phenotypic expression in exactly the same way. Thus, both the mutations *vermillion* and *cinnabar* result in bright red eyes in *Drosophila melanogaster* instead of the usual dull red eyes. These mutations affect different genetic loci producing different enzymes, both of which are active in the same pathway leading from tryptophan to the brown pigment ommatin and apparently inactivate these enzymes or slow their catalytic rates drastically. Again, at least three nonallelic mutations are known to cause hemophilia in man by inhibiting the normal production of three of the many different components needed for normal blood-clotting. One importance of this to the evolutionist is that there is more than one way to achieve a single end. Should it become useful for *D. melanogaster* to have bright red eyes instead of dull red eyes because of some change in the environment, natural selection will opportunistically seize upon the first convenient mutation (one that does not also produce untoward pleiotropic effects) that comes along. Over the entire range of a common species this could in fact result in several new alleles being favored in different localities for exactly the same reason. In very closely related species as well, different alleles at different loci may serve the same adaptive purpose. For this reason should two groups of organisms become genetically separate (see Chapter 14), the most probable ensuing genetic process will be divergence. The chances that two independent populations will "adopt" the same alteration of their genomes in response to the same environmental stimulus are very small, even though their adaptations to that response may be virtually identical and may even involve modification of an identical biochemical pathway (see Chapters 4 and 5—parallel evolution). This consideration goes part way at least to explaining the apparent large amount of variation found in the characteristics of primary gene products when comparing different populations composing a species or when comparing closely related species.

NEOTENY

The timing of the appearance of relatively fixed traits during individual ontogeny is a function of both the information encoded in the genome and of specific environmental stimuli. It is also partly the result of the nature of epigenesis; for example, the vertebrate eye could not form if the brain had not formed first. But the kinds of processes that determine exactly *when* the eye shall form if a normal brain is present are of some evolutionary interest. This is because the relative rates of formation of organs and organ systems has been evolutionarily modified during the history of many lineages. This kind of evolutionary process is usually detected and described when some organ or system has its development delayed in comparison with the rest of the organism—a process known as paedomorphosis or neoteny. Several examples can be found in man; man is more like primate fetuses and embryos than like most adult primates in the flatness of the face, lack of hair over most of the body, shape of the foot, relatively large size of the brain, relatively ventral position of the vagina, and in a number of other features. Thus, all primates have a man-shaped foot at a certain stage of fetal development, but soon the foot and toes become relatively elongated and that shape is lost. In man that shape is retained during subsequent development, presumably because selection has favored that shape for plantigrade walking. One way to describe this is by saying that in man some aspects of differentiation stop at an early stage in the development of the foot, whereas other aspects of differentiation, say, those involving the teeth or ears, continue to later stages characteristic of primates in general. Another way to interpret neoteny is to see it as a slowing of the rate of development of the foot in respect to other organs that develop at rates more comparable to those of primates in general. In the example of hairlessness, apparently some enzymes involved in the development of hair follicle cells have been so modified in human evolution that their full development does not occur over most of the body surface. Other examples abound in almost every lineage of higher organisms.

It is doubtful that neoteny should be viewed as a special evolutionary process. It is better viewed as a description of the specific route (or means) taken (or exploited) during the evolution of specific lineages. The shape of the primate fetal foot undoubtedly has nothing to do with plantigrade walking, but it is *preadapted* to promote such a function. Natural selection simply seized on the first convenient way to improve plantigrade walking in the lineage leading to man, and it happened to be some change in the rate or timing of foot development. The reason this mode of evolution came to be given a special name has to do with both the frequency with which it has occurred and the relative ease with which it can be detected. After all, both embryos and adults of each kind of living higher organism are available, and it was inevitable that they should be compared in detail.

PAEDOGENESIS

Another related evolutionary mode of more interest in the present context is paedogenesis, the coming to sexual maturity of larval forms and elimination of the adult habitus. The adult genome never achieves expression. This has been arrived at in a number of different ways in different lineages, including modification of the hormones responsible for bringing on metamorphosis (as in the salamander genus *Ambystoma*), or by a modification of the ability of the tissues to respond to the appropriate hormones (as in the salamander genus *Necturus*). These changes do not, however, interfere with maturation of the germ cells and continued development of the primary sexual organs. Perhaps this mode of evolution can be considered to be a special case of another mode, gerontomorphosis, in which the rates of development of some features are speeded up relative to those of the rest of the organism, the particular features here being the sexual apparatus. It is, however, necessary in this case to consider some further changes preventing metamorphosis, such as the two suggested previously. The case involving the modification of the hormones that normally initiate and coordinate the various changes that occur rapidly during metamorphosis can be taken as an excellent example of modification of loci with very widespread pleiotropic effects. Usually, modification of such loci would result in effects so drastic as to be completely without adaptive value. In this case, however, the larval forms are already functioning in the external environment and are adapted to it. Preventing them from achieving imaginal status is not as drastic an event as would be one that prevented further development of a gastrula, provided that they could still reproduce. One should be aware of the interesting possibilities for what amounts to quantum evolution in this area. Tadpoles ecologically and anatomically show more variety than do frogs. In the far-fetched event of eliminating the imaginal stage (paedogenesis is not known in frogs), the genetic and taxonomic distances between genera and species of the members of this class would increase drastically in a (geologically speaking) sudden fashion. Moreover, this would have been achieved by modification of relatively few genetic loci.

PHYSIOLOGICAL TRAITS

So far we have been referring to traits that develop once during ontogeny and are not subsequently modified (although at least some of what has been described could apply to other kinds of traits as well), to relatively "fixed" traits. Also inherited are certain continuing ranges of response—the machinery underlying both physiological homeostasis and individual adaptability (see Chapter 3). What is fixed here are the limits beyond which homeostatic mechanisms cannot restore a certain level of adequate functioning to the system. These tolerances have evolved in conjunction with the environments previously experienced by generations of successful parents. Typically, an organism adapted to function in boreal regions cannot adjust to keep its life processes moving at an adequate rate in, say, very warm temperatures. Data exist, for instance, showing that some of the physicochemical characteristics of proteins from conforming organisms differ according to the temperatures at which the organisms usually live. Procaryotes from hot springs have proteins with temperature stabilities that differ from those of organisms living in the supercooled waters below the ice caps at the poles. Dromedaries undoubtedly do not possess the coordinated responses and reflexes needed to traverse rugged, high alpine terrain or the ability to recover from an attempt to do so in an atmosphere containing so little oxygen. Even if they were anatomically adapted to living in such places (which they are not), they would not be so physiologically and behaviorally.

The molecular basis for an inherited range of response is, of course, not different from that of a fixed anatomical trait; they both involve enzymes encoded in the genome. In the case of fixed traits, enzymes that were active only during a short period of ontogeny or special transient combinations of always-present enzymes participated in endergonic, anabolic pathways to build up a permanent structure one time. In homeostatic responses enzymes continue to operate at different rates, depending on changes in pH, temperature (if any), substrate concentration, coenzyme concentration, ionic strength, and so on. For example, the evolutionary process has favored an enzyme whose change in rate of activity with low-ered pH, say, is such that the organism as a whole functions better at the low pH than it would have if this change in rate of enzyme function had not been able to occur, or whose change in activity is such as to help to restore the pH back to normal. All enzymes known undergo alterations of catalytic rate with changes in the environment (see Figure 7-6); the question is whether the changes will have appropriate magnitudes and occur at adequate rates. A typical homeostatic response will, of course, involve considerably more than one enzyme, and each one has to have properties that are coadapted with those of the others. What comes under the regulation of natural selection, of course, is the adequacy of the final, orchestral response, not the performance of an individual kind of enzyme. This means that one enzyme can compensate for a slight deficiency in an-other, giving us in addition a molecular explanation for the ability of more than a single genotype to achieve a single phenotype and also for the phenomena of suppressor mutations and complementation whereby a previously unfit genotype caused by a mutation at one locus can be restored to normal by a second mutation at another locus (see also the discussion of physiological traits in Chapter 6).

MAGNIFICATION OF VARIABILITY AT THE GENE PRODUCT LEVEL

The ability of enzymes to undergo significant changes in rates of activ-ity is conferred in part by being composed of more than one subunit, as most of them are. This is what allows them physiochemically to assume different tertiary and quaternary structures, a process known as allosteric transition. A model of this situation is as follows: an enzyme is acting at a certain rate. An effector substance (any kind of usually small molecular weight chemical, often a by-product or end product of another or the same pathway, sometimes from outside the system) arrives and contacts the en-zyme and combines with a subunit (sometimes at a specific site) in such a way as to alter its folding and its relationship to the other subunits, with the final effect that the catalytic activity of the multimer itself is changed. The effect is transitory and depends on the concentration of the effector molecules.

Being composed of more than one subunit allows enzymes to be made of the products of different genetic loci (some are and some are not). This has allowed a kind of amplification of the variability present in a single genome. For instance, lactate dehydrogenase in vertebrates is composed of four subunits (it is a tetramer). There are two genetic loci producing subunits that can form this enzyme. In a cell where both loci are actively transcribing, the two kinds of subunits can combine with each other at random, giving some enzymes composed of four identical subunits (1111), some composed of four of the other sort of subunit (2222), and three classes of combinations, 1112, 1122, and 1222. Thus, being composed of tetramers has allowed this enzyme to exist in five different molecular forms (see also Figure 7-9). These forms are known to produce different rates of catalysis. If we now have a diploid organism heterozygous at any one of these loci, random combination would produce 15 different kinds of lactate dehydrogenase isoenzymes, whereas if the organism is heterozygous at both loci, the total number of different kinds of lactate dehydrogenase isoenzymes would be 35—not allowing for positioning effects which would give more than a thousand but for which there is no evidence as yet (see Figure 8-6). Each of these enzymes could show somewhat different catalytic rates. Perhaps one of them would be more efficient in one type of extreme condition and another more efficient under another kind of extreme condition. This may be a factor in homeostatic responses. In addition to this genetically controlled variability, there is with enzymes a

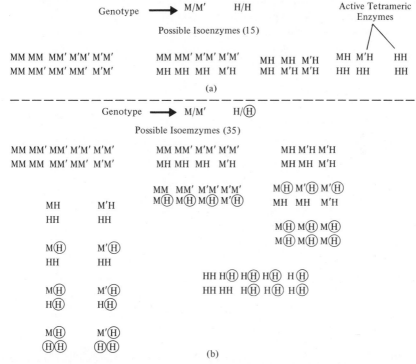

Figure 8-6 Different possible sorts of tetrameric isoenzymes given (a) heterozygous at one of the lactate dehydrogenase loci; (b) heterozygous at both lactate dehydrogenase loci.

possible kind of phenotypic plasticity summed up under the name of conformers. There is some evidence that a single form of genetically determined enzyme multimer may exist in an undetermined number of different, permanently altered forms. The effect would be produced by having the enzyme, say, acetylated or decarboxylated or iodinated, and so on. Many such enzymes may be inactive, but others may have a certain amount of activity in the cell. In this way, the total number of different molecular forms of a given enzyme may be only dimly foreshadowed by the number of loci involved in its coding or by the degree of heterozygosity (Figure 8-7). Because of these effects, there may be a very large number of different molecular forms of an enzyme at a given time in a given organism. The large number itself could be an important factor in explaining homeostatis and, combined with allosteric effects, makes that phenomenon almost predictable on the molecular level.

Another molecular homeostatic mechanism that has been shown in single-cell organisms and may possibly exist in higher forms is known as enzyme induction. Here, a locus is not actively transcribing until stimulated by an effector molecule, which may be a substrate or hormone or other relatively small-sized molecule that combines reversibly in some way with the gene. Loci of this type would be intermittently active, depending on the concentration of effector substances in the system. Perhaps loci of this type would transcribe only under extreme or stressful situations. The op-

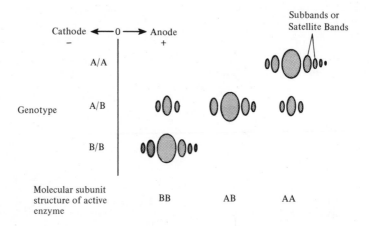

Figure 8-7 Electrophoretic picture (zymogram) of a genetically understood dimeric enzyme, showing the presence of more active enzyme types than would be predicted from the structure of the system. The satellite bands may represent conformers in some cases.

posite effect has also been observed in microorganisms; that is, a locus being intermittently repressed by high concentrations of an end product of a pathway in which the enzyme it helps to code for is involved. Thus, effector molecules can be involved in two conceptually different kinds of homeostatic reactions at the molecular level: (1) allosteric effects involving changed affinities of enzymes for substrates and (2) synthetic effects involving either induction or repression of transcription, and perhaps translation. The concept of effector molecules acting in these ways was proposed by and has in part been developed by the French molecular biologists François Jacob and Jacques Monod.

The temporary physiological adaptability of individuals to temporary extreme conditions can be considered to be simply a special case of homeostasis occurring near the limits of the range regulation in a given case. This range may be wider or narrower, depending upon the physiological response and the species in question. It is doubtful whether this ability in itself can be considered to be a special adaptation except when the abilities are extraordinary. It is probably a function of the complexity of organisms per se and as such was one of the preconditions for evolution on this inconstant planet. The limits or range of the ability, however, has no doubt been tailored to each species by natural selection.

INHERITANCE AND OPEN-ENDED POTENTIALITIES

There is still another kind of trait that can be inherited—the open-ended potentiality. This trait cannot be completely programmed in the genome in the sense of being entirely predictable given total knowledge of the environment during ontogeny. We have already mentioned one such trait—the enzyme conformers—although the internal milieux of the cells of a given organism are probably to a large extent predictable in principle. If that is so, then the numbers of different kinds of conformers would also be predictable. What we are referring to here are traits not completely predictable (programmed) even *in principle*. In practice detailed predictions about many kinds of traits of *individual organisms* are not possible. Examples would be the exact pattern of veins in an individual vertebrate or the exact interconnections between its neurons. However, these are in principle predictable given total knowledge of genome and environment, and therefore are to be considered completely coded for in the genome *and* subject directly to natural selection. A truly open-ended trait would not be subject to natural selection concerning its detailed structure, since only its potentiality is part of the genome. The exact pattern of spots on a frog's back appears to be one such trait inasmuch as isogenic siblings develop widely different patterns (Figure 8-8). Since every human being has a unique fingerprint pattern, this trait too, is probably not coded for as such. An interesting example of such a trait is the human capacity to

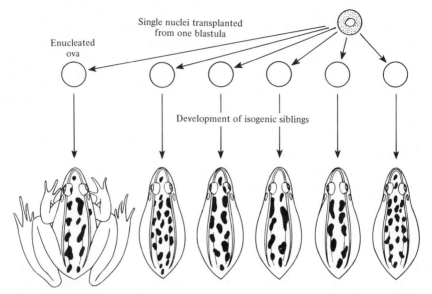

Figure 8-8   Dorsal spotting patterns of isogenic clones of leopard frogs. Each pattern is different, but the spotting density is the same in all members of a clone, and so appears to be coded for genetically. (Data from Casler, 1967.)

symbolize or to construct a symbolic language (see Chapter 15). Natural selection can be said to have favored the evolution of the capacity to symbolize but could not determine the details of linguistic structure. We know this because different languages differ even in their principles of structure. This example will be taken up again in Chapter 15.

Another candidate for an open-ended trait is culture (again see Chapter 15) or learned tradition. Different troups of gorillas living in proximity (sympatry) to each other do not feed on exactly the same plants. Each group has its own tradition in matters of this kind. The range of possibilities here is, however, subject to natural selection in the sense that poisonous items are excluded. Another restriction can be seen to arise if a favored food plant becomes extinct locally. Individuals persisting in a search for these could be put at reproductive disadvantage, but such a temperamental quirk really has no connection with the ability to acquire a learned tradition (the ability to acquire conditioned responses). Flexibility must be a part of the ability to learn. Instances in which organisms can learn only predetermined patterns are not true learning (see Chapter 15), and are more closely analogous to embryonic induction.

At present there is no generally accepted, general theory of learning at the cellular or molecular level. The mechanism of neuron function seems to be limited to the passage, in varying frequency, of a single kind of signal (a single amplitude). The frequency is varied in a number of ways, all involving differential thresholds at the synapses. The latter mechanism is also responsible for the multiplication of the signal into many different channels and gives rise to the possibility of feedback inhibition or feedback facilitation. It is generally felt that if the system simply gets more complex—that is, if more neurons are added—the phenomena of mind and spirit arise spontaneously, perhaps at some critical level of complexity. It is not clear how a very complex system involving a single kind of signal and different channels varying in threshold could give rise to open-ended phenomena, such as true learning and symbolic language. It is therefore not clear just how the genome is represented in an open-ended system; neither is it clear how natural selection could promote such a system nor what kinds of preadaptations were subsequently refined in the new adaptive zone (that of what may be called conditionable responses). There may be an analogous problem in the synthesis of antibodies. The possibilities of unrestricted primary structural modification of proteins are being examined in connection with antibody synthesis and may well shed light indirectly on the biological basis of the mind. It may be, of course, that proteins need not be modified as such in order to invent or learn a new word (see discussion of tooth development in Chapter 6).

SUMMARY

In summary, the heritable variability of populations is stored in the gene pool. In the process of expressing itself, the interaction between the genomes of individuals with specific environments generates further variability. This is because the phenotype is not completely determined by the genotype. Different genetic loci express themselves at different times during the life cycle of individuals and come under different selection pressures. Because of this, characteristics of different parts of the life cycle can evolve independently. There are relatively more restrictions on the evolution of genetic loci that have large-scale pleiotropic effects, and these include many genes that act early in development. Most genes probably have some pleiotropic effects, and most traits of organisms are polygenic or multigenetically determined. For this reason no locus evolves independently of other loci acting during the same life history stage. On the contrary, in producing the fittest genotypes, natural selection produces coadapted genomes. Thus, although one can describe the history of a trait, it is perhaps too loose a use of the word to say that traits "evolve." Lineages of organisms evolve in fact, traits evolve as a convenient descriptive fiction. Some traits are the fixed results of the actions of gene products at some fleeting point in the life history of organisms; others are the actions themselves; still others are similar actions that take place over long periods of the organism's life (metabolic events, for instance). Any property of an organism not involved in one or another such trait (that is, that cannot be influenced by natural selection) is not a part of the phenotype and may vary with time in a random fashion.

REFERENCES

Bonner, J. T., *The Molecular Biology of Development.* Oxford University Press, London, 1965, 155 pp.

deBeer, G., *Embryos and Ancestors.* Oxford University Press, London, 1958. 197 pp.

Haddorn, E., *Developmental Genetics and Lethal Factors.* Wiley, New York, 1961, 350 pp.

Ingram, V. M., *The Hemoglobins in Genetics and Evolution.* Columbia, New York, 1963, 165 pp.

Jacob, F., and J. Monod, "Genetic Regulatory Mechanisms in the Synthesis of Proteins," *Journal of Molecular Biology,* **3:**318–356, 1961.

Markert, C. L., "Mechanisms of Cellular Differentiation," in J. A. Moore, ed., *Ideas in Modern Biology,* XVI International Congress of Zoology, 1965, pp. 229–258.

Milkman, R. D., "The Genetic Basis of Natural Variation," I. *Genetics,* **45:**35–48, 1960; II. *Genetics,* **45:**377–391, 1960; III. *Genetics,* **46:**25–38, 1961.

Monod, J., J. P. Changeux, and F. Jacob, "Allosteric Proteins and Cellular Control Systems," *Journal of Molecular Biology,* **6:**306–329, 1963.

Prosser, G. L., "Comparative Physiology in Relation to Evolutionary Theory," in S. Tax, ed., *Evolution after Darwin: The Evolution of Life.* The University of Chicago Press, Chicago, 1960, pp. 569–594.

Rendel, J. M., *Canalization and Gene Control.* Academic, New York, 1967, 166 pp.

Stebbins, G. L., *Variation and Evolution in Plants.* Columbia, New York, 1950, 643 pp.

Strickberger, M. W., *Genetics.* Macmillan, New York, 1968, 886 pp.

Vessel, E. S., ed., "Multiple Molecular Forms of Enzymes (a Symposium)," *Annals of the New York Academy of Sciences,* **151:**1–689, 1968.

Waddington, C. H., "The Genetic Assimilation of an Acquired Character," *Evolution,* **7:**118–126, 1952.

Waddington, C. H., *The Strategy of the Genes.* George Allen and Unwin, London, 1957, 262 pp.

CHAPTER 9

Natural selection in the absence
of environmental change

Living systems on earth evolve in continually changing environments. Many of the features of these systems reflect this
fact, including the basic strategy for maintaining adaptedness—the sacrifice of individuals for the continued adaptedness of the group. Any system that did not develop this
strategy became extinct. Because of its fundamental importance, this strategy is perhaps best viewed as a necessary
precondition for continued survival on earth rather than as
an evolved adaptation.

In the absence of continued environmental change there
would be no need for evolutionary change, although that
might still occur for a time in terms of continued niche subdivision (*vide* the wood warblers in Chapter 3) produced
by character displacement (Chapter 6) until the ecosystem
reached a level of complexity with every organism relatively
highly specialized (Chapter 3). Selection under such circumstances would be primarily concerned with increased efficiency at all levels unopposed by counterselection arising
from the necessity for adaptation to changing, and therefore
rigorous, climates. Even after reaching some level of complexity beyond which selection could produce no further
evolutionary change (were that possible), it must be realized that natural selection would still continue to operate.
Environmental change is necessary for long-continued
evolutionary change but not for the process of natural
selection.

Given that there are differences among individuals in a population (Darwin's major point), some will perforce be better adapted to a given environment than others. These will leave more offspring than those not so well adapted. Genotypic differences will, however, be maintained (or continually regenerated) by recombination of the components of the gene pool and by mutations (Chapter 7). The variability of phenotypic traits in a population tends to be distributed either normally (for continuously varying traits) or binomially (for discontinuously varying traits), with the great majority of individuals having measurements close to the means (Figure 6-5). This chapter will be concerned with the maintenance of genetic variability by natural selection itself.

NORMALIZING OR STABILIZING SELECTION

The bell-shaped distributions of measurements of phenotypic traits are in fact produced by natural selection working specifically in the absence of environmental change, or what is the same thing, over short periods of time (few generations). Thus individuals with extreme measurements of any trait are at a disadvantage relative to individuals with the more usual or more common measurements, because individuals with average measurements have been in the recent past the most successful in producing offspring. This we know because most of the current population has measurements close to the mean (in a normal distribution some 62 percent of the population is within one standard deviation from the mean), thereby demonstrating which traits were successful in the last generation.

An interesting example of selection working in this way was described in 1899 by the American biologist H. C. Bumpus. A severe snow and sleet storm accompanied by high winds resulted in unusually large mortality among the house sparrows at Woods Hole. Bumpus collected as many injured birds as he could find. Of 136 birds, 64 subsequently died, giving him two groups—those killed by the storm and those that survived it. He then measured a randomly chosen series of traits, such as tarsus length, wing length, and so on, in the two groups, and, upon comparison, found that those that were killed by the storm tended to be those with measurements falling nearer the ends of the bell-shaped curve. Thus, individuals too different from the average (which has been favored by selection in the past) tend to be eliminated during catastrophic events or in stressful situations. This is in fact how the current mean measurements have been selected for. (This example, and the following one, are interesting also because, in the absence of doing selection experiments (Chapter 10), such observations would be the only way to demonstrate natural selection at work).

Another example of natural selection acting in this way was described by the British biologist W. F. R. Weldon in 1901. Exact measurements were made of the shapes and angles formed by the inner whorls of the shells of individuals in a population of land snails. The inner whorls are formed during growth in the youngest ages and, once formed, they do not change, thus providing a permanent record of youthful growth rates and patterns. Weldon then compared the innermost coils of all individuals, thereby comparing, in individuals of all ages, the pattern of growth at a certain young age. He discovered that surviving cohorts showed decreasing variability in the inner whorl measurements from year to year, while the mean measurement did not change. Thus, individuals that survive for longer periods of time are those with measurements nearest the mean for the population, whereas the most abnormal individuals tend to be eliminated from the population first. Similar processes have been demonstrated for other systems, for example, the pattern of scales on the heads of lizards.

The Bumpus experiment is relatively easily interpreted. Selection had in the recent past favored birds that could survive environmental catastrophes of all sorts (only those survived to breed). These were birds that, given the ecological niche of the house sparrow, had body proportions within certain limits; thus, the birds most easily blown down were those that had wings too long for their weight and so presented too large a surface area to the wind, or those that had wings too short for their size and so could not make adequate headway in a gust of wind. The range of individual measurements providing adequate adaptation in the absence of environmental catastrophe is larger than the range that can survive the more stringent requirements of extreme stress. Every catastrophe will take its toll of the phenotypic variability of a population, thereby weeding out suboptimal genotypes (defined as suboptimal precisely because they did not produce phenotypes adequate to survive the normal flux of environmental conditions).

The Weldon observations can be thought of as extending in time the one-shot observations of Bumpus. As time goes on, one catastrophe succeeds another, gradually chipping away at the variability of a cohort of individuals of the same age. (The variability of the entire population will not necessarily change, however, since variability is continually regenerated during sexual reproduction.) Each catastrophe may differ somewhat from the others, each probing a different set of phenotypic traits; or, with passing time, the probability of a more severe catastrophe qualitatively like that of a previously survived one increases, and the number of survivors in our cohort decreases until, finally, the last one succumbs. In this view, no individual is perfectly able to survive any and all catastrophic events even though there are always some who can survive a given one (unless the population becomes extinct). Even an organism with optimum measurements for most of its traits will finally be carried off by a chance event (such as a landslide, or a chance encounter with a predator after exhaustion). Such an organism, however, tends to survive longer than others, and, what is crucial, tends therefore to reproduce more times. In those organisms that show senescence there is a further built-in decrease in the probability of surviving after some threshold age is reached, apparently involving a breakdown in physiological homeostasis (Figure 6-4).

One might wonder in the Weldon example what sort of selective forces would be working on the products of growth rates that were concluded before the selective event. Put another way, why would the shapes of the innermost coils of a large snail affect its ability to survive long after many new coils had been added? There are at least two kinds of answers. First, the early growth rates produced a structure that has remained as part of the phenotype of the individual; it was not obliterated. Even an evanescent enzyme activity can have lasting, if indirect, effects on the phenotype. If an event or structure A is abnormal, other events or structures that grow out of it or are built upon it; even if they are closer to normal in themselves, they will tend to appear further from the mean because of A's influence. A second point is that the gene products that produced A may still be functioning (although in other contexts, that is, pleiotropically) and, having produced a slight abnormality once, might do so again. Thus, suppose that a snail with relatively tight initial coils in its shell has this trait because some biochemical pathway is working faster than normal. Suppose this pathway continues to do so and that, while some compensating processes prevent that phenotype from appearing in later coils, another result of this fast rate is a thinner shell. This might not affect the very smallest snails, but would begin increasingly to be a handicap as a snail grew larger. In this sense, an early abnormality, even if it has no effect then or at later developmental stages, is a sign of a genetic abnormality that may have other, perhaps multiple, pleiotropic effects in these later stages. Put another way, a phenotypic trait may be selectively neutral at one stage of life, but the gene products that produced it, if they are still operating, will not be.

Natural selection acting in these ways, in the absence of large-scale environmental change, to produce bell-shaped curves of variability (Figure 9-1) is referred to as stabilizing selection or normalizing selection (see Glossary for distinctions here). For some examples discussing the mechanism whereby selection produces an optimum (mean) measurement in a population, see the discussion of clutch size in Chapters 6 and 12, of sexual selection in Chapter 12, and of antler size in Chapter 5. Here we may point out that the principle involved in the explanation is that of counter-selection working to produce a system whose attributes are coadapted with each other as well as being adapted to the total environment. Thus, suppose there is a spectrum of seed sizes produced in an environment, with relatively fewer large seeds, common medium-sized seeds, and very common small-sized seeds. In a seed-eating organism, selection for food parti-

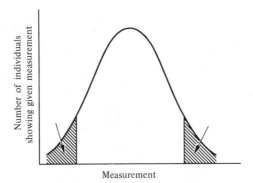

Figure 9-1 Normalizing or stabilizing selection. The arrows point to the measurements of those individuals relatively most heavily selected against (shaded portion of curve).

cle size will be complex. On the one hand, the use of the largest seeds entails that less energy need be expended per seed eaten or per calorie of energy gained, and so selection would favor organisms specialized in the direction of taking larger seeds. But in the environment postulated, large seeds are scarce, and so an organism specializing in them would have to spend more energy hunting for them, thereby decreasing the energy gain. This condition will establish a counterselection against specializing in the very largest seeds, and the species will finally home in on some optimum seed size given the environment in question. This will involve achieving an optimum measurement in some phenotypic trait such as, for example, beak size in birds. Another kind of example would be where a predator could, by increasing its size, prey upon more different kinds of prey, thereby broadening its niche and allowing it to survive years of scarcity better. If this predator was a burrow-living form, however, selection would begin to oppose increased size beyond some maximum allowable in terms of its entire mode of life, thereby favoring some intermediate, optimum body size. Individuals having these intermediate body sizes will leave more offspring than those that are too large or too small.

Another aspect of stabilizing selection should be mentioned here. In stipulating that this sort of selection would be characteristic of an unchanging environment, there is no implication that real environments ever actually endure without change. However, portions or aspects of an environment relevant to certain traits may undergo little or no change over fairly long periods of time so that such traits may experience only or largely stabilizing selection during this time. On the other hand, certain features of organisms themselves may be so fundamental to their functioning as to experience primarily stabilizing selection in any environment. For example, it has been found that the basic protein histone IV is practically identical in beans and in cattle, there being only two (conservative) sequence differences out of 102. The last common ancestor of these forms cannot have existed less than 600 million years ago, and so we may say that this locus has essentially not changed since that time in many different sorts of organisms. What this in effect means is that this locus has undergone intense stabilizing selection during all of this time, with any variant arising by mutation being eliminated as unfit. Presumably the functions of histones are crucial to basic cell homeostasis and involve all of the sequence equally, and presumably these functions have not been altered since the time of the common ancestors of plants and animals.

HETEROSIS, BALANCING SELECTION, AND SEGREGATIONAL LOAD

Natural selection will, then, maintain optimal forms of traits. These optima will correspond to the means of the binomial and normal curves of variability for these traits. Only when the environment changes significantly will these means shift to new values (Chapter 10). Selection acting thus at the ends of the measurement distribution curves acts against genotypes producing the most extreme measurements; it will in fact weed out any alleles at various genetic loci that tend overwhelmingly to produce such extreme measurements. Alleles will remain in the gene pool only insofar as they can combine to produce the desired optimal measurements (see Figure 9-2). Notice that in the simple model shown in this figure that optimal or near optimal genotypes are mostly heterozygotes (six out of seven) and that the two genotypes least favored by selection are both homozygotes. The superiority of heterozygotes over homozygotes, for which there is ample evidence in nature (see below), is not fully explained in Figure 9-2. In this figure the b/b homozygote is as good as any of the heterozygotes and, indeed, as the model is set forth, allele b would replace a and c completely with time, because both a and c enter into one genotype each which is not favored, whereas allele b is involved in no such genotypes. Indeed, it has been shown that in a uniform environment in the laboratory, selection for some intermediate measurement will lead to homozygosity. In nature, however, it frequently (or always) appears that, other factors being equal, heterozygotes are superior to homozygotes, a phenomenon referred to as heterosis.

Originally this concept grew out of observations whereby different species, when crossed, would frequently produce individual offspring of great phenotypic and/or ecological vigor; this phenomenon was early known as hybrid vigor. Good examples are found among various plants where the hybrid offspring of different species are frequently larger and more productive of seeds than either parental species. In animals similar observations can be made, but frequently the hybrids are sterile, therefore having adaptive values of zero. In order for the hybrid to be fertile, the

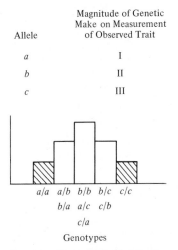

Allele	Magnitude of Genetic Make on Measurement of Observed Trait
a	I
b	II
c	III

a/a	a/b	b/b	b/c	c/c
	b/a	a/c	c/b	
		c/a		

Genotypes

Figure 9-2   Histogram showing the actual distribution of measurements in a natural population of observed traits and responsible genotypes at a given locus, indicating which genotypes have been least favored by selection (shaded portion).

two parental forms must have, at a minimum, the same chromosome numbers, or else meiosis is not possible and so gametes cannot be formed. The major problem facing hybrids is the fact that, although the parental forms are each specialized and well adapted for some ecological niche, the hybrid, being phenotypically intermediate in most cases, is well adapted for neither niche, and, unless some third niche is available into which it can move, will not survive competition with the parental forms. In plants it does happen that such new niche possibilities are frequently available and natural hybrids have been found thriving in nature. Data concerning animals are only now beginning to become available, and it is not clear whether this phenomenon is as common among them as among plants.

About 50 percent of genetic loci in sibling species of *Drosophila* flies are represented by alleles producing electrophoretically different gene products. We may then estimate that any two species closely related enough to be able to form hybrids carry different alleles at about 50 percent of their loci. The hybrids, then, will automatically be heterozygous at a minimum of half of their loci and, by normal recombination, probably at a great many more. The demonstrable vigor of hybrids thus correlates with a high degree of heterozygosity.

There are many classical examples of heterosis involving single traits. Thus, there is a mutation in *Drosophila melanogaster* that segregates as a recessive and causes a very dark body color in the double recessive associated with markedly decreased viability. It has been found in several laboratories that this recessive trait (known as ebony) is maintained in laboratory stocks at rather high frequencies (higher than predicted on the basis of the high selection coefficients associated with the double recessive genotype), the frequencies varying with different environmental conditions. Since the heterozygotes cannot be distinguished from the dominant homozygotes in this case, there has been only a suspicion that heterosis is involved. For an example where the heterozygote can be detected, we may examine a situation in a marine copepod, *Tisbe reticulata*. This animal has a number of different color phases in each population, three of which have been experimented with as follows: V^v/V^v, violacea; V^m/V^m, maculata, V^v/V^m, intermediate. The heterozygote was found to be markedly more common in all examined environmental conditions, but was increasingly so as crowding increased. That is, there were many more heterozygotes present than predicted by the Hardy-Weinberg formulation. The proportions of the genotypes found were not compatible with Hardy-Weinberg proportions, and the heterozygote values deviated most from such predictions. For a discussion of still another example, see the section on sickle-cell anemia in man in Chapter 11.

In terms of the Hardy-Weinberg formulation, the genotype frequencies in such situations can be derived from the following:

$$AA \qquad\qquad Aa \qquad\qquad aa$$
$$p^2(1 - s_1) \qquad\qquad 2pq(1) \qquad\qquad q^2(1 - s_2)$$

Examination of this shows that even if s_1 or s_2 or both equal unity, neither allele can ever be eliminated from the gene pool. Instead, they will reach some equilibrium proportion depending upon the respective values of their selection coefficients, and at equilibrium can be shown to be

$$p(A) = \frac{s_1}{s_1 + s_2} \qquad \text{and} \qquad q(a) = \frac{s_2}{s_1 + s_2}$$

(Sometimes the Darwinian fitnesses in a situation of heterosis are calculated on the basis of the most favorable homozygote being set at unity, and that of the even more favored heterozygote being therefore greater than unity. This was the result of historical accident and goes back to a time when all favored alleles were considered dominant and all less favorable alleles recessive. Faced with the phenomenon of heterosis, the best homozygote was still considered to have an adaptive value of 1.00, while the heterozygote was considered to be favored over and above this, and to be therefore "overdominant.")

Two different major theories have been put forward to explain heterozygote superiority. P. M. Sheppard, of the University of Liverpool, has suggested an explanation involving pleiotropy. The products of most genes are involved in more than one physiological or ontogenetic process. It is most likely that such a product will be the best possible allotype in only some of these processes, and that selection favors allotypes that have the best all-around effects. This will often involve compromises, with the more important processes being scanned by selection most intensively. This means that some of the processes in which a gene product is involved will exert more influence concerning which allele will be favored at a given locus than will others. However, it is most unlikely that major selective intensity in process A will produce the allotype best fit for the less crucial process B. Under these circumstances the genome will evolve in such a way as to convert the favorable effects of our allotype in process A into dominant effects and its less favorable effects in process B into recessive effects (see Chapter 7). The result of this is that the gene product of a given allele will be dominant in one process and recessive in another. In a heterozygote the favorable, dominant effects of this allotype will be expressed, but its less favorable, recessive effects will be suppressed. In a homozygote, even for an allele with favorable activity in several processes, some unfavorable recessive traits will appear.

Can this idea be reconciled with the heterotic properties of many interspecies hybrids where it is not possible that the genotypes of the parental species would be coadapted with each other in the necessary way to produce dominance? One could argue that the overall fitness of such hybrids is usually in fact less than that of their parents, citing the fact that heterosis applies (almost by definition) to viability and fertility features and not to traits that can be measured as such. Thus, it is irrelevant that the mule is stronger than a donkey or a horse; what is relevant is that it is sterile.

Another idea that has been used to explain heterosis is that heterozygotes are biochemically more versatile than homozygotes, an idea first suggested by Haldane. We have already noted that selection for some intermediate measurement (say, an enzyme rate) would in a uniform environment eventually produce homozygosity for alleles coding for allotypes producing just that intermediate condition. Normal environments, however, are nonuniform, both in space and over even short periods of time. During the course of its life one individual usually meets, and must successfully cope with, more than one external environment, and, indeed, in most eucaryotes, passes through an ontogenetic series of different internal environments as well. Thus, we can compare the environment of a hibernating frog in midwinter with that of the active frog in summer. Or we can compare the internal milieu of a prospective neural crest cell in an amphibian gastrula with that of a liver cell in an adult frog. If there is some enzyme activity that must be present under all these conditions, it almost certainly is unlikely that a given single allotype would be the best possible allotype under all of these conditions. Given two allotypes that are slightly different kinetically, a heterozygote producing two of them is probably better off than a homozygote. One might function a little better in winter, the other somewhat better during gastrulation, and so on. An individual might just survive a sequence of environmental catastrophes because, say, it had at one time an enzyme with a slightly higher rate of activity in the cold and, later, another with slightly higher activity under acid conditions in the brain. Having survived two such catastrophes, an individual might just succeed in breeding. For further considerations of this idea see Chapter 8.

It would appear, then, that heterozygotes are more fit than homozygotes. We might suppose, further, that individuals heterozygous for more loci would be more fit than those heterozygous for fewer loci. If Darwinian fitness were determined mainly by heterosis, it would be possible to conceive of superfit individuals that are heterozygous at most loci. It has been shown mathematically, however, that such individuals would be extremely rare in a population, and that most individuals would be heterozygous at only about half their loci (Figure 9-3). In nature the meaning of such heterotic superfitness is in any case dubious. Such fitness would accrue if adaptive value were determined purely by the genotype, which is clearly not true except in the special case of genetic disease. Fitness in a real population (all genetic diseases already weeded out) is a relative matter, one individual being relatively superior to another. Since so few superfit heterotic individuals are generated, the carrying capacity of a given environment will be made up largely by individuals somewhat less fit than these, and if these still do not saturate the environment to the carrying capacity, then the next most fit class of individuals will survive, and so

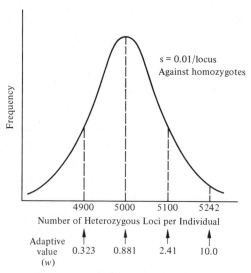

Figure 9-3 The frequency and adaptive value of individuals with different numbers of heterozygous loci (total loci—10,000). (Modified from Sved, Reed, and Bodmer, 1967.)

on. It has been found, in fact, that the average number of polymorphic loci in individuals as determined by electrophoresis in man is 16 percent, in *Drosophila pseudoobscura,* 12 percent, and in the house mouse 8.5 percent. These are lower limits since gene products may be different without being electrophoretically different. Furthermore, in a growing population (the carrying capacity increasing or the population range expanding) more genotypes will survive this culling, finding themselves with adaptive values above the necessary threshold value. Thus, even though a superheterotic individual can be calculated to have a relatively very large adaptive value, in actuality he has no more adaptive value than an individual who is just barely heterotic enough (has just enough heterotic loci) to pass in the given environment. Perhaps, however, individuals tending to survive environmental catastrophes (periodic drastic lowerings of the carrying capacity) would be those heterotic for more loci.

To sum up, although it is advantageous to be heterotic at as many loci as possible, it is usually sufficient to be so at an intermediate number of loci (10 to 50 percent?), because (1) the carrying capacity allows for a certain number of individuals regardless of their genotypes and (2) so few individuals are generated that are heterotic at more than half their loci that their effect on the gene pool is minimal; that is, they will be so diluted in the population that few "normal" individuals will have occasion to compete with them. Also, upon breeding, their offspring will segregate into a spectrum of differently heterozygous individuals with very few as heterozygous as they were; heterozygosity is not heritable, as such. The super heterozygote is the accidental result of there being a large degree of polymorphism in the gene pool. A large degree of genetic polymorphism will be maintained in a population because an average degree of heterozygosity can be and is favored by selection.

Selection favoring heterozygotes has the paradoxical effect of maintaining at fairly high frequencies in the population, alleles that form phenotypes with very low adaptive values when homozygotic. In one sense this statement is formalistic only; if there are three adequate genotypes AA, Aa, and aa, with Aa somewhat more favored by selection, this does not mean that genotype aa or allele a are in any sense bad. Indeed, a certain proportion of a alleles must be maintained in the population in order to achieve the favored Aa genotype. However, it can and does happen that alleles are maintained in this way at high frequencies that produce outright genetic diseases in homozygotic combination. Thus, in man, sickle-cell anemia ($w = \pm 0.07$ for the homozygote) is a condition caused by homozygosity of a certain allele for the hemoglobin beta chain. As discussed in Chapter 11, this allele can go as high as 30 to 40 percent in some populations in Africa because the heterozygote combination of this allele is favored above the usual wild-type beta chain homozygote. In this case heterosis is maintaining a "bad" allele in the population at a frequency it could never achieve on the basis of its physiological effects (which are detrimental in some cases to some degree even in the heterozygote).

Situations of this kind, where lethal, semilethal, and subvital alleles are kept as part of a gene pool at fairly high frequencies because of their ability to confer selective advantage in heterozygote combination give rise to a kind of genetic load different from that already discussed in Chapter 7. This concept is usually referred to as segregational load. To illustrate with an extreme example, supposing both homozygotes were lethal ($s_1 = s_2 = 1$), then only half of the offspring in a population would survive just based on this locus alone. If there were two such loci, then 0.5^2 offspring, or 25 percent of the offspring in any generation would be eliminated on the basis of their not being heterozygous at both loci. If there were five such loci, 0.5^5 or $1\frac{1}{2}$ percent of zygotes would survive, the rest being weeded out because they were not heterozygous at all five loci. That is, these are part of the genetic load of the population—that part generated by heterosis. Reflecting upon this situation, it soon becomes apparent that not many such loci could be maintained in the population if it were not to become extinct given the limited fecundity of most organisms. Supplanting much more reasonable selection coefficients (0.01 or so) does not change the picture very much, there then being a maximum of only some 100 to 200 loci that could be maintained polymorphic in a given population on the basis of heterosis. This would represent only some 10 to 20 percent of all loci. Recent electrophoretic studies of primary gene products in *Drosophila,* in oats, and in the house mouse have revealed, on the contrary, that populations are polymorphic at a minimum of 30 percent of their loci.

There are at least three ways to resolve the contradiction between theory and observation here. First, the genetic model here (and usually) makes the simplifying assumption that genes are inherited independently. If significant degrees of linkage were involved, the genetic load generated by heterosis would be alleviated. Second, the genetic model also makes the simplifying assumption of independent action of the gene products, leading to a multiplicative model of adaptive values. Thus, assuming independent action, it can be shown that the effects on total fitness of all loci will be cumulative in a multiplicative fashion (w locus $1 \times w$ locus $2 \times w$ locus 3 . . .), leading to calculations such as those used above. This assumption is most probably wrong in view of the fact that most gene products act as members of biochemical pathways, the actions of which, in addition, have repercussive effects on all other biochemical pathways. Probably the adaptive value of a total genotype will not be simply the product of the adaptive values of each of its genetic loci; that is, adaptive value is a phenomenon arising out of higher levels of organization than that of single gene products. This is clearly demonstrated in sickle-cell anemia, where, even though the heterozygote is physiologically less good in an absolute sense than the wild-type homozygote (Hb A), under conditions of chronic blood parasitism it is favored by selection. Third, a significant proportion of the polymorphisms observed in nature may have a basis other than heterosis. For example, some of them may be selectively neutral, with various alleles drifting in and out of the population via mutation and gene flow (see Chapter 13). Also, some of the polymorphisms may reflect the process of one allele replacing another, but this would not actually help resolve the problem because such polymorphisms would be part of a substitutional load analogous to the segregational load (see Chapter 10).

A possible mechanism for maintaining polymorphisms without heterosis could arise if the environment of a population were such that it allowed several different subniches within its confines. Then, in a single population, some alleles may be superior in one microhabitat and others in others. If mating in such a population is nevertheless random, heterozygotes would be generated. If, further, there is habitat selection in this model, each genotype could seek out the subniche to which it is best suited and spend most of its time there, heterozygotes possibly choosing intermediate or other microhabitats. This, however, would actually be a special case of selective neutrality. Further possibilities are taken up below.

A certain number of individuals in a population, then, suffer genetic death as a result of the generation of homozygotic genotypes during the reproduction of previously successful heterozygotes. These form part of the individuals excluded because the carrying capacity is limited and cannot support as many individuals as can be produced. The proportion of all genetic deaths caused by this factor makes up the segregational load on the population, but its magnitude is unclear.

Selection favoring heterozygotes is one form of what is termed balancing selection. Balancing selection is selection that leads to a balanced polymorphism in a population (as opposed to directional selection leading to transient polymorphism—Chapter 11). A balanced polymorphism is one that is maintained for a significant number of generations unchanged. At the genetic level a balanced polymorphism exists when two or more alleles are maintained at more or less the same frequencies for many generations without having one or another increase significantly in frequency, thereby supplanting the other(s). Heterosis is clearly one way by which that can happen. Other kinds of mechanisms for maintaining balanced polymorphisms should be examined here as well, together with the means by which selection maintains them.

HETEROSTYLY

The maintenance of various forms of mating systems affords a number of interesting examples of balancing selection. In many plants we find the phenomenon of heterostyly. This system has been extensively investigated in the primroses (*Primula*), a convenient example for description. There are two kinds of flowers: one, called pin, has a long style, raising the stigma far above the anthers which are perched low on short stamens (Figure 9-4). The other kind (thrum) has a very short style, and the stamens arise from the corolla tube high up, placing the anthers far above the stigma. In a population we find roughly 50 percent of each. Since these flowers are bee-pollinated, pin flowers can only fertilize thrum flowers and vice-versa, as shown in the figure. The resultant progeny of a normal pin-thrum fertilization show a 50:50 ratio of pin to thrum. Since experimental fertilization of a pin plant by another pin gives a few progeny and they are all pin, and since such an experimental cross between thrums results in a few progeny in a 3:1 ration of thrum to pin, the normal 50:50 ratio from a pin-thrum mating is due to thrum being dominant. Thus, the polymorphism is maintained by the combined nature of the genetic underpinnings and of the pollenating agent. The system does not break down because, even should pin pollen land by accident on a pin stigma (or thrum on thrum), there are further genetic determinants such that the growth of the pollen tube will be inhibited. That is why so few offspring are obtained in experimental crosses of these types. In addition, the shapes of the pin and thrum pollen grains are such that they tend not to stick to their respective stigmas. Such a system, in which a number of traits obviously controlled by more than a single gene are yet inherited as a block, has been termed a supergene, and we will need to explore the possible origins of such genetic structures below. In primroses it has been possible to obtain experimentally some recombination within the supergenes we have been discussing, and so to prove that the different traits are in fact coded for by different loci.

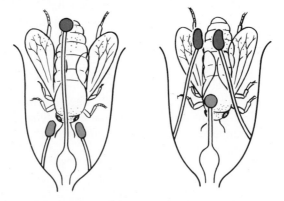

Figure 9-4   Heterostyly in primroses, showing how pin pollen can only pollinate thrum flowers, and *vice versa*.

FREQUENCY-DEPENDENT SELECTION

Another example of a mating system based on a balanced polymorphism is animal sexuality and its concomitant sexual dimorphism. Individuals intermediate between the two sexes will be selectively disfavored. This works to maintain the phenotypic distinctness of the sexes. In addition, one sex does not replace the other, because as soon as there comes to be a scarcity, say, of females, the chances of the remaining females reproducing are increased; that is, their adaptive value increases. This is an example of frequency-dependent selection, another means by which balanced polymorphisms can be maintained.

Frequency-dependent selection has been demonstrated at the level of primary gene products in *Drosophila*. Thus, looking at experimental populations with two different alleles for an esterase or for alcohol dehydrogenase, it was found that the egg to adult (larval) viability of homozygous genotypes was indirectly proportional to the gene frequency of an allele (Figure 9-5). One explanation for this effect could be that the environment occupied by the population is heterogeneous enough to provide subniches within a given ecological niche. Such differences might be different categories of food substances or distinctly different microclimates (temperature, humidity), such that one phenotype is somewhat better able to cope with one than another. Each of these subniches is capable of being generated by only a limited number of individuals because any environment is finite (the population size must be at carrying capacity in this model). Thus, a given environment might be capable of supporting x number of homozygotes AA. If an experiment is begun with a gene frequency of A such that more than x AA homozygotes are generated by random breeding, selection will disfavor A until such time as its gene frequency approaches closely to the value that will generate just x of the AA genotype. It will not drop below that number because then, in the

Percentage F at Start of Experiment	Adaptive Value (w) of Genotypes		
	F/F	F/S	S/S
70	0.44	0.68	1.00
50	0.77	0.94	1.00
30	1.00	0.89	0.83
15	0.97	1.00	0.40

Figure 9-5 Frequency-dependent selection on an esterase locus in *Drosophilia melanogaster*. The adaptive value of the genotypes varies according to the gene frequency at the beginning of the experiment. (Modified from Kojima and Yarborough, 1967.)

same way, there will come to be more opportunities for AA homozygotes and so their Darwinian fitness will increase; there will also be relatively too many *aa* homozygotes for the given environment, and *they* will be selected against until the appropriate gene frequencies are again attained. Different environments would be characterized by different gene frequencies of the various alleles in the system. Such a system could stabilize regardless of whether the heterozygote was favored or not if mating was random between the two groups.

As the gene frequencies approach those that are appropriate to the environment in question, the selection coefficients diminish in value, and at equilibrium the alleles are selectively equivalent; that is, they are selectively neutral or very close to it. Once selection has achieved an appropriate balance of alleles, it effectively stops operating at the locus in question. See Chapter 13 for further considerations of selective neutrality.

Disruptive selection

Before continuing with another example of frequency-dependent selection, it will be convenient here to make a digression concerning disruptive selection, another process that may result in balanced polymorphisms based on frequency-dependent adjustments to heterogeneous environments. Disruptive selection would be a process whereby a trait with a unimodal distribution of values of some measurement in young individuals gave rise later to a bi- or trimodal distribution in the adults, intermediate organisms having been in the meantime selected away (a kind of converse of normalizing selection). One such example may have been discovered concerning molar tooth sizes in house mouse populations in England. The adult measurements were more variable than those of the young at the time of first acquiring their adult set of teeth. Such a phenomenon could occur if the food supply were heterogeneous for two extremely different sorts of particles (large and tough versus soft and small, for example), with individuals selected on the basis of their efficiency of assimilating food. Presumably the proportions of the different phenotypes present after disruptive selection has occurred would bear a direct relationship to the relative frequency of each of the subniches involved as in frequency-dependent selection. Random mating would return the variability of the trait to a unimodal distribution in the young each generation. Thus, in this case, the genetic load is borne by individuals with genotypes that generated intermediate phenotypes. (Disruptive selection is a term that has been used for many different processes by different authors—see Glossary.)

Batesian mimicry

Another sort of frequency-dependent selection is exemplified by Batesian mimicry (named after H. W. Bates, a nineteenth century British naturalist). This is defined as the sort of mimicry that occurs when a harmless, palatable organism mimics a harmful or noxious one (compare with Mullerian mimicry, Chapter 12). There have been a number of cases studied, all involving visual mimicry (Figures 9-6 and 9-8) and predators that use primarily vision in their search for prey. It is known that mimicry exists also in connection with other senses—smell and hearing, for example—but we know little of these cases being, as we are, visually oriented ourselves. Not surprisingly, the majority of well-known cases involve birds as the predators. It has been possible to demonstrate in laboratory experiments that both birds and frogs can and do learn to avoid noxious prey organisms (with stingers, bad taste or other chemical defenses) after three or four attempts to eat them if these organisms have some easily seen, bold or bright color pattern. These patterns (aposematic coloration; see Chapter 12) act as negative reinforcement stimuli in establishing conditioned avoidance responses in the predators. The avoidance behavior will be maintained, once learned, by occasional interactions with the unpleasant prey as the individual predators begin to "forget." Once programmed to avoid the specific noxious organisms, both birds and frogs will now avoid other, quite palatable organisms that have been disguised to look like the unpleasant ones. These processes are the bases for the evolution of Batesian mimicry, wherein a palatable prey organism gradually comes to resemble, often quite astonishingly closely, some sympatric noxious organism.

There is one stricture imposed upon the mimetic population right from the start, and it is this restriction that is of interest to us here. The

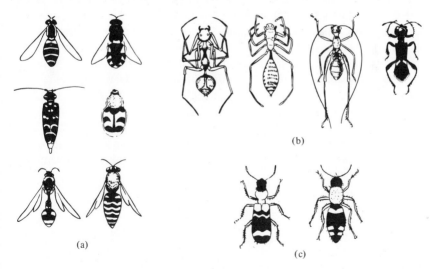

Figure 9-6   Various noxious insects and their mimics. (a) wasp mimics: flies on top, then beetles, then wasps. (b) ant mimics: (*l. to r.*) ant, spider, larval grasshopper, and beetle. (c) beetle on left mimicks wingless wasp. (Portman, 1959. By permission of University of Michigan Press.)

models must be many times more common than the mimics or the mimetic system breaks down. The exact number of models necessary is that number required to keep reinforcing the avoidance behavior in the predators. If the mimics become too common, the conditioned stimulus (aposematic coloration) will no longer be very effective, since three or four palatable forms may be encountered between every noxious one. The adult predators will "forget" or the young will not learn. This will result in increased predation upon the mimics (their bold colors, in fact, now attracting attention while the noxious traits of the models will still protect *them*) until their numbers are reduced below the point at which the proportion of models is sufficient to reestablish the conditioned responses in the predators. (Apparently the predator-prey interaction does not usually result in the destruction of noxious prey individuals, the predator quickly relinquishing its hold and the prey able to continue its life only somewhat battered.) It has been demonstrated in the field (Figure 9-7) that Batesian mimicry breaks down where models are relatively few. In addition, it has been shown in laboratory experiments that predators will reduce the numbers of mimics to a low level without eliminating them because they again learn avoidance at low mimic density.

Batesian mimics, then, have severe limitations set on their population sizes, and this is a factor of major importance in establishing Batesian mimicry as a polymorphic situation. Thus, it frequently occurs (for example, in butterflies) that the mimetic population is polymorphic, with some normally colored individuals and some mimetic individuals. In butterflies it is considered that the females are the mimetic sex because, apparently, the males respond sexually to odors rather than colors. For the females, on the other hand, the colors of the males are important, and so the latter must show species-specific color patterns or be rejected. In this way only half the individuals of a species can be mimetic. Hence there is some chance for successful mimicry in that the models are likely to be at least twice as frequent as the mimics, assuming similar carrying capacities for both species of butterflies. (It may be pointed out that butterfly models are noxious because they taste bad, apparently because they or the caterpillars feed upon plants, such as milkweeds, that contain foul-tasting sub-

| | No. of Individuals of Models (Noxious Species with Aposematic Coloring) | ♀ Papilio Dardanas Mimics | |
		Totals	Imperfects
Locality No. 1	1949	111	4%
Locality No. 2	32	133	32%

Figure 9-7   Data showing the breakdown of Batesian mimicry where models become relatively rare. (Data from Ford, 1953.)

stances.) In some species not all the females are mimetic either, possibly (proximately) because there is some slight preference by males for normally colored females or, possibly (ultimately) because the carrying capacity for the species is so large in respect to that of the model that there would be too many mimics if all the females of the palatable species were mimetic. The mimicry then becomes less effective, and since relatively more mimics die, selection favors nonmimetic morphs, possibly in a frequency-dependent fashion.

The latter seems to be the case in a famous example of butterfly mimicry from Africa, where a population of one species (*Papilio dardanus*) may contain females mimicking several different sympatric noxious models as well as nonmimetic males. Since the males remain nonmimetic, the color pattern is sex-controlled, perhaps with sex hormones acting as inducers or repressors. In Madagascar and Ethiopia the species is nonmimetic, there being no noxious models present. In other places in Africa, such noxious models are present and the females are mimetic, with as many as three or four morphs being found in a population (Figure 9-8). The species as a whole has been found to mimic at least six species of models. P. M. Sheppard and C. A. Clarke, of the University of Liverpool, found that a common situation was one with three morphs present, one a very common form, a rare form dominant to the latter, and a form of

(a)

(c)

(b)

(d)

Figure 9-8. Batesian mimicry in some African butterflies. Two noxious species *Amauris carawshayi* (a), and *Amauris niavius* (b), both mimicked by forms of a third edible species, *Hypolimnas dubius,* form *mima* (c), and form *wahlbergi* (d). (From *Mimicry in Plants and Animals* by Wolfgang Wickler. Copyright 1968. Used with permission of McGraw-Hill Book Co.)

intermediate frequency that showed no dominance in respect to either of the others. As in heterostyly, the complex color patterns appear to be coded for by a number of different genes inherited together as supergenes. In the case where individual females are born heterozygous for "alleles" of the supergene that show no dominance and are, therefore, intermediate in color pattern, they no doubt have low adaptive values inasmuch as they are not protected by mimicry. Evidently the r of these populations and the K of their environments are such that the populations can carry this genetic load. The common recessive form mimics a very common model, while the others mimic rarer models, the frequency of each determined via frequency-dependent selection, by the frequencies of the respective models. The population, therefore, can maintain large numbers of mimetic females because they are not all in the same subniche.

Richard Levins, of the University of Chicago, has examined theoretically the conditions under which different sorts of polymorphisms will be maintained in a population, representing these as different strategies for survival. Thus, if relevant environmental features change during the life of an individual (encountering different microclimates, traversing different sorts of terrain, eating different sorts of food), and if the range of variation of these environmental features is less than what can be tolerated by an individual, the optimal strategy is seen to be an intermediate phenotype adapted to an intermediate value of the environmental variable. This intermediate phenotype could be generated by a heterozygous genotype, and a genetic polymorphism will be maintained on the basis of heterosis (superiority of intermediate phenotype). On the other hand, if relevant portions of the environment change during the life of an individual as above, but the range of change is much larger than an individual can accommodate, the optimal strategy will be to have an extreme phenotype, specialized for one small range of environment, the specific one depending on what is the most frequent or accessible to the population. Here there will be no polymorphism maintained. If, now, the individuals do not experience a change in an environmental variable during their lives (no change in soil type for a plant or in the pattern of host substrate concentration changes for an internal parasite), but the environment does differ from place to place and different individuals in the population live in different places (on different soils for plants and in different individual hosts for parasites), then the optimal strategy for the population (assuming panmixis) is to maintain several different extreme phenotypes in a polymorphism maintained as hypothesized above in connection with frequency-dependent selection.

SEASONAL SELECTION

One situation where the environment does not change during the life of an individual, yet where its condition is quite different for different individuals, is the one where generation times are much shorter than a year in seasonal climates, resulting in seasonal selection. This is found frequently among insects in temperate climates and a case has been described concerning European ladybugs. Here there are several generations produced during the spring and summer, and then the individuals hibernate during the winter in crevices and crannies. Those that survive begin reproducing again in the spring. Some populations are found to be polymorphic for color, with some individuals being red with black spots and others being a very dark red so that they are effectively black or dark brown, the spots not showing. During the breeding season, dark individuals increase in frequency, from about 40 to 65 percent, while during the winter these are apparently relatively less able to survive and so decrease again to 30 to 45 percent by the next spring. Thus, red, spotted individuals are selectively favored during the winter because they are more viable under these stress situations, whereas the dark individuals are better adapted to conditions during the breeding season (perhaps they are more fecund, for example). Here a polymorphism is maintained by fluctuating selection coefficients or alternating adaptive values. Presumably if the environment did not undergo seasonal changes one form would entirely replace the other, this situation then being seen as a case of directional selection (Chapter 11) frequently or cyclically alternating its direction.

This example is somewhat like the third of Levins' categories in that the population in autumn and winter is composed of individuals experiencing one sort of environment (gradually deteriorating temperature regime) while the individuals in the population in summer experience another (increasingly crowded) environment. Seasonal selection, however, is conceptually different from Levins' third category in that it does not involve different individuals of the population simultaneously adapting to different portions of a heterogeneous environment. In fact, it is really closer to Levins' second category, but with cyclic fluctuations in the direction of selection such that no phenotype ever actually replaces the other. It has also been referred to as cyclical selection.

Levins' third category, however, can be exemplified both by disruptive selection and by frequency-dependent selection. Together with heterosis, these and seasonal selection are the forms of balancing selection. Interestingly, heterosis often involves a frequency-dependent component of selection. Thus, in all cases of heterosis where both homozygotes are not lethal, the actual gene frequencies will be determined by the relative fitnesses of the two homozygotes in the environment in a frequency-dependent fashion. This is shown in the example of esterase polymorphism shown in Figure 9-6 and is also well-exemplified by the extensive and multifaceted work of Theodosius Dobzhansky and his collaborators formerly at Columbia University and the Rockefeller University, on chromosome inversions in *Drosophila pseudoobscura*. This work provides a further example of seasonal selection and also provide an example of density-dependent selection—a process fitting, with seasonal selection, into Levins' second category.

CHROMOSOMAL POLYMORPHISM IN
DROSOPHILA PSEUDOOBSCURA

A much-studied polymorphism in some kinds of flies can be detected by examining the giant salivary chromosomes of the larvae. These chromosomes exhibit bands that have been shown by indirect means to represent different genetic loci, or groups of them. Frequently flies were found with somewhat different gene arrangements on a given arm of a given chromosome. Upon close examination, these variant gene arrangements were disclosed to be caused by a segment of the chromosome having been inverted from what was later determined to have been the ancestral gene arrangements at a given chromosome arm. Although there is theoretically no limit to the number of different gene arrangements one could have on a chromosome, the number of different ones found in any given species or population has turned out to be finite, ranging from none to six or seven. This suggests either that the inversion process is not frequent or that selection has disfavored all but a few possible arrangements. Different populations of one species are characterized by different frequencies of the different inversions at any given time of the year, and so this trait, like any other, shows geographic variation (Figure 9-9). The frequencies found at a given locality at one time of the year have been found to be constant from year to year over a period of some ten to twenty years of observation, suggesting that selection is maintaining a given frequency in a given en-

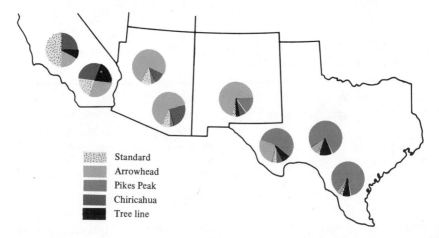

Figure 9-9 Geographic variation of five third chromosome inversions of *Drosophila pseudoobscura*. (Data from Dobzhansky and Epling, 1944.)

vironment. This is further suggested by repeatable observation of seasonal changes in the frequencies of the inversions at a given locality (Figure 9-10). This is a clear demonstration of seasonal selection. Furthermore, laboratory experiments with population cages have demonstrated that under environmental conditions mimicking those of various seasons at a given locality, the frequencies of the different inversions approaches those found in nature at the appropriate season (see the following for details).

These same laboratory experiments also demonstrated heterosis under some conditions. Thus, at 0 to 4°C St/Ch is more viable than either homozygote, whereas at 28 to 30°C the St/St homozygote is most viable. Heterosis was also suggested by the inversion frequencies found in nature at some localities in some seasons, that is, by the fact that an excess of heterozygotes (compared with that expected from the Hardy-Weinberg formula) could be demonstrated. It was found that heterosis could only be demonstrated in the laboratory if the different gene arrangements were derived from the same population, suggesting that, although specific gene arrangements in different populations may be related by descent, subsequent divergence of the genetic information has occurred. This is a further demonstration of the coadapted genome principle (Chapter 7). Under a given set of laboratory conditions adequate to elicit heterosis, one can begin with any proportion of the different inversions and always end up ultimately with the same proportion (Figure 9-11), thereby demonstrating frequency-dependent selection as well. That is, under conditions where genetic equilibrium is attained at 55 percent St, St at frequencies greater than this are selected against, while at lower frequencies it is selectively favored—until the equilibrium frequency is reached.

Density-dependent selection has also been shown in this system. Thus,

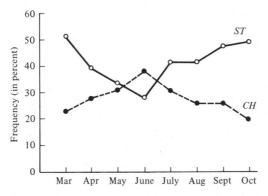

Figure 9-10 Changes in frequency of chromosomes carrying two inversion types in natural populations of *Drosophila pseudoobscura* in the San Jacinto Mts. during the advance of the season from March to October. (From *The Origin of Adaptations* by Verne Grant. Copyright 1963. By permission of Columbia University Press.)

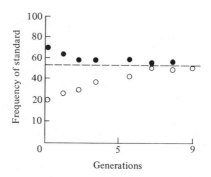

Figure 9-11 Frequency-dependent selection on chromosome inversions in *Drosophila pseudoobscura* with Standard and Arrowhead configurations segregating in the population cages, one of which (black circle) was started with a high frequency of *ST* and the other (white circle) with a low frequency of that inversion. Both cages were kept under identical conditions. (Modified from Dobzhansky, 1948.)

in one locality, Pinon Flats, California, the *St* gene arrangement increases rapidly during the summer while the *Ch* arrangement decreases correspondingly. It was shown in a population cage experiment that in warm temperatures the *St* arrangement from this locality would increase at the expense of *Ch* from that locality. In nature there is no change in the frequencies of these inversions in the winter and, correspondingly, no selective advantage accrued to either in population cages kept at cool temperatures. So far we are dealing only with seasonal selection. In nature in the spring, the *Ch* arrangement increases rapidly at the expense of *St*, and it was found that this was due to a density-dependent selection, as follows: Duplicate population cages were kept at warm temperatures; in one, maintained as usual, the *St* inversion increased at the expense of *Ch*; in the other, where a superabundant supply of larval food was provided so that competition at the larval stage was more or less eliminated, the *Ch* arrangement increased at the expense of *St*. This experiment indicated that in the spring in nature, while the temperatures are such as to favor *St*, the density of flies is so low that larval competition for food is not a factor, and that this is why *Ch* increases in frequency at this time. Other experiments, including some using primary gene product markers, have demonstrated this kind of effect of population density on the direction of selection. More will be said about this later.

K-SELECTION AND *r*-SELECTION

Here it might be pointed out that very often in experiments demonstrating heterosis, the effect is obtained only under crowded conditions, whereas, if the populations are artificially depleted, the heterotic effect does not turn up. This is in general due to the fact that heterotic individuals are superior in terms of viability, but under conditions where the environment is not saturated with individuals (below K) there is room for all genotypes to survive (except for those producing genetic diseases—including those that are inadaptive to the external environment). Thus, relatively less viability selection comes into play until the population has reached the carrying capacity of its particular environment. Selection below the carrying capacity will be primarily in terms of genetic diseases and also will emphasize fertility, since if all individuals survive, some will nevertheless still produce more offspring (more gametes, zygotes or propagules per unit time) than others and their genes will preferentially accumulate in the next generations—since other (viability) factors have become more equal.

These considerations are the basis for describing the dichotomy between K-selection and r-selection. The former is felt to occur under conditions where the environment is more or less saturated with individuals, and is primarily directed at increased viability and ecological intraspecific competitive ability. On the contrary, r-selection is found in places (often marginal) where the populations do not saturate their environment, and the main thrust of selection is to increase fertility and to maintain or acquire adaptedness to the external environment (as in inhospitable climates). This too will be taken up again below.

DURATION OF POLYMORPHISMS IN TIME

Before leaving the topic of polymorphism and its maintenance in populations something should be said concerning the durability of polymorphisms in time. It was pointed out in Chapter 7 that there are a number of instances where allelic differences involve more than a single codon difference. Here the alleles had to have been maintained as such for a long enough period of time so that more than the presumably single original difference between them would have had time to accumulate. Indeed, it was pointed out in that chapter that there are cases of allelic differences probably antedating species differences. The conditions for such long duration of polymorphisms are not known or whether they involve anything other than the various balancing and cyclical selection processes just discussed.

PATTERNS OF VARIABILITY IN NATURAL
POPULATIONS

Although almost any kind of detailed pattern of variability can be expected in natural populations, there are two major patterns so commonly met with that they deserve specific citation. These are central population-peripheral population differences in the amount of variability and clines or clinal variation.

VARIABILITY IN CENTRAL VERSUS PERIPHERAL POPULATIONS

Haldane apparently originally suggested that one could expect natural selection to act differently in central populations from the way it acted in peripheral ones. He based his reasoning on the idea that central populations probably occupy an external environment to which the species is well adapted, whereas peripheral populations represent expansion of the species range outward from this center into less hospitable climates or, at least, to new communities for which the already existing species adaptations are not optimally fit. Under these conditions, he suggested that selection at the periphery of the species range would tend to emphasize adaptations to the rigors of the external environment, while conditions at the center, which he described as tending to overcrowding, would elicit selection based instead on the ability of individuals to compete ecologically with conspecifics. These ideas have been incorporated previously into the parallel ideas of r-selection and K-selection based mainly on the difference between selection for fertility and selection for viability, again characteristic of sparsely populated (peripheral) populations and densely populated (central) populations, respectively. (It may be noted here that peripheral populations are not invariably smaller and sparser than central ones.)

An interesting example of such a dichotomy in the pattern of variability again concerns *Drosophila* chromosome inversions, this time in *D. robusta* as described by Hampton Carson of Washington University in St. Louis. Figure 9-12 shows the geographic distribution of the number of gene arrangements found in a given population. This pattern, with more variants found in the center of the range and fewer in peripheral populations, is a very commonly encountered situation at all levels from that of primary gene products to that of multigenetic factors. Dobzhansky, and also Ernst Mayr of Harvard University, have explained this pattern as follows: Assuming that that the different inversions represent different supergenes, each specialized for a different subniche, the situation in the center of the species range is one where, because the populations have been there for a long time, they have had ample time to increase the numbers of subniches in which they can interact with the environment. Every time a new subniche was created, it was accomplished by the production of a new supergene. Thus, there might be a supergene for acid food in damp situations, one for cooler temperatures in less humid situations, one for living in garbage cans, and so on. The populations can "play" these ecological games largely because the external or physical environment in general is one that is "'balmy" because the species has long ago adapted to it. In peripheral populations, on the other hand, this is not so; the more peripheral they are the more rigorous (to them) is the climate. As the external

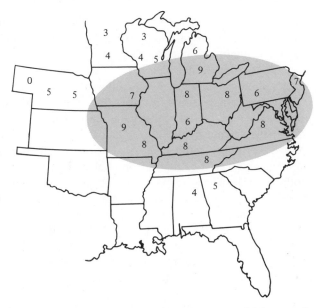

Figure 9-12   Number of different inversions in addition to Standard in 24 populations of *Drosophila robusta*. The central populations are found in the shaded area. (Based on Carson, 1958.)

environment becomes more and more foreign to their fundamental adaptations, the flies must give up more and more of their subniches, becoming increasingly restricted to earlier-attained subniches for which they are most fit. That is, marginal subniches will be dropped as the climate becomes increasingly rigorous. Finally, in the most rigorous climates, the flies are only able to exist in one or two subniches.

Carson has presented another, actually complementary, viewpoint. He noted that if inversions represent supergenes protected from being broken down by crossing over (Chapter 7), each one a specialization for a different subniche, then the very fact that so many are found in central populations means that very much of the potential variability of such populations is tied up in inversions; thus, although there is a great deal of variability expressed, there is correspondingly less potential variability; the populations are specialized in a few different directions. Peripheral populations, on the other hand, are not, and could not be, so specialized, since they must continually adapt to new, rigorous climates and unusual (to them) habitats, and so selection has favored a breakdown of older specializations (inversion supergenes) in favor of potential variability. Also, there has not been time yet in these populations to invent new specializations since peripheral populations are, by definition, those more recently established. This probes deeper into the nature of variability, pointing out that actual variability might well be obtained at the expense of potential variability.

Indeed, any time a series of supergenes is created by coadaptation of different loci in the genome, potential variability is sacrificed for the sake of specializations. Any real situation must be a balance of these needs.

Mayr has contributed a further idea here—that gene flow from different peripheral populations specialized for different environments could carry alleles or gene arrangements toward the center, and so some of the greater variability in the central populations could be due to this factor as well. No gene arrangement that originated, say, in a warm, humid environment at one periphery could be carried by gene flow into another extreme environment (say, cold or dry) at another periphery, but arrangements from both of these places might prove useful in the center of the range by forming raw material for experimentation with new subniches there. In this view, the basic subniches represented by the inversions were created in peripheral populations, each specialized in a different way, and then found their way to central populations and served as the basis of adaptations to new subniches there. Quite possibly all three kinds of explanations are valid—they do not contradict each other.

CLINAL VARIATION

A cline is, quite simply, a gradient in the value of some measurement of a trait. Very frequently a trait will be found to change gradually in a single direction across a geographic transect. As an example, a study was made of various traits in 16 species of Scandinavian butterflies. Six species were migratory and showed no clinal variation at all, whereas of 59 characters studied in the rest, 79 percent of them showed north-south clinal variation along the Scandinavian peninsula. Most factors of the physical environment change gradually with distance. Obvious examples are the changes in light quality and temperature with latitude or altitude. On a smaller scale one could move from a humid forest through a drier savannah to a desert, with gradual changes in climate. However, not all features of the environment change in a gradual fashion. Soil types may change extremely suddenly, resulting in sharp ecological boundaries; rain shadows in mountains may result in there being a desert on the lee side of a mountain and a rain forest on the side facing the ocean. In cases like these the ecotone is very sharply defined instead of a very gradual transition from one community to another. Thus we may expect some features of organisms to show clinal geographic variation and some not, depending upon the relationships of the traits to various geographic features.

Even without gradual changes in environmental parameters it would be possible to observe clinal variation if the populations observed showed some degree of gene flow. For example, suppose certain alleles are selected in a marginal population in a warm climate, whereas others are selected in a marginal population in a cold climate (for example, at different altitudes in the same geographic region). It could be possible that these alleles, selected in the extreme climates, could exist in the intermediate populations, not necessarily for their own qualities but in connection with heterosis, having gotten to these intermediate populations via gene flow. There are, therefore, two quite different theories of clinal variation; one holds that it represents direct selection by the environmental parameters in question at every point along the transect, whereas the other holds that selection is working on the trait as such in only some points along the transect, with the genotypic differences in regions between these points smoothed out, as it were, by gene flow.

A few clines have attracted enough attention to have been given names, and they should be mentioned here. Bergmann's rule, originally thought to have been confined to homeotherms (birds and mammals), has now been discovered to hold for some cold-blooded vertebrates and invertebrates as well. This rubric describes the condition where individuals living in colder climates (usually further away from the equator) tend to be larger than individuals living in warmer climates. This observation has been made with individuals from different populations of a single species and also when comparing individuals from different but very closely related species. This phenomenon is clearly related to the need to conserve metabolic heat in colder regions and/or to lose it in tropical regions. The volume (of cells producing metabolic heat) increases as the cube of the linear dimensions, whereas the surface (from which the heat can be radiated) increases only as the square of the linear dimensions.

A related clinal phenomenon is Allen's rule, which describes the almost universally observed phenomenon of enlarged and elongated appendages (ears, legs, fingers, antennae) in tropical regions, and short, stubby appendages in boreal regions. This again is related to heat radiation—necessary in the tropics, undesirable in cold regions. Thus, in man it has frequently been cited that races characteristic of tropical regions tend to be tall and thin, whereas Eskimos, for example, are short and stocky with small ears and relatively stubby fingers.

Gloger's rule (apparently restricted to homeotherms) records the observation that individuals from populations living in colder regions tend to be lighter colored than individuals from tropical populations. There has yet to be proposed a satisfactory explanation for this phenomenon. It has been shown that dark pigment (as in frog eggs) does afford protection from ultraviolet radiation by screening it out before it reaches the cells. But it is not clear that tropical populations are necessarily exposed for longer periods to stronger sunlight. Indeed, many tropical mammals live in dense rain forests where sunlight rarely penetrates, and there is a further correlation between living in humid tropical regions and dark pigmentation as opposed to living in dry (desertlike) tropical regions, where light pigmentation prevails among mammalian pelages. Gloger's rule, then, is restricted to moist environments. Also, animals living in tropical regions where the sun is a factor in their lives often are crepuscular or nocturnal, and so do not appear in the open during the day. A recent attempt at explanation notes that in tropical savannah regions nonburrowing animals tend to spend much of the hottest part of the day in the shadows under clumps of trees, and suggests that dark pigmentation was selected as a kind of cryptic coloration for this situation. Another plausible suggestion involves vitamin D, formed in the skin when exposed to sunlight. Populations living in cold climates tend to remain hidden from the sun (having thick fur or living in caves) so that selection favors a decreased skin or pelage pigmentation so that more of the (weaker) sunlight available in northern winters can pentrate the skin sufficiently to help produce adequate amounts of vitamin D. In this explanation light skin is seen as the evolved specialization.

SUMMARY

In the absence of relevant environmental change, stabilizing selection will continue to weed out less fit genotypes continually produced by sexual recombination and mutations. This sort of selection is capable of maintaining genetic loci unchanged for hundreds of millions of generations. Polymorphisms can also be maintained for many generations by the different kinds of balancing selection, notably heterosis and/or frequency-dependent selection and disruptive selection, and also by the various kinds of cyclical selection, such as density-dependent selection and seasonal selection. The general directions of selection in sparsely populated and dense populations are different, r-selection, working mainly on fertility in the one and K-selection, stressing relative viability, in the other. Furthermore, marginal populations will tend to be adapting to extremes of tolerable climatic conditions, and will have less polymorphic traits and loci than will central populations, where selection is mainly concerned with intraspecific competition. This difference in variability between peripheral and central populations is fairly generally observed, as is clinal variation.

REFERENCES

Baker, H. G., and G. L. Stebbins, eds., *The Genetics of Colonizing Species*. Academic, New York, 1965, 588 pp.

Bumpus, H. C., "The Elimination of the Unfit as Illustrated by the Introduced Sparrow," *Biological Lectures, Woods Hole* **1897**:209–215, 1899.

Dobzhansky, T., *Genetics and the Origin of Species*. Columbia, New York, 1951, 354 pp.

Ford, E. B., *Ecological Genetics*. Wiley, New York, 1964, 335 pp.

Ford, E. B., *Genetic Polymorphism*. M.I.T. Press, Cambridge, Mass., 1965, 101 pp.

Grant, V., *The Origin of Adaptations*. Columbia, New York, 1963, 606 pp.

Kojima, K., and K. M. Yarborough, "Frequency-Dependent Selection at the Esterase 6 Locus in Drosophila Melanogaster," *Proceedings of the U.S. Academy of Sciences,* **57:**645–649, 1967.

Levine, R., *Evolution in Changing Environments.* Princeton University Press, Princeton, N.J., 1968, 120 pp.

Lewontin, R. C., ed., *Population Biology and Evolution.* Syracuse University Press, Syracuse, New York, 1968, 205 pp.

MacArthur, R. H., and E. C. Wilson, *The Theory of Island Biogeography.* Princeton University Press, Princeton, N.J., 1967, 203 pp.

Mayr, R., *Animal Species and Evolution.* Harvard, Cambridge, Mass., 1963, 797 pp.

Maynard Smith, J., *The Theory of Evolution.* (2nd ed.) Pelican, Baltimore, 1966, 336 pp.

Mettler, L. E., and T. G. Gregg, *Population Genetics and Evolution.* Prentice-Hall, Englewood Cliffs, N.J., 1969, 212 pp.

Sheppard, P. M., *Natural Selection and Heredity.* Harper & Row, 1960, 209 pp.

Sved, J. A., T. E. Reed, and W. F. Bodmer, "The Number of Balanced Polymorphisms That Can Be Maintained in a Normal Population," *Genetics,* **55:**469–481, 1967.

Waddington, C. H., *The Strategy of the Genes.* George Allen and Unwin, London, 1957, 262 pp.

Wallace, B., *Chromosomes, Giant Molecules, and Evolution.* Norton, New York, 1966, 171 pp.

Wallace, B., *Topics in Population Genetics.* Norton, New York, 1968, 481 pp.

Weldon, W. F. R., "A First Study of Natural Selection in *Clausilla Laminata,*" *Biometrica,* **1:**109–120, 1901.

CHAPTER 10

Environmental change
and natural selection

Although natural selection will operate in the absence of environmental change, it is obvious that it seldom has occasion to do so. The geological record informs us that, in general, the environment has always been changing, and one might reasonably assume that it has been continually changing everywhere since the origin of the earth. This continual change is what imposes upon living systems the need to change, in order to adapt to the continuously generated new environments. This necessity no doubt imposed upon the lineage(s) that gave rise to present-day living systems the precondition strategy of sacrificing individual units after only a short time, because only with such a strategy was it possible (at first probably only by mutation) to generate individuals somewhat different from those adapted to the previous environment. If enough such altered individuals are produced, some will by chance be better adapted to the altered environment than the previous sorts of individuals, and so the population will survive by maintaining its adaptedness. The means of producing new individuals is reproduction (implying a system of heredity), an obvious part of the preconditional strategy of those living systems that were successful.

DIRECTIONAL OR PROGRESSIVE SELECTION

The mode of selection associated with environmental change has been termed directional or progressive selection (as opposed to stabilizing or normalizing selection—Chapter 9). Examining this notion first at the level of the phenotype, we can begin by referring to Figure 10-1. In an unchanging environment, stabilizing selection weeds away, every generation, individuals with "unusual" forms or measurements of traits from both sides of the normal or binomial curve of frequency distribution. In an altered environment the previous mean for some measurement may no longer correspond to the optimally adapted form of the trait. Suppose the environment is gradually becoming cooler and that we are examining the relative length of some appendage (ear pinna of a mammal or limbs of an arthropod). All other things being equal, we would predict that those individuals with relatively shorter appendages will tend to leave relatively more offspring (see Allen's rule, Chapter 9). In such a situation, stabilizing selection will select away individuals from both ends of the normal curve unequally. More individuals will be selected out from those at the larger end of the curve than from individuals at the smaller end. Clearly, this will begin in a short time to affect the population mean, notably by decreasing it (assuming the usual sort of hereditary situation). Individuals with relatively longer appendages will tend not to survive passing catastrophes, and their offspring, also having relatively longer appendages, will suffer the same fate. Thus in retrospect it will turn out that parents with longer appendages will have left fewer breeding offspring than will those with shorter appendages. Note that stabilizing selection is not conceived of as being absent in this situation; it simply does not work evenly at both ends of the distribution curve.

The reason why stabilizing selection must continue to occur is because it is necessary to maintain a coadapted genome (Chapter 9). Genetic

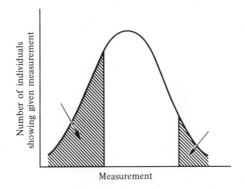

Figure 10-1 Directional selection. The arrows point to the measurements of those individuals (shaded portion of the curve) most heavily selected against. Note that stabilizing selection is occurring, but unequally at each end of the curve. The net effect will be to shift the mean to the right.

changes can presumably accumulate only slowly in a population if typical recombinants are to be fit in an all-around sense. Even though smaller appendages are favored in the case under consideration, an individual with the smallest appendages it is possible to produce by recombination of available alleles will almost certainly not have a spectrum of alleles at different loci that work well together to produce a harmonious phenotype. It takes time to develop a spectrum of coadapted alleles. Shortened limbs would be a liability in the absence of appropriate behavioral changes, and the probability of mutations and recombination producing such coordinated changes in a single individual is very low. In the absence of this low probability simultaneous change, the length of appendages can change only slightly in each generation so as not to unbalance the harmonious phenotype.

A more drastic situation can perhaps exist in which stabilizing selection at the favored end of the distribution curve could be almost eliminated. This might be so in a case of character displacement (Chapter 6) for one or two traits after the catastrophic bringing together of similarly adapted forms from different communities. Interspecific competition might be so intense that individuals with unusual phenotypes, *because* they differ most from individuals of the other species, would be overwhelmingly favored by selection (provided only that they are not too inviable in their adaptions to the physical environment) since the more normal individuals have what virtually amounts to a genetic disease in the presence of the other species. The difference here from the more usually conceived directional selection is the relatively greater intensity of selection envisioned.

It may be worth noting that the degree to which an organism can regulate (such as developmentally, behaviorally, physiologically), that is, its degree of phenotypic flexibility, can materially affect its evolutionary response to altered environmental conditions. Regulators, as opposed to conformers, should be capable of experiencing greater environmental change before there is an evolutionary response in their genomes. If individuals of a species typically can regulate to develop normally over a temperature range of ten degrees, a change in, say extreme low temperatures during the breeding season over a period of time of less than this magnitude, would not result in an evolutionary change, the organisms being able to regulate within the framework of the older genomes. Because most organisms are regulators in some systems and conformers in others, most would fit into either category, depending on which traits are being examined; hence during a given period of environmental change one would expect some aspects of the phenotype of some species to change while others do not, as in mosaic evolution (Chapter 5). This, indeed, can serve as another explanation for that phenomenon.

INDUSTRIAL MELANISM

This is an opportune point at which to review a famous case of directional selection from nature that has been observed by man sufficiently long so that the story can be pieced together with reasonable confidence. The case is that of industrial melanism in the peppered moth, *Biston betularia*. The early part of the nineteenth century saw the dramatic rise of industrialization in Europe. With the burning by factories of cheap fuel, such as soft coal, the smokestacks produced tons of black, sooty smoke that fell upon the countryside leeward of the factories, covering forests and fields with various amounts of soot. In the vicinity of large manufacturing centers, this effect was quite extensive, materially altering neighboring communities. In forests, tree trunks became covered with so much soot that lichens normally growing on them could not do so, and this, together with the color of the soot, changed the usual color of the trunks from a mottled greenish-gray to black. It is this environmental change that we will be concerned with here.

The wing color of the typical peppered moth is a mottled gray that blends perfectly with lichen-covered tree trunks, rendering the moths cryptically colored when resting on such a trunk (Figure 10-2). Around 1850 the first dark-colored peppered moth was taken by a butterfly collector in the region east of Manchester, England. This is a variant that was to become common enough to merit a variety name, notably *Biston betularia carbonaria*. During the next 100 years the frequency of dark individuals gradually increased from less than 1 to about 90 percent or more in the vicinity of industrial areas. This we know from many well-preserved butterfly collections, especially in England, and by population surveys today. (Many other sorts of organisms showed an increase in the frequency of melanistic individuals during this same period in industrial Europe, but the record for this moth is especially good and much work has been done with it.)

The causes of this striking increase in numbers of melanic individuals have been worked out by several British biologists, most notably by E. B. Ford and H. B. D. Kettlewell, both of Oxford University. Ford found that caterpillars that later metamorphosed into melanics were generally much more vigorous and viable than were the usual wild-type caterpillars; they could survive environmental hardships, such as cold and lack of food much better than the latter. This being so, one might have expected that *carbonaria* would long ago have replaced the peppered form. Kettlewell, however, showed why this event did not occur until the advent of soot pollution. Using high-speed photography, he first proved that various sorts of birds (nuthatches, and so on) feed upon the resting moths. Since birds locate their prey by sight, it was probable that, in a nonsooty forest, no

Figure 10-2   Peppered moths, *Biston betularia,* showing how the two morphs, *typica* and *carbonaria,* are cryptically colored on different backgrounds. (By permission of H. B. D. Kettlewell, Oxford University.)

matter how many more melanics transform, they would be the first to be devoured by the birds because they are not cryptically colored on lichens (Figure 10-2). On the other hand, with the elimination of lichens in some forest communities the *carbonarias* are cryptically colored, whereas the older, wild types, are not. Kettlewell performed the following experiments to test this notion. He released equal numbers of typical and *carbonaria* males in two localities, one polluted with soot, the other a normal forest not on the lee of an industrial region. Both sites had *Biston betularia* populations. The males were marked so that when recaptured they could be recognized. After a few days traps were set up containing female phero-mones, capable of attracting the males from rather long distances. In the normal forest many more peppered marked males were returned than melanics; just the opposite was found in the soot-polluted site, as expected (Figure 10-3).

Thus, the trigger that released the greater competitive ability of the melanic form was the alleviation of the major aspect of the environ-ment that selected against them—bird predation. This example is a nice demonstration in that it shows well that it is the total organism interacting

		Typical Form	Carbonaria
Unpolluted Woodland	Released	496	473
	Recaptured	62(12.5%)	30(6.3%)
Soot-Polluted Woodland	Released	137	447
	Recaptured	34(16)%	154(34%)

Figure 10-3 The results of marking and recapture studies on the color morphs of *Biston betularia* in two different sorts of environments, suggestive of relative differences in survival of the two morphs. (Based on Ford, 1964.)

with its total environment that is scrutinized by selection. Superior viability alone is not enough; nothing alone is enough; the phenotype must work together as a whole in its entire environment.

Modern breeding experiments have shown that in any given population *carbonaria* segregates genetically as if it were a dominant at a single genetic locus. There is, however, historical evidence that at one locality this form was not always dominant. First, the *carbonarias* of today are noticeably darker than those preserved from the past; second, one breeding experiment done at the turn of the century resulted in progeny not in the expected proportions if *carbonaria* were then a dominant. It has been suggested that we have here an example of the gradual evolution of dominance (Chapter 7). This idea is supported by modern breeding experiments showing that, in two populations, both containing a low percentage of typical forms, *carbonaria* from one population did not behave as dominant when crossed with a typical individual from the other population. Again, crossing a *carbonaria* with typical forms from nonsooty habitats shows that the dark color is not fully dominant in such a cross. In short, the *carbonaria* allele appears to have evolved dominance independently in different isolated populations subjected to the same environmental change. Or, to put it another way, the genomes of different populations became reorganized around the production of dark-wing phenotypes in different ways. The question can then reasonably be raised as to whether the original mutations selected for in each polluted community were in fact identical chemically (the same actual alleles). It is most likely that they were not. On this score, there is in England another dark grey melanic, *Biston betularia insularia,* not as dark as *carbonaria* and tending to be restricted to populations on the peripheries of polluted areas, that behaves genetically as if it were located at a separate locus from that carrying *carbonaria.*

Haldane has, by making certain assumptions (including that *carbonaria* has been dominant ever since its appearance), calculated, in a way that will be shown below, that the selection coefficient against the previous wild-type allele averages out at 0.70 over the 100 year period—extraordinarily high compared to what had been considered reasonable before that calculation. First, bird predation is extremely efficient; birds have relatively high metabolic rates and exceedingly keen eyesight. Also important here is the fact that *carbonaria* caterpillars are demonstably more viable than the wild type in adversity. In other words, not only was there an advantage in the negative sense of being more cryptically colored on soot but also the positive advantage of more vigorous caterpillars. These factors might explain how such a high selection coefficient might appear outside a genetics laboratory. In fact, selection was probably less intense at first (absence of dominance, less dark coloration of melanics) and gradually increased until now *s* must be higher than 0.70, but it can for simplicity's sake be averaged out over the whole time as 0.70.

If a population is sampled a short time after a large-scale environmental change has initiated a burst of directional selection acting on a trait that tends to be polymorphic (either dark or light, either one allele or another), the population will be found to be polymorphic for that trait, because there has not yet been time for one morph to replace the other. This transient polymorphism cannot be distinguished from a balanced polymorphism by observations over only one or a few generations. Thus, a study of peppered moth populations around Manchester in 1900 would no doubt have disclosed a polymorphic situation with high frequencies of both morphs. Finding such a situation, the population biologist of today would probably initially guess that the polymorphism was balanced rather than transient, since the latter situation is intellectually trivial and relatively uninteresting. Yet, clearly one must not assume that one is always observing equilibrium conditions for all traits; some observed traits of almost any sort of organism must be undergoing directional change in response to changed or changing environments at the time of observation. It is often assumed that a trait that is disfavored by selection would disappear so fast from a population that one could only rarely observe it. But we have seen that in the peppered moth, rather intense selection pressure did not effectively eliminate (reduce to less than 1 percent) a formerly wild-type morph in less than 100 years. Such a period of time is large enough compared to the time over which observations are made so that different degrees of polymorphism should have been clearly observable all through it.

It is of some interest that there are examples of increased melanism that occurred during this same period but in which industrial wastes can have played no part. A case in point is that of the black hamster. Hamsters were (and are) sought for fur in Siberia. Throughout the nineteenth century there was steady increase in black pelts listed in the fur trapping records in Russia. The case is interesting because of the great probability of irrelevant bias in the data. We may assume that in any luxury item rarity is sought. Suppose that early in the nineteenth century a trapper, having fewer of the kinds of furs he usually sent to Moscow, padded the shipment with a few black ones—kinds he had been catching at a certain rate for some time but had never shipped. Upon arrival in Moscow, these pelts created a sensation and the trapper was ordered to search for black pelts. His usual methods initially return him the few he had always caught, but in time would he modify his methods in order to catch more of the valuable animals. In a situation like this we can reasonably expect to see increased numbers of black pelts arriving at headquarters as more and more trappers begin searching for them. This record, then, is not a valid one in our present sense, possibly being an artefact of nonrandom searching, and we do not in fact know whether there was or was not an increase in the frequency of dark-furred hamsters in nature during this time.

LABORATORY EXPERIMENTS WITH *DROSOPHILA* POLYGENIC TRAITS

Moving now from observations in nature and experiments on natural populations to the more controlled world of the genetics laboratory, it would be worthwhile to cite some classical experiments demonstrating directional selection, those of Kenneth Mather and associates of the University of Birmingham on bristle number in *Drosophila melanogaster* being a good case in point. The object of the artificial selection in the experiments to be discussed here was to increase (in some, "high" lines) and decrease (in others, "low" lines) the numbers of bristles located on the ventral portion of the fourth and fifth abdominal segments. Figure 10-4 shows that, beginning with a population mean of 36 bristles, the mean could be lowered to 25 in 30 generations and increased to 56 by 20 generations. In general, it has been found that numerical traits (measurements, counts, and so on) can be doubled or halved in value fairly rapidly by

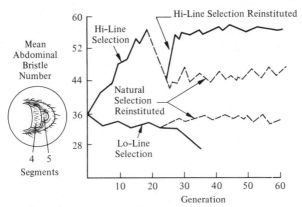

Figure 10-4 Artificial selection for abdominal bristle number in *Drosophila melanogaster*. Inset shows a ventral view of the posterior portion of the abdomen and the bristles in question. (Based on Mather and Harrison, 1949.)

intense artificial selection of the kind maintained in these experiments—notably to discard (and so prevent from breeding) any individuals with more than the mean value in the low line and less than the mean in the high line each generation.

The limits imposed on this kind of intense selection appear as severely impaired fertility and greatly lowered viability, so that continued selection beyond certain limits can lead to the extinction of the population being subjected to it. There are two classes of explanation for this phenomenon, both of which may be operative in any given case. First, such intense selection, occurring over relatively few generations, must have a negative effect on the number of different alleles present at the various genes involved in some way with bristle number. There is ample evidence that quantitative traits of this sort are generally coded for by more than a single gene—by what Mather called polygenes (see Chapters 7 and 8), each gene having a small effect on the trait being measured, each of which can vary independently of the others; the final size phenomenon is the cumulative effect of the action of the products of all the polygenes. In the system we are here examining, for example, it was found in other experiments that genes affecting bristle number are located on all four chromosomes and at many loci on each. Using the model in Figure 9-2, we can see that selection favoring high bristle number would tend to eliminate from the population alleles at the various polygenes that produce gene products whose action tends to decrease bristle number. And there simply cannot be enough time in a few generations to experience enough new mutations so that a few can be selected to replace the lost alleles. This being so, the population must come to be increasingly homozygous at more and more loci, thereby uncovering previously undetected recessive lethal, semilethal, and subvital alleles and also leaving fewer possibilities for heterosis. At the same time the trait being selected will come to show decreased variability. This is the phenomenon referred to as inbreeding depression. Another aspect of this sort of explanation is that as the genetic variability gets to be less and less with the progress of the experiment, the gene pool is capable of generating by recombination fewer and fewer different sorts of genotypes and so the population cannot respond by continuing to alter its mean bristle number; it simply cannot generate genotypes coding for even more (or fewer) bristles.

The other kind of explanation invokes a breakdown of the previously coadapted genome by the intensity of selection for a minor trait. In this idea the polygenes are seen, quite reasonably, to have each of them various pleiotropic effects on other phenotypic features, including, for example, fertility and various aspects of viability. Of interest for this notion is the fact that in the bristle number experiments, many anatomical traits were altered simultaneously with bristle number, although selection was not directed at them. There were changes in pigmentation, eye form, and in the numbers of spermathecae (for holding sperm after mating) in the females, for example, as well as in fertility, mating behavior, fecundity, and larval survival. The idea here is that natural selection, working on the genome as a whole, has produced well-balanced genotypes capable of generating harmonious phenotypes and that in this picture bristle number is a very minor feature indeed. Bristle number might even, in natural populations, be whatever happens when the more important effects of the genes have been established by selection. In the laboratory, however, the investigator chooses to make this relatively minor trait the *sine qua non* for survival, and does so regardless of the effects this fixation might have on the rest of the phenotype. Connected with this explanation is the observation that as selection continued in this experiment, other phenotypic features become more variable. At first this was considered to argue against the first type of explanation—the decreasing variability in the gene pool—but is better seen to represent a breakdown of canalization (Chapter 8) due to the disruption of the coadapted genome. Previously well-buffered developmental phenomena become disrupted, and development becomes less predictable. Thus, in females there are normally two spermathecae, but in these experiments many females began to turn up in the late generations with none, one, three, four, or five such organs. Clearly it is illogical to suppose that this was caused by an increase in the numbers of alleles in genes coding for spermatheca number.

Figure 10-4 shows that when selection was lifted from some populations, bristle number tended to return to the normal mean value. The fact that it could not get there argues in favor of the idea of the gene pool having been depleted, especially of alleles coding for the artificially determined unwanted bristle numbers. That there was a tendency, with artificial selection lifted, to return to the phenotypic condition that had been produced by natural selection (one might argue that such lifting of artificial selection is equivalent to reinstating natural selection) suggests that combinations of alleles giving genotypes for viable phenotypes tend to produce the original, wild-type bristle number. Whatever was left of the original gene pool had a preferred way of combining to produce the most viable and fertile phenotypes, and those combinations tended in turn to produce the wild-type bristle number.

Clearly, natural selection rarely works on a single trait with such intensity, although the previously discussed case of industrial melanism in *Biston betularia* perhaps comes close (which may be why it stands out as so clear a case of directional selection). If natural selection were for some reason to favor increased bristle number, it is unlikely that this could become an objective of such urgency as to render other features involved with viability and fertility relatively less important. It is more likely that a compromise situation would exist in which bristle number would gradually increase over many thousands of generations, never going beyond what might disrupt the harmonious phenotype in any generation. During this long period new mutations would become incorporated into the gene pool that tend to favor higher bristle number without inflicting damage on other aspects of the phenotype, so that depletion of genetic variability would not be a part of the phenomenon as it is of artificial selection; as alleles for low bristle number leave the gene pool, new ones for higher bristle number would enter it.

Thus, many of the phenomena observed during artificial selection experiments are considered from this point of view to be artifacts of the abnormal intensity of such selection. These include predominantly the decrease in genetic variability and its attendant inbreeding depression and decrease in variability of the selected trait, coupled with an increase in variability of unselected traits and loss of fertility and viability. Such phenomena do not arise out of natural selection as a rule; otherwise the process probably would not work.

It is interesting that in the bristle number experiments, when new high lines were started six generations after natural selection had been reestablished in lines that had been subjected to high-line selection for 20 generations, an increase in bristle number was again obtained, this time without loss of viability and fertility, but bristle number could still not be driven beyond a mean of 56. If our view of what is going on here is correct, it is surprising that enough reorganization of the genome could occur in ten or so generations (Figure 10-4) to allow the new mean bristle number to occur without impairment of viability and fertility. It would be interesting to know if such flies could have competed with more "normal" ones, however, in a given ecosystem. For example, an attempt by a state game commission to breed some of the meat qualities of domesticated turkeys into their wild turkey populations met with complete failure. The domesticated turkeys released into the wild populations disappeared without issue in a single generation, presumably having been rendered unfit for their ancestral environment by the kinds of selection experienced under domestication. The flies maintained in the population cages were capable of breeding in that artificial environment, but probably many more generations would be needed for enough genetic reorganization to take place so that adaptively competent flies would be produced. This, however, is purely speculative, and the seemingly quick recovery of viability shown in these experiments remains unexplained. It may be noted also that when natural selection was reinstated in some lines developed from this new, viable, high line six generations later, they did not show a tendency for bristle number to return to the previous wild-type mean, indicating that in 16 or so generations the genome had apparently become sufficiently reorganized around this new bristle number so that it had become, to some degree at least, canalized. It may be that genome reorganization can occur at a much faster rate than suggested here.

SUBSTITUTIONAL LOAD, COST OF EVOLUTION, POPULATION FITNESS

Recall that with a selection coefficient of 1 (Darwinian fitness equals zero) a recessive allele could be reduced by directional selection from 99 percent to less than 1 percent frequency in some 100 generations (Figure 6-6), and that Haldane, using similar calculations but with various other assumptions (smaller selection coefficients, and so on), calculated that on the average it would take some 300 generations to replace one allele by another in natural populations. He also showed that while the number of generations needed to accomplish this is less with more intense selection pressures, the total number of individuals that must suffer genetic death or impairment of their Darwinian fitness is the same in all cases. If fewer generations are needed (higher selection coefficients), more individuals per generation must be selected adversely in each generation than if more generations are needed for the process to take place, in which case fewer individuals in each generation must suffer genetic death due to the trait in question. Thus, the total number of genetic deaths per locus is independent of the intensity of selection. Haldane also estimated that the total number of these individuals is often of the order of magnitude of 30 times the number of individuals in a generation. This number of selected individuals he referred to as the "cost of natural selection."

This "cost" was later seen by Kimura as an aspect of genetic load, which he dubbed substitutional load; that is, a genetic load imposed upon a population in order for it to remain adapted by having one allele replaced by another. The load would be maintained until the substitution was effectively accomplished. A number of workers have realized that there is in this formulation an unnecessary negative connotation, as though this "cost" or "load" were something bad and, more importantly, that this is also a very restrictive view of how fast evolutionary change can occur. They have pointed out that contrary to being bad, this process is the way populations remain adapted. This controversy is worth exploring here, because it allows us to refine our definitions and make more precise our view of what selection is and does.

Alice M. Brues, of the University of Colorado, Bruce Wallace, of Cornell University, and J. A. Sved, of the Stanford Medical Center, have in particular pointed to two important facts that make the concept of substitutional load somewhat pointless, notably that (1) by and large death rates are set by density-dependent factors emanating from the size of the carrying capacity and (2) Darwinian fitness in cases other than in outright genetic diseases is defined as a relative concept. The argument is as follows: All organisms usually overproduce offspring every generation, so that, unless there is an increase in the carrying capacity of the environment, only enough of them will survive to replace older individuals lost by death.

The rest are excess population and are removed, regardless of how many there are, and to some extent, regardless of their genotypes. Natural selection is the agency that "decides" which individuals will replace those lost. It does so by first culling from the population all individuals suffering from genetic diseases, then individuals that carry important semilethal traits, then individuals carrying subvital traits, then individuals that have less stamina or agility than others, then individuals that have difficulty mating or rearing offspring, and so on, the distinctions made by selection becoming increasingly more subtle and refined as the ratio of offspring produced by the population to the carrying capacity is increased. Conversely, if this ratio is small, as in an expanding population or one recovering after a severe environmental catastrophe has wiped out most of its members, there is more room for more individuals to survive until a new carrying capacity is reached (or the old one regained), and so selection will make fewer distinctions between individuals—in the limiting case differentiating only those with outright genetic diseases from those without, allowing all of the latter to survive. The Darwinian fitnesses of all these latter individuals will be unity *because* there is room for them (or, strictly, their offspring) to survive; that is, the magnitude of Darwinian fitness, except in genetic diseases, is relative to the ratio of individuals produced to individuals that can survive, as determined ultimately by the environment.

There is then no need for concepts such as the "fitness of a population." Kimura has proposed this term and defined it in terms of the proportion of individuals in the population carrying the fittest genotypes. In the case envisioned by Haldane, when an environmental catastrophe occurs such that the previously best allele at some locus is no longer the best, the fitness of the population is seen to plummet, and to be slowly regained as a gene substitution takes place at the given locus over some period of time. According to the argument just given, it can be seen that unless the previously best allele becomes associated with a genetic disease ($s = 1$), this sort of idea has no meaning. If a previously rare allele becomes the best in the new environmental conditions, that does not mean that individuals carrying the original best allele now must all suffer genetic death simultaneously. In other words, the most usual sort of case would probably be where a previously slightly better allele becomes slightly worse, and, indeed, this only because there is a better contender in the ring, as it were. Darwinian fitness is a *relative* concept.

An interesting correlate of the relativeness of adaptive value is the idea that regardless of what allele is present as the wild type of some population, there can always be one that might be better in the given environment. This one may not appear simply by chance or, having appeared, it might be carried out of the gene pool by the accidental death of its possessor. Actually, this notion that there can always be a better allele in any given environment may be erroneous if the "best" is determined to any large degree by coadaptation of the genome. In this case, there may well be only a single best at any given time in any real population in any given environment and this became "best" by simply being there from the start, its gene product happening to be the one that formed part of the physiological background for all recent major evolutionary occurrences in the populations. In this case almost any alternative allele would code for a gene product producing a genetic disease.

Thus, supposing that Haldane's catastrophe occurs, as, say, in the case of industrial melanism in *Biston* (a case Haldane had in mind), is the fitness of the population lowered? Birds that had presumably previously fed upon other sorts of invertebrates will now begin to capture more peppered moths because their cryptic coloration has been neutralized. The very few melanics, on the other hand, fare better than before and are not captured by the birds. The size of the peppered moth population may decrease somewhat initially, depending upon the intensity of the predation. One might say that the carrying capacity has been decreased because of the environmental change initiated by air pollution. In some localities the populations might become wiped out, in which case one could perhaps say that the fitness of the population had decreased. But in other localities, every generation sees an increased proportion of melanics making up the shrinking population until, with time, enough melanics are surviving so that the population size can begin to grow again. When the old carrying capacity is reached, the population stops growing, being limited, as before, by various features of the environment. Now the populations will be composed mostly of melanics instead of peppered forms. As long as a population continued to survive, what need is there to consider that its "fitness" had declined? Even at its lowest point individuals were surviving and reproducing up to the point allowed by the altered environment. The state of the environment is crucial, and any surviving population is, *because it survives,* fit enough and, as fit as it was in another environment where the carrying capacity was greater.

A situation in which the notion of population fitness may be more appropriate is one where a population, because of the spread in its gene pool of an allele that happened by chance to be better than the original wild-type allele at some locus, begins to increase its geographic range and/or its niche breadth without any major environmental change. It is widely held that this process must happen in nature. Is the population more fit if its population size has been increased? Not unless, as in this case, the increase was allowed by an increase in carrying capacity or range extension initiated by a change in the gene pool rather than by a change in the external environment. This distinction, however, is perhaps a bit overfine, and one wonders whether any purpose is served by making it. Any population that survives is fit and any that does not, or is reduced below the critical population size (Chapter 4), is not. There does not seem to be any use for the notion of different degrees of fitness in terms of populations. It may be objected that in ecological catastrophes, when two very similar organisms are brought together in the same community, and one survives while the other does not, having been "outcompeted" by the first, that the notion of relative population fitness would be useful. When, however, it is realized that individuals and populations of different species cannot compete reproductively (Chapter 6), it is clear that members of another species are effective only as portions of the external environment and not other "individuals" at all. Their effect on the individuals of a population is to reduce the carrying capacity in that community because they are utilizing some requisite needed by the species in question. Again, the external environment, in its total sense, is of overriding importance, and a population cannot be more fit than some component of that environment since fitness is not a concept appropriate to the environment (except in a poetic or metaphysical sense).

Another criticism must be leveled at the notion of population fitness, and that is the artificiality of the viewpoint of environmental change implicit in it. Thus, in the classical Haldane situation, a sudden catastrophic change lowers the fitness of certain predominant genotypes in the population. Then it is considered that the environment stops changing, and the population slowly regains fitness as some alleles replace others. Some, particularly drastic, environmental changes may act like this (perhaps in some places, industrial pollution—that is, is it present or absent), but as a general rule the environment is continually changing at all levels. If that is so, no population would have high fitness because they can never come to equilibrium with an environment that has always just changed again after the gene pool had almost become adjusted to the last change. Again, a population having achieved high "fitness" at one locus, may not have it at another.

ADAPTIVE VALUE AND GENETIC LOAD

Another problem with the Haldane-Kimura formulation of allele replacement arises out of the way they viewed the interactions of several different loci. More specifically, they saw the Darwinian fitnesses of different loci interacting multiplicatively. That is, if an organism has w at locus $1 = 0.67$, w at locus $2 = 0.32$, and w at locus $3 = .71$, its total adaptive value is $0.67 \times 0.32 \times 0.71 = 0.15$, that is, very low indeed. Thus, each locus is considered to contribute to the adaptive value of an individual independently, rather than in an interacting sense. Although this may be true in some cases, it does not seem reasonable to suppose that it is the general situation. Indeed, it seems much more likely that selection generally scans the phenotype as a whole since a number of different forces impinge upon organisms at a given time. Some of the important consequences of the Haldane-Kimura viewpoint render it even less likely to be generally true.

First, if total adaptive value is multiplicative and there is more than one locus undergoing directional selection, the number of such loci would have to be restricted to one hundred or so, and even then the adaptive values of much of the population would be so low that the population would become extinct. It hardly seems likely that evolution can take place at only less than a hundred loci at a time. Instead, one can suppose that total adaptive value is somehow an average of those found at different loci; then our organism above would have

$$w = \frac{0.67 + 0.32 + 0.71}{3} = 0.56$$

Or, as John Maynard Smith, of the University of Sussex, has suggested, there may be "threshold selection," that is, individuals will survive that have more than a threshold number of favored alleles, up to the point allowed by the carrying capacity. Of course, a lethal allele at any locus is still best seen in a multiplicative sense; when $s = 0$ at any locus, the total adaptive value is affected by this in a multiplicative sense ($\Sigma w \cdot 0 = 0$).

These ideas bear importantly on rates of evolution as well. As noted, Haldane suggested that an average figure for a gene replacement might be $1/300$ generations. This was calculated, using the assumption of a multiplicative total Darwinian fitness, on the basis of an average of 10 percent selective deaths per generation for all loci together. Higher intensities of selection could not be conceived, because then too few loci could undergo evolution at the same time because of the multiplicative restriction. But if the multiplicative restriction is lifted, any locus could evolve much more rapidly than one replacement per 300 generations. This is because now we can have larger selection coefficients at each or any locus without worrying about eliminating the population in one generation. In fact, the multiplicative model of adaptive value forced those accepting it to consider that in nature selection coefficients seldom rise above figures such as 0.03, 0.007, and so on) despite the evidence of the peppered moth example. With only such selection coefficients, evolution would be slow indeed. But, eliminating the concept of multiplicative total Darwinian fitness, we need not stipulate that selection coefficients are so low as a rule, and consequently evolution can be conceived of as moving more rapidly than suggested by Haldane.

SUMMARY

Environmental change elicits directional selection in natural populations. The more drastic environmental changes (as in laboratory experiments) elicit more dramatic evolutionary responses, but have consequences in the disruption of coadapted genomes. Natural selection (as distinct from artificial selection) rarely focuses only upon some single trait, but is concerned with overall viability and fertility in all situations—even when some obvious phenotypic trait is being modified. When considering genetic load caused by allele substitution at most genetic loci, it is best considered as an average over all the loci, with lethal genotypes being treated multiplicatively.

REFERENCES

Brues, A. M., "The Cost of Evolution versus the Cost of Not Evolving," *Evolution,* **18:**379–383, 1964.

Dobzhansky, T., *Genetics and the Origin of Species.* Columbia, New York, 1951, 364 pp.

Grant, V., *The Origin of Adaptations.* Columbia, New York, 1963, 606 pp.

Haldane, J. B. S., "The Cost of Natural Selection," *Genetics,* **55:**511–524, 1957.

Hamilton, T. H., *Process and Pattern in Evolution.* Macmillan, New York, 1967, 118 pp.

Kimura, M., "Optimum Mutation Rate and Degree of Dominance as Determined by the Principle of Minimum Genetic Load," *Journal of Genetics,* **57:**21–34, 1960.

Kimura, M. and T. Ohta, *Theoretical Aspects of Population Genetics.* Princeton University Press, Princeton N.J., 1971, 219 pp.

Li, C. C., *Population Genetics.* The University of Chicago Press, Chicago, 1955, 366 pp.

Mather, K., and J. L. Jinks, *Biometrical Genetics.* Cornell, Ithaca, N.Y., 1971, 382 pp.

Mayr, E., *Animal Species and Evolution.* Harvard, Cambridge, Mass., 1963, 797 pp.

Mettler, L. E., and T. G. Gregg, *Population Genetics and Evolution.* Prentice-Hall, Englewood Cliffs, New Jersey, 1969, 212 pp.

Sheppard, P. M., *Natural Selection and Heredity.* Harper & Row, New York, 1959, 209 pp.

Smith, J. Maynard, *The Theory of Evolution.* (2nd ed.) Penguin, Baltimore, 1966, 336 pp.

Smith, J. Maynard, "Haldane's Dilemma' and the Rate of Evolution," *Nature,* **219:**1114–1116, 1968.

Sved, J. A., "Possible Rates of Gene Substitution in Evolution," *American Naturalist,* **102:**283–293, 1968.

Wallace, B., "Polymorphism, Population Size and Genetic Load," in R. C. Lewontin, ed., *Population Biology and Evolution,* Syracuse University Press, Syracuse, New York, 1968, pp. 87–108.

CHAPTER 11

The best of all possible worlds

"PERFECTION" IN BIOLOGICAL ADAPTATION

It is often supposed that biological adaptations are "perfect."
It is obvious that organisms do show various teleological (or
useful) characteristics, but closer examination of the situation
reveals that these can hardly be characterized as "perfect."
Some adaptations may, indeed, be more intricate than others,
and must have involved more actual evolutionary change in
order to come about (say, the vertebrate eye as compared to
a tooth, or the molecule chlorophyll *a* as compared to vertebrate collagen). It takes a more improbable or highly-ordered
configuration of structures and events to focus light into an
image than to afford a hard surface for grinding, or to interact in a specific way with photons of light than to form part
of a generalized structural material.

If a lineage moves into an adaptive zone where more
complex adaptations would enhance the chances of the reproductive success of its members, and if by chance the
necessary preadaptations are present, then there is no reason
not to assume that selection would gradually improve the
preadapted trait as individuals in given populations compete
with each other in terms of which has the more efficient
chlorophyll or eye. Given long periods of time, such structures could become increasingly complex—if they continue
to remain preadapted for further new functions. Thus, an
early version of the eye might simply detect light and dark.

A later version, for which the first was preadapted, might detect movement in a more precise way. A still later stage might involve the ability to focus images, and later the ability to change focus and accommodate distance. Each step requires the previous one as a preadaptation.

Structures like the eye are more meaningfully thought of as complex than as perfect. It is highly probable that an engineer could design a "better" vertebrate eye than now exists. This is because natural selection has never been "bringing about the evolution of the eye" as such (see Chapter 4), but has been simply choosing the individuals from each population in a series that would leave relatively more reproductive survivors. It happened that in some lineages those that left the most successful offspring were those that could "see" better, and it was this that resulted in the gradual complication of eye structure sometimes referred to as "improvement."

We sometimes obtain a hint simply from observing the traits of organisms that their adaptions are less than perfect. Thus, man's vertebral column is an adequate structure certainly, but the number of ailments involving it (such as slipped discs) suggest that it is not a "perfect" structure. In fact, medical persons frequently approach the organism as a rather poorly put together but usually adequate mess, a viewpoint gained from their experience with the malformations and maladjustments they see every day. In this chapter an attempt will be made to demonstrate by several different sorts of examples that natural selection is not a cunning architectural force, but a force better described as being involved with expedience and adequacy, or with the *first* (not the best) solution for any given challenge. It always works only on short-term advantage.

Notice that the point here is not that natural selection is not the agent responsible for the evolution of a structure like the vertebrate eye, but that it is not capable of planning such an evolutionary sequence. It will be the aim of this chapter to demonstrate that selection cannot and does not produce perfect structures as such. Demonstrating that natural selection often produces less than admirable adaptations perhaps leaves open the possibility that some other force was responsible for those adaptations we *do* admire. It is the task of the rest of this book to present the argument that this was not so, and that natural selection provides an adequate explanation for any adaptation.

SELECTION AT DIFFERENT LIFE HISTORY STAGES

SENESCENCE

Since most organisms undergo ontogenetic changes, it is probable that the internal milieux of the cells also undergo changes with time. Perhaps this was the condition that led to the invention of isotypic forms of gene products, with ontogenetic switching from one to another during the course of differentiation (Chapter 8). However, it is clear that some gene products remain in some cells during the entire life of the organism. Such proteins must be capable of adequate function in many different sorts of cells of all ages. It is unlikely, however, that such a protein would be equally well-adapted to all sorts of cell environments. One theory of aging, especially developed by George C. Williams of the State University of New York at Stony Brook, is based on this consideration.

First, selection can only act on prereproductive and reproductive life history stages (Chapter 12). If a particular cytochrome c was selected in a certain population, it was because it functioned well with other gene products in the cells of "young" organisms. If, however, there is a continual change in intracellular conditions beyond sexual maturity (it is not clear why this should be, but it does appear to be so—see below), then conditions will probably sooner or later change from those to which the particular cytochrome c is optimally adapted. This will result in malfunctioning that will contribute to the alteration of the system away from its youthful condition. Multiplying this example by all the gene products active in the cells of postreproductive individuals results in a gradually accelerating departure from the homeostatic conditions typical of prereproductive organisms, to which the gene products are adapted. In this view, aging is the result of a gradual breakdown of homeostasis because of the limitation on natural selection that it can only detect traits in young organisms because it is the result of differential reproduction. Since this limitation exists, natural selection can never mold a perfect organism.

A word should be said concerning the possible reasons why intracellular conditions should continue to change after sexual maturity has been reached. Development is partly epigenetic—that is, processes present at one moment alter the conditions in which they occur simply by operating, leading to new conditions (or to the next stage) in a developmental sequence. Primary gene products are adapted to the sequence of such conditions as they occur up to and including the typical age at reproduction. But there is no reason why processes operating at reproductive maturity should not also modify the internal conditions so as to bring about new conditions—in this case conditions never "seen" by natural selection. It must be remembered that living systems are always dynamic systems and cannot fail to continue changing.

PREZYGOTIC SELECTION

In the house mouse there is a series of *tailless* (or t) alleles, all located in one region of the ninth chromosome. There are now upwards of 21 such mutations known. They are defined as t-alleles because they all, on heterozygote combination with the *brachyury* allele (T) (for example, T/t^3, T/t^{12}), result in a tailless phenotype. In homozygote combination (t^5/t^5) they are lethal (or cause sterility), the organisms dying at various stages of development depending upon which t-allele is involved, but in most cases the lethal homozygotes abort before birth. The heterozygote with the wild type allele $(+/t)$ has a tail of normal length, t being recessive. Usually the combination t^x/t^y results in a shortened tail, and in some of these the males are sterile. Despite the obvious liabilities of any of these t-alleles, most house mouse populations carry one or another of them, often at rather high frequencies. Indeed, an experiment carried out by Paul K. Anderson, L. C. Dunn and A. B. Beasley, in which they introduced t^{11} (in a single heterozygote male) into an island population that had no t-alleles, showed that the allele reached quite high frequency in the local population in two years despite the fact that selection could not have favored its increase as such.

Not only is the homozygote recessive lethal, but, if there are other ts present, or T, a shortened tail is also selected against since that organ is used in steering while running. There is, in addition, no evidence for heterozygote advantage for the $+/t$ genotypes. The answer to this puzzle was partially found when it was discovered by L. C. Dunn and Howard Levine, of Columbia University, that some 90 percent of the offspring of heterozygous males carry the t allele, indicating that the effective sperm pool is not a $50:50$ mixture of $t:+$. For some reason the sperms carrying t are far and away more effective at fertilizing ova than are $+$ sperms, this being a case of prezygotic selection.

Although the presence of t confers a high adaptive value on a mouse spermatozoan, it produces one or another sort of genetic disease in many of the zygotes thus preferentially formed. The direction of selection on this locus at one life history stage is quite different than at another, the locus being widely pleiotropic. The abnormal segregation ratios of t-alleles result in high frequencies in local family groups (each a territorial male with several females and their offspring) despite strong postzygotic selection against them in homozygotes. It is not clear, however, why such an allele should spread in a local population beyond the family group of the original carrier male. It may be noted here that at high densities young females will disperse to neighboring territories, and these could carry the allele by chance.

Since males accept their own daughters (undispersed in crowded conditions) as mates, it is clear that in the family of a carrier male many lethal homozygotes will begin to be produced in a generation or so. One effect of this is to cut down on the number of offspring actually weaned by a heterozygote mother since in many cases death is prenatal or immediately postnatal. In effect, the male carrying this trait begins soon to produce fewer offspring than a neighboring, noncarrier male. This is not to say, however, that he produces relatively fewer reproductive individuals in the next generation. It is perhaps possible that the fewer individuals (all heterozygote carriers) weaned in a litter where many were lethal homozygotes are actually in some respects more physically fit simply as a result of having developed in a less crowded uterus or nest. It should be noted that mice have quite high reproductive potential and yet frequently live in very restricted territories with definitely limited carrying capacities. Any family group must reach saturation numbers quickly. In such a situation it may be advantageous, or, at least not definitely disadvantageous, to produce relatively fewer but stronger offspring. It is possible, then, that the t-allele system is a density-dependent population control mechanism, the numbers of heterozygous females mating with heterozygous males increasing with density and resulting in smaller litter sizes containing more—physically—fit offspring.

Ultimately, however, the allele could spread very widely in the population until present in every family. At that point the reproductive potential of the population is reduced, and catastrophic events (such as attack from farmers or homeowners) could result in local extinction. Thus, a trait that may at one time be adaptive is at another disadvantageous, the adaptation in question being far from "perfect". Indeed, survival gained by means of t-alleles may seal the ultimate doom of the family, and even of the population (although descendants of a family apparently often succeed in dispersing).

Another curious example of a trait being adaptive in one situation and not in another is that reported by P. S. Dawson, of the University of Illinois, in the flour beetle *Tribolium castaneum*. Here, during a competition experiment between a population of these and another species, *T. confusum*, a recessive eye color mutant appeared that alters eye color from black to light brown. This mutant condition spread in the population rather rapidly, presumably being selectively favored, but, as the gene spread in the population the latter began to decrease in size while the population of *T. confusum* increased proportionately. Thus, the spread of the new mutant gene in the population seems to have resulted in a loss of competitive ability vis-à-vis *T. confusum*. The conditions used in this experiment were unfavorable to *T. castaneum* but the line used had been selected for fast development, and such lines typically outcompeted *T. confusum* in similar situations. Presumably the eye color mutant was favorable in the stressful environment as such, but its presence hindered the ability of its possessors to compete with members of the other species ecologically, or hindered their reproduction in some way. The case has not yet been fully analyzed.

GENETIC LOAD

The concept of genetic load provides another example of how "imperfect" the products of natural selection are, especially in its form of segregational load. The other forms of load, substitutional (Chapter 10) and mutational (Chapter 7), are so much a matter of precondition and mathematical necessity as to be of less interest in the current context. Segregational load arises out of sexual recombination in connection with adaptations that are either coded for on a multigenetic basis or involve heterosis. Adaptively adequate genotypes always give rise to some individuals with extreme measurements that are destined to succumb to passing environmental catastrophes. The situation as seen in heterozygote advantage is especially clear, and we will examine one such case, that of sickle-cell anemia, in more detail at this point.

Sickle-cell anemia is a clinical condition in man arising out of the presence in the blood of some proportion of hemoglobin that is constructed of beta chains containing valine at the sixth position from the amino end instead of glutamic acid as in the wild type, hemoglobin A. Such hemoglobin precipitates in the deoxygenated state causing the shape of the red cells to be deformed by its elongated crystals. This deformation (the "sickle" shape) somehow disrupts the integrity of the red cell membrane and the cell lyses, initiating a loss of active hemoglobin in the blood and, therefore, causing anemia. In heterozygotes, some proportion of the red cells lyse periodically, but evidently not enough of them at any one time to render the individuals unfit in normal environments. However, even a small loss of hemoglobin cannot be taken as a favorable condition. As such the trait would be eliminated by natural selection, all other factors being equal. Further, the homozygote is effectively lethal, although individuals may survive up to the teen ages before dying; they almost never breed. Nevertheless, the frequency of this trait in various parts of Africa is quite large—up to 40 pecent of a population may be carrying it—and so it can be described as forming a polymorphism with Hb A.

Several lines of evidence have been adduced to show that heterozygotes carrying Hb S are more resistant to falciparum malaria. First, there is the correlation between the geographic distribution of hemoglobin polymorphisms in general and the malaria belt (Figure 11-1). Second, statistical studies of children in hospitals indicate that only very rarely does a sickler contract falciparum malaria symptoms. Also, a study was carried out using volunteer convicts that indicated that heterozygote sicklers tended not to contract falciparum malaria when bitten by carrier mosquitos, while nonsicklers did contract the disease. It was found that in newborn infants the genotype frequencies fitted exactly the Hardy-Weinberg prediction using the gene frequencies found in the adult population, and so apparently there is no selection in the uterus on this locus. However, after birth the Hb S/Hb S genotype is heavily selected against, and the Hb A/Hb A genotype is at a slight disadvantage with respect to Hb S/Hb A—there comes to be an excess of heterozygotes. The adaptive values of these genotypes in one locality were found to be, respectively, 0.169, 0.860, and 1.00.

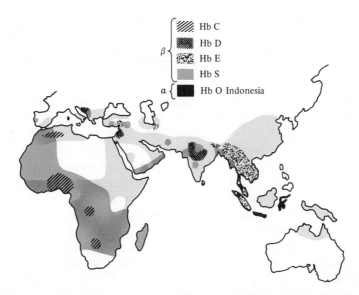

Figure 11-1 Map showing the distribution of four beta chain and one alpha chain human hemoglobin variants that form major polymorphisms with the wild type (Hb A). The gray shading indicates the distribution of falciparum malaria.

Now, in West Africa there is another hemoglobin allele found at rather high frequencies in some populations, Hb C, a mutation at the same codon—but this time from glutamic acid to lysine. It was discovered that the adaptive meanings of Hb S and Hb C were probably the same inasmuch as they replace each other in the populations—that is, there is an inverse correlation between the frequencies of these two in populations in West Africa (Figure 11-2). The adaptive values of the genotypes involving Hb C in one locality, where it was along with Hb S, were: Hb A/Hb C, 0.970; Hb C/Hb C, 0.485; Hb C/Hb S, 0.358. That is, Hb S protects better against falciparum malaria than does Hb C (one suggestion is that Hb C evolved in connection with quartan malaria found endemic in West Africa, and so is not as effective against falciparum as is Hb S). But notice that the Hb S/Hb S homozygote is more severely ill than is the Hb C/Hb C homozygote ($w = 0.169$ instead of 0.485 in the

	$\dfrac{\text{Hb A}}{\text{Hb S}}$	$\dfrac{\text{Hb A}}{\text{Hb C}}$	$\dfrac{\text{Hb A}}{\text{Hb A}}$	$\dfrac{\text{Hb C}}{\text{Hb C}}$	$\dfrac{\text{Hb C}}{\text{Hb S}}$	$\dfrac{\text{Hb S}}{\text{Hb S}}$
w	1.00	0.970	0.860	0.484	0.358	0.169

Figure 11-2 The adaptive values (w) of various hemoglobin genotypes in a population in western Africa. (Data from Allison, 1956.)

locality in question). Thus, this sort of adaptation to living in malarious regions is more effective the more severe the genetic load; the heterozygotes are protected to the degree that the homozygotes are ill. This observation has suggested the following mechanism for the protection given against the malaria plasmodium.

During contact between man and the mosquito, the sporozoite stage of the plasmodium transfers from the latter to man. These sporozoites enter red cells and undergo a series of developmental stages until they reproduce asexually inside the cells to form many merozoites. These soon burst through the cells and infect many other red cells. Many of these simultaneously undergo still further asexual reproduction in 48 hours, again bursting out of the large number of red cells they occupy, causing anemia and releasing toxins that cause fever and chills. Now, if the host has his own periodic large loss of red cells because he carries a hemoglobin variant, that loss will not be synchronized with the life cycle of the parasite and the random destruction of red cells will destroy many of the parasites because they are not at the proper (merozoite) stage to survive in the blood stream. As a consequence, significantly fewer sporozoites will successfully develop into merozoites in an individual carrying a hemoglobin mutant—often so few as to suppress the symptoms of the disease. In this view the degree of protection from malaria (protection from symptoms only) is directly related to the severity of the anemia caused by the structure of the hemoglobin. Hemoglobin mutants would not be selected for if their physiological effects were so severe that the genetic disease was as bad as the infection disease. They will be selected for if they cause a condition less bad physiologically than the infection. And they will be selected specifically because they do cause anemia. In other words, protection from malaria has been won at the price of genetic disease, and the greater the protection, the greater the price—hardly a "perfect" adaptation. (A second theory of why Hb S protects from malaria is that the plasmodia have difficulty feeding upon the altered hemoglobin because of its higher viscosity. Again, the severity of the phenotypic results of the mutation would be correlated with its ability to protect from malaria.)

OVERSPECIALIZATION

Another view of the imperfection of adaptations arises out of the very fact that they are specific adaptations (serving one or a few specialized functions). Thus, having become adapted to seed-eating itself effectively prohibits the possibility of feeding on flesh. In this context all specialization is *over*specialization since it would be very handy to be generalized enough to pursue most sorts of possible food. Ecological competition, however, on both the inter- and intraspecific levels ensures specialization in that the gain in efficiency thereby acquired allows the specialist to reproduce at a favorable balance vis-à-vis the more unspecialized, who must devote more energy simply to existing and, therefore, less to reproduction. However, specialization to the degree, for example, of becoming restricted to a single food item, probably inevitably leads to extinction with the first major ecological catastrophe. Thus, the most "perfect" food-getting adaptation is simply a guarantee of early extinction for the lineage acquiring it. Clearly, the idea of "perfect" adaptations is inappropriate to organic nature—ultimately because the external environment will not stop changing. Any adaptation may become a liability in a new environment.

CHANCE EVENTS

In order to emphasize that not all observable biological situations have been molded by natural selection, and, therefore, that not all observable biological phenomena are adaptive, we may examine a peculiar situation in salamanders, described by Thomas M. Uzzell, of Yale University, and also by Henry Wilbur, of the University of Michigan. The blue-spotted salamander and the Jefferson's salamander are very closely related, the latter being found in the southeastern United States, the former in the northeastern parts of North America. It has been suggested that they arose as separate species from a common ancestor during the last glaciation during a period when they were in different glacial refuges (see Chapter 14 for mechanisms of speciation).

All along a broad zone in which the two species come into contact, running from Michigan through Maine (Figure 11-3) there are found mixed in the populations of each, populations of very similar-looking, all-female, triploid salamanders. Examination of several primary gene products electrophoretically in all these salamanders suggests that the triploids carry two sets of genes from one species and one from the other, indicating that their origin somehow derived from hybridization between the two diploid species. Currently the most probable event seems to have been an impairment of gametogenesis in hybrids by suppression of the first meiotic division because the genomes of the two parental species were too dissimilar to synapse successfully. That event would result in the production of unreduced (diploid) ova that were subsequently fertilized by a normal haploid sperm from one of the diploid species. This gave rise to triploids, of which only the females were fertile because of an unknown mechanism whereby they can regularly produce triploid ova. The most probable mechanism here may be an endomitosis prior to the reduction division of meiosis so that at synapsis each of the three chromosomes pairs with a copy of itself (Figure 11-4).

The mode of reproduction of the triploid females involves obligatory mating with diploid males because the eggs will not develop unless activated by a spermatozoon, and because there are virtually no triploid males.

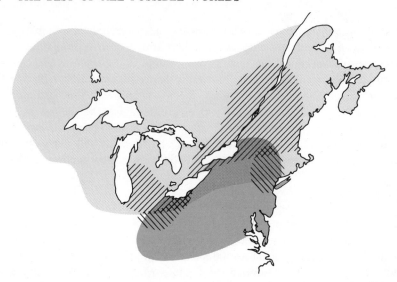

Figure 11-3   Geographical ranges of the blue-spotted salamander [ ] ,
and Jefferson's salamander [ ]   , and the two types of triploid female
populations—those that resemble the blue-spotted salamander phenotypically
[ ] , and those that resemble the Jefferson's salamander phenotypically
[ ] . (Data from Uzzell, 1964, and Uzzell and Goldblatt, 1967.)

The spermatozoan contributes no genetic material to the zygote, however,
and so the triploid female gives rise only to daughters (if the mother was
X,X,X, or X,X,Y, each ovum will carry that genotype too). The mode of
reproduction of triploid females, therefore, is gynogenetic (genetic material
deriving only from the mother).

   If the triploid females survive as well as do the diploids, and if the
males mate with them equally frequently as with diploids, it can be shown

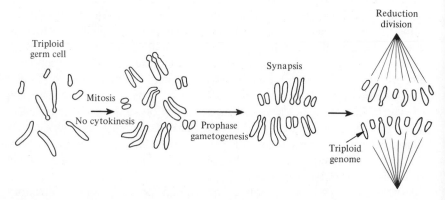

Figure 11-4   Possible mechanism whereby triploid females could give rise to
triploid ova.

that there will be a very rapid increase in the proportion of triploid to diploid females accompanied by an increasing deviation of the sex ratio from the 50:50 male:female ratio typical of sexual organisms. Unchecked, this process should reduce the number of available males to a point where diploid females will have difficulty in finding one to mate with, the few present being monopolized by the more common triploid females. It should be noted that it has been calculated that a typical male of this complex of salamanders could mate with only from seven to ten females in any given season. If the number of diploid females goes below 10 percent, there will be a real problem connected with their finding the very rare males. At this point the deme in question could become extinct, and the triploids will go the same way since they have need of spermatozoa in order to reproduce. Before this happens the size of the population would have begun to shrink as well since any increasing rarity of males would result also in less triploid reproduction. Thus, the gynogenetic triploids are in the position of stealing sperm from the diploid population with ever increasing frequency until there is simply not enough left to propagate the sexual diploid species.

Is this process actually occurring in the populations? First, some triploid-diploid populations have been under observation for some 40 years (13 to 20 generations) without apparent change in status (even though it has been calculated that only 13 generations would be needed to carry the triploid proportion from 1 percent to 54 percent of the available females). Second, modern hybrids between the two diploid species do not appear to occur in nature, the males of one never courting females of the other, and so the original hybrids must have occurred some time ago—perhaps when the species first met after the last glaciation. These two observations suggest that the triploid-diploid relationship has been present for a long time, and is not, therefore, unstable in the sense postulated above.

Examination of this situation has disclosed several mechanisms operating to the disadvantage of triploid females, and these (and perhaps others) must be involved in stabilizing the situation short of extinction. First, and perhaps most important, diploid males *do* discriminate against triploid females, mating with them less intensely (thereby conserving sperm) or, in the presence of a choice, choosing diploids. Many triploids deposit unfertile egg clutches as a consequence. This has resulted in the triploids evolving so that they more closely resemble the diploid females (only those that most closely resembled these left offspring), giving rise to two quite different sorts of triploids, one associated with each species. On the other hand, male discrimination has no doubt evolved as well since males that tended to waste their sperm on triploids have left no diploid descendants. So far the triploid females have not become as good at fooling the males as the latter have become at detecting the fraud.

As other factors, it has been found that the triploids are less fecund (have smaller clutch sizes) than are the diploid females. This may be a result of their having larger cell sizes, since their ova are larger than the diploid ones. This feature would slow down the processes postulated above leading to extinction. The eggs of triploids take longer to hatch (and, therefore, to feed). This gives diploid larvae a head start in feeding and growing that could well be significant since food is limited in the ponds where the larvae develop. It has been found that in some ponds there is clearly a lower carrying capacity for triploids than for diploid larvae; this is a result perhaps of character displacement (Chapter 6) such that the ecological niches of the two sorts of larvae are no longer identical (if they ever were) even though they do overlap. This would result in there always being more diploids metamorphosing from the pond regardless of how many more triploids actually started out as eggs. Another factor is that triploids take a significantly longer time before metamorphosing. This is of extreme importance in temporary ponds such as are typically utilized for breeding by these animals, because animals with longer larval life stand a greater chance of drying up with the pond before they get a chance to transform into adults.

Even though mechanisms such as those just enumerated are capable of explaining why the Jefferson's and blue-spotted salamanders have not become extinct along their zone of contact, it is clear that the triploid-diploid relationship is not an adaptation—that is, it is not good for either individuals or populations of these species nor for the species themselves. The generation of this triploid parasitization was a purely chance event such as may happen often, but which usually would not be preserved. Perhaps some extinctions were caused by such freak accidents. Indeed, since some portion of the carrying capacity for some populations of the diploid species in this case must be shared with triploids, it may be that various populations of these species would not be capable of generating enough genetic variability per unit time to survive extreme environmental catastrophes.

SUMMARY

Organic nature is not "perfect" in the sense that man can conceive of perfect design—as, for example, in the design of a watch. Many structures and relationships in organic nature are engendered and preserved by chance even though they are of no benefit to the individuals or lineages that harbor them (triploid salamanders). The products of natural selection are "flawed" from the point of view of elegant design. First, selection cannot operate on postreproductive organisms, thereby guaranteeing senescence by its very nature. The operation of selection always involves a "wasteful" genetic load, and the more effectively it operates the larger is the load (hemoglobin mutants). Even when an adaptation is very nearly "perfect" from the point of view of design (for instance, an efficient beak) this very perfection is one aspect of an overspecialization that may prevent the lineage possessing it from surviving serious environmental change. The very fact of ontogeny prevents "perfect" adaptations inasmuch as selection for some trait in the gamete stage might result in genetic disease at the gastrula stage (*t*-alleles) or the other way around. An organism must simply be an *adequate* compromise between the needs of the various stages it must live through up to breeding (or, in some cases, the breeding of its offspring). This will involve a balance between the past payoffs on both specialization and generalization, and an accommodation to various chance events imposed upon the history of its ancestors. Natural selection does not select the best possible individuals, only the best available.

REFERENCES

Allison, A. C., "The Sickle-cell and Haemoglobin C Genes in Some African Populations," *Annals of Human Genetics,* **21:**67–89, 1956.

Allison, A. C., "Polymorphism and Natural Selection in Human Populations," *Cold Spring Harbor Symposium on Quantitative Biology,* **29:**137–149, 1964.

Anderson, P. K., "Lethal Alleles in *Mus musculus:* Local Distribution and Evidence for Isolation of Demes," *Science,* **145:**177–78, 1964.

Anderson, P. K., L. C. Dunn, and A. B. Beasley, "Introduction of a Lethal Allele into a Feral House Mouse Population," *American Naturalist,* **98:**57–64, 1964.

Ayala, F. J., "Biology as an Autonomous Science," *American Scientist,* **56:**207–221, 1968.

Comfort, A., *Ageing, The Biology of Senescence.* Holt, Rinehart and Winston, New York, 1964, 365 pp.

Dawson, P. S., "A Conflict between Darwinian Fitness and Population Fitness in *Tribolium* 'Competition' Experiments," *Genetics,* **62:**413–419, 1969.

Lewontin, R. C., "Interdeme Selection Controlling a Polymorphism in the House Mouse," *American Naturalist,* **94:**65–78, 1962.

Lewontin, R. C., and L. C. Dunn, "The Evolutionary Dynamics of a Polymorphism in the House Mouse," *Genetics,* **45:**705–722, 1960.

Livingstone, F. B., *Abnormal Hemoglobins in Human Populations.* Aldine, Chicago, 1967, 470 pp.

Murayama, M., "Molecular Mechanism of Red Cell 'Sickling'," *Science,* **153:**145–149, 1966.

Raper, A. B., "Sickling in Relation of Morbidity from Malaria and Other Diseases," *British Medical Journal,* **1:**965–966, 1956.

Selander, R. K., "Behavior and Genetic Variation in Natural Populations," *American Zoologist,* **10:**53–66, 1970.

Uzzell, T. M., "Relations of the Diploid and Triploid Species of the *Ambystoma jeffersonianum* Complex," *Copeia,* 1964, 257–300.

Uzzell, T. M., "Notes on Spermatophore Production by Salamanders of the *Ambystoma jeffersonianum* Complex," *Copeia,* 1969, 602–612.

Uzzell, T. M., and S. M. Goldblatt, "Serum Proteins of Salamanders of the *Ambystoma jeffersonianum* Complex and the Origin of the Triploid Species of this Group," *Evolution*, **21**:345–354, 1967.

Wallace, B., *Topics in Population Genetics*. Norton, New York, 1968, 481 pp.

Williams, G. C., "Pleiotropy, Natural Selection, and the Evolution of Senescence," *Evolution*, **11**:398–411, 1957.

Williams, G. C., *Adaptation and Natural Selection*. Princeton University Press, Princeton, N.J., 1966, 307 pp.

CHAPTER 12

Altruism

The unit of natural selection is the individual organism. Natural selection distinguishes between the phenotypes of individual prospective parents in such a way as to modify the gene frequencies of the gene pool of the next generation. Each individual organism competes with each other individual in its population (with the exception of its mate or mates) to contribute a larger proportion of the next generation's individuals, or to spread more widely in the next generation's gene pool genes from its own genotype. This ultimate competition, of course, may involve many proximate, specific, physical competitive acts vis-à-vis other individuals, such as competition for food, for nesting materials, and for mates. It is in the context of these interactions, as well as in connection with escaping from predators, that most specific traits of organisms evolve as adaptations, because it is in these contexts that one individual can be superior to another. An individual with a genetic disease such as hemophilia is so little able to compete with other individuals that his abilities to obtain food and mates are irrelevant; his adaptive value is close to zero. Individuals free of major genetic diseases, however, do compete with each other in reference to the prerogatives from the external environment. For this proximate competition to result in an evolutionary event it must ultimately be translated into differential reproduction. Intraspecific competition in areas not related to relative reproductive ability (if any such exists) will have no evolutionary effect. Alleles producing enzymes with effects that result in some curtailment of reproductive ability will tend to have large selection coefficients associated with them, and will either never come to form a significant part of the gene pool, or will be eliminated as major factors in it.

SEXUAL SELECTION, COADAPTATION FOR REPRODUCTION

While the traits found in a population of organisms are present because individuals in the past who had them were relatively more successful at intraspecific competition than were individuals that did not have them, it is not true that *all* of these traits are connected with proximate competitive ability, and some of them are even connected with what appear to be cooperative interactions between members of a population. The production of offspring in sexual organisms, for example, must involve some cooperation between males and females. This, of course, is self-evident, but it must be produced and maintained by natural selection. Cooperation between the sexes involves many adaptive (and coadaptive) traits. Males and females must have specific attractants to bring them into close contact. These may be visual signs such as the colored feathers of cock birds, chemical stimuli such as the sex pheromones of female moths, or behavioral signals such as those involved in the courtship rituals of many kinds of animals. Along with these signals or *releasers,* behavioral traits must develop in the opposite sex so that appropriate responses can be elicited by the sexual stimuli, resulting finally in mating behavior, which, too, must involve many coadapted anatomical and behavioral traits. Individuals deficient in any aspect of attracting and mating with a member of the opposite sex obviously will not project their genes into the next generation, and whatever alleles at whatever loci were involved in the maladjustment to signals from the opposite sex will be selected against. We may call this process *sexual selection.*

Sexual selection has from time to time been accused of producing inadaptive individuals, and even of leading to extinction. One sort of case that has been cited is the development of elaborate tail feathers in pheasants and of elaborate wing feathers in birds-of-paradise. It has been shown that birds will often respond preferentially to signals of supernormal magnitude. If length of tail feathers is important in producing the display that arouses the female's responses, it could happen that males with longer tail feathers would tend to produce more offspring than those with shorter tail feathers, thus spreading alleles for long feathers in the population. At some point, however, it becomes relatively more difficult for an individual to escape from predators because it possesses a tail of such length as to interfere with flight. The argument thus far can be granted; alleles for long tail feathers (in the context of male-level sex hormones) spread in the population and soon recombination among the polygenes responsible begins to produce some individuals with dangerously long tails. But this is not enough to insure extinction! What will occur at this point is a counter-selection brought on by the relative inability of birds with overly long tails to flee from predators, maintaining male tail feather length at some optimum mean size. Sexual selection alone cannot drive a species to extinction because there will always be some individuals present with less than most extreme measurements for whatever trait is involved. It could perhaps happen that sexual selection would compromise a population that is already barely surviving for other reasons, and tip the scale, as it were, into extinction. In this case we still could not ascribe extinction to sexual selection alone.

Returning to the main point, one can note that in sexual organisms a precondition for any evolution at all (that is, for the continued maintenance of adaptiveness) is coadaptation between the sexes involving a certain amount of cooperation between them. Thus, individuals may share food with their mates, for example, but not with anyone else. In plants, of course, the coadaptation between sexes or hermaphroditic individuals is only structural, there being no behavior involved (other than that of insects or the wind).

It can be shown that the energy an individual organism expends on the reproductive functions is, as it were, deducted from the total energy available to it. Reproduction exerts a high price in viability. Developing eggs draw energy from the female that could otherwise contribute, say, to somatic growth. Mating is a time of great danger vis-à-vis predators, Nesting ties an individual to a place more easily tracked by a predator. Defense of the young involves some degree of risk to an individual's life. It has been demonstrated experimentally that individual flies prevented from reproducing lived longer than siblings that did reproduce. However, one cannot say that reproduction is good for the species only and not for the individuals making up the species. What is at stake in these considerations is *not* the lives of individual organisms but the survival of their genes into the next generation. If an individual did not reproduce and thereby gained many years of continued life, its adaptive value would be zero.

SOCIAL COADAPTATION

Other cooperative interactions have been cited by various authors, (particularly by V. C. Wynne-Edwards of Marischal College in Aberdeen, Scotland) along with statements of disbelief that ordinary natural selection, as summarized so far in this chapter, could be responsible for their evolutionary development. An example would be the utterance of warning calls by members of a population at the approach of a predator. The individual that gives the call first runs the risk of attracting the attention of the predator to itself. If this is so, then individuals in the past who tended to give warning cries at the first sign of a predator should have produced fewer offspring than those, say, that slunk away when they first detected the predator, leaving their neighbors to be captured. This process should have resulted in the elimination of alleles that affected the nervous system in such a way as to elicit cries at fairly low levels of stimulation. But it was pointed out by Wynne-Edwards that this statement of the problem is not complete, and that it should be realized that such cooperative inter-actions are found in the context of a *social system*. It will be seen that all such examples of cooperation between individual organisms occur in a social context, even though the animals involved need not necessarily be fully social animals.

ALTRUISM AND GROUP SELECTION

At this point Wynne-Edwards raised serious controversy by invoking the notion of *altruistic traits* to explain such cooperative interactions as described above, and by postulating *group selection* as the process whereby alleles favoring such traits can spread in a population. A paraphrase of his argument is as follows: There are some traits involved in cooperative interactions between members of a social group that are clearly good for the group itself, but which are detrimental to the individual lives of the members of the social group. Ordinary natural selection could not select such traits since the unit of natural selection is the individual. Thus, we need to postulate a different force involving the group as the unit of selec-tion rather than the individual; this we can call group selection.

Now, there is no reason why one may not postulate anything one wishes. There is nothing *wrong* with the idea of groups being selected as such. It is a conceivable possibility. The problem is that as scientists we wish to have the simplest possible explanations for all phenomena; we wish to unify as much data as possible under a single explanatory hypothesis. If we had a separate explanation for every separate event we would have no explanations, only chaos. The notions of altruistic traits and group se-lection may be superfluous. We would like to be rid of these ideas if at all possible because they seriously complicate our picture of organic evolu-tion. If, on the other hand, the concept of natural selection is truly incapa-ble of explaining certain social phenomena, then we must be prepared to accept further complications.

The fundamental question here is whether or not there really are al-truistic traits—that is, traits that are good for the population, deme, or species, and not for the individual, and therefore could not have been pro-duced by natural selection. George C. Williams, of the State University of New York at Stony Brook, has devoted himself to this question exten-sively, and, along with the present author, remains unconvinced. The fol-lowing discussion will be based in part on the more extensive ones in Williams' books, and has a serious weakness in that it consists of raising possible examples of altruistic traits and then knocking them down again. The problem, however, demands this kind of treatment in that it is a question of whether there are any such traits existing in nature, and the only way one can proceed is to examine the possibilities that have been raised by the proponents of the idea. At any moment someone might discover a true altruistic trait that cannot be explained away. This does not appear to have been done yet. There is another reason for going through these examples that has nothing specific to do with the problem at hand. It will afford a good opportunity for refining our concept of just what natural selection is by examining cases where it is not perhaps immediately apparent that it has been operating.

BIOTIC ADAPTATIONS

PARENTAL BEHAVIOR AND KIN SELECTION

In order to get away from words with highly charged connotations, Williams has suggested that adaptations that are beneficial for a group and deleterious for individuals be called biotic adaptations, as opposed to the usual kinds of adaptations of individuals (organic adaptations). We can now examine some of the proposed biotic adaptations.

First we can examine some possibilities from nonsocial contexts. One whole class of possibilities here is involved with the reproductive process. If natural selection operates only in terms of direct intraspecific competition for leaving the most offspring in the next generation, then it is not clear why some organisms do not maximize their intrinsic rates of increase during the operation of selection. Birds, for example, are capable of laying considerably more eggs than they normally do. If one robs a bird's nest, the hen will deposit a new clutch, and if one steals that too, she will again deposit a new clutch, and so on. Why, if she is competing with other birds to leave more offspring in the next generation, has selection produced a behavioral response inhibiting the production of eggs at a very much smaller total than she is physiologically capable of producing? Why has selection not maximized the intrinsic rate of increase in birds? Surely that should be the natural outcome of intraspecific competition in terms of reproductive success. David Lack, of Oxford University, has studied this situation in a number of species of birds and has found it to be a case where selection is producing an *optimum* clutch size, in the following way. Birds that deposit less than the optimum number of eggs simply do not produce as many offspring as those that deposit more eggs. Birds that produce clutches larger than the optimum, however, do not actually contribute a larger proportion of the individuals of the next generation. First, the parents have trouble feeding the numerous nestlings; they simply have not the energy needed to catch all the insects needed by their offspring. Many of the nestlings die before fledging. Once fledged, marking and recapture studies indicated that fledglings from large clutches do not survive as well as those from smaller clutches, presumably because they are weaker due to their insufficient early diets. In this way polygenic systems would be built up favoring the deposition of only as many eggs as can be fed by the parents once hatched (see also Chapter 6). This example shows that natural selection need not be expected to favor automatically the super-ficially obvious means for maximizing reproductive capacity. It has been cited, not so much as an example of a possible biotic adaptation, but as a warning against thinking too simplistically in terms of reproductive ca-pacity when considering intraspecific competition. That will be an impor-tant moral to keep in mind when thinking about biotic adaptations.

Among many ground-dwelling birds with precocious chicks, the hen will lead a predator away from her chicks by feigning injury and flopping about directly in front of the predator but never letting it get close enough to attack. Thus, it seems as if the hen were throwing herself into a danger-ous position in order to aid other individuals. Two observations can be made about this example which eliminate it as a biotic adaptation. First, the hen is not usually captured during the time she leads the predator away; she is in a hyperaware state in which it is *less* likely that she could be captured than if she were calmly feeding, unaware of the approach of a predator. Therefore the hen is not actually placing herself in serious dan-ger in this situation. Second, and more crucial is the fact that individuals as such are not what is important from the point of view being explored here. In protecting her chicks, the hen is protecting her genetic investment in the next generation, and is materially increasing the probability of spreading alleles for chick protection by herself contributing a greater pro-portion of the individuals in that generation. Natural selection occurs via differential reproduction. A behavior pattern such as that described here is not materially different from the feeding of nestlings discussed above; in each case certain behavioral traits have evolved in connection with re-production which aid in its realization. In fact, energy expended in these ways is not materially different in effect from that expended by mammals in producing milk and suckling the young, or from that expended by male spiders and insects when in some cases these are eaten by the female after mating, thus providing energy for the development of yolk substances in the eggs that will produce their own offspring. In all these cases a certain amount of energy is expended and personal risk of life is undergone as part of the effort to insure that the individual's alleles will form part of the gene pool of the next generation. There is no altruism here *in a genetic sense.*

We are now in a position to state a general proposition, that when-ever it is competitively disadvantageous to simply drop eggs into the environment (as do some insects, frogs, or fishes) various behavioral mechanisms (nest-building, care and feeding of the young, defense of the young, and other parental behavior) will evolve as integral parts of the reproductive function. These will sometimes involve behaviors that are altruistic in the sense of one individual aiding another. There is no problem with conceiving of ordinary natural selection producing these traits—that is, nothing beyond selection working at the level of individuals is needed to explain their evolution. The crucial point is that the individuals aided by another individual are genetically his close relatives. Statistically they tend to have the same alleles as he does at most genetic loci. Alleles that tend to favor altruistic behavior patterns can only spread in a population if the individuals that are aided by the altruistic behavior themselves possess those alleles. An individual that aided unrelated individuals to survive and breed could only do so at the expense of its own offspring (or its own alleles, including those that enhanced the altruistic behavior) if, that is, the population is at carrying capacity, as most populations seem to be. Thus, we can see that within the context of small primary family groups there is no problem evolving certain kinds of altruistic behaviors by natural selection. This process has been called kin selection by several authors. While kin selection can produce altruistic behavior (and a whole range of emotional states associated with affection on which the behavior is based), there is no genetic altruism involved.

SOCIAL CONTEXT

Passing on now to possible biotic adaptations in the context of social situations, we should note in advance that the social groups of many animals consist of closely related family groups, so that kin selection must often have played a part in the evolution of any altruistic behaviors found in such groups. Some supposed examples of altruistic behavior result from confusing individuals with populations, as in the famous case of the honeybee, where a worker, because it has a barb on its stinger, frequently dies after stinging, when the stinger is ripped out of its abdomen. Such an individual sacrifices its life in defense of the hive. It is important to realize that the worker bee is not a bona fide individual in the sense that most organisms are. The workers are nonreproductives; the queen in the center of the hive is the only reproductive individual. The workers are in a sense robots constructed by the queen to perform certain functions, among which is defending the hive. In fact the whole structure of the social system of many social insects is more closely analogous to that of an individual organism than it is to a population of social animals. The unit of selection is the hive, not the individual bee (unless that bee is a queen or a drone), but the hive is not a real group because it is made up of nonreproductive individuals mostly having identical or very similar genotypes.

Phenomena often cited as altruistic traits are the various warning signals produced by individual social animals at the approach of a predator.

As mentioned above, the proposition is made that if the warning signal attracts the attention of the predator, the giver of the first signal would be at a selective disadvantage in respect to the other members of a population. First, such an adaptation could evolve by natural selection in a social system composed of closely related kin. Second, warning signals of some kinds, such as calls in primate troops or in flocks of birds in forest situations can be seen as adaptations equally for distracting and misleading a predator. Many types of warning calls have ventriloqual qualities and do not lead to easy location of the caller. Frequently the first call immediately elicits calls by all members of the group from all directions resulting in a tumultuous din, as with chimpanzees. In other words, it is probably not true that the giver of the first warning call places itself in any special danger for having uttered the call. Another type of warning signal that is more easily interpreted as dangerous to the signaller is that found in the white-tailed deer and cottontail rabbits of eastern North America. Here, as an individual turns to flee, it raises its tail exposing a white rump and a white ventral portion of the tail. This appears to act as a danger signal for other members of the population, and could certainly attract the attention of the predator as well. Since many predators have evolved hunting methods involving fixation on one member of the prey population as a means of avoiding being distracted by defensive adaptations of the prey, being the first individual in the group to become visually obvious could indeed be dangerous. In connection with this example we may point out that while a trait may currently function in some manner, it is not necessary to consider that it was originally selected for in connection with that function. On this basis one can consider, for example, the tail flashing as originally a by-product of nervous tension, while the white rump region could be explained as having been favored by sexual selection. If this were the case, but it was also true that individuals that tended to flee first were more often taken by predators, could there not arise a counterselection against white rumps? Suppose, as is certainly reasonable, that the speed and evasive abilities of the individual prey organisms enter importantly into this picture. If all individuals tend equally frequently over a period of time to give the signal first, then selection will be primarily focused on the ability to run and dodge. If an individual tends to give the signal more frequently than others, he will incur a selective disadvantage on *that* basis, not on the basis of whether the signal is given or not; that is, such a situation would not select against the signal itself, but against giving it as the lowest intensities of stimulation. The discussion so far is in the context of the signal being simply a byproduct of sexual selection; in that context it could be maintained because sexual selection continues to favor it, while the disadvantage incurred in terms of attracting predators is not of a kind that differentiates between individuals on the basis of the presence of the signal per se. An interesting possibility is that the warning signal is partly a device to place running prey between the giver of the signal and the predator inasmuch as the other members of the prey group do not know where the predator is when they are startled by the signal. By chance some of them may rush headlong towards the predator (say, toward their burrows if they are rabbits), saving the giver of the signal. Of course, if the signal were really selected for originally as a warning device, it could easily have been produced by natural selection in a kin group situation—but it is hard to imagine how this could happen in a group of unrelated individuals. Summing up on warning signals, then, two points can be made: (1) natural selection in the form of kin selection can favor such signals even if their production entails risk for the individual giving the signal and (2) it is not at all clear whether it is necessary to consider the issuance of warning signals to entail any special risk. Risk must be demonstrated in each case. Thus, there is no need yet to invoke group selection as a process different from natural selection in this connection.

Another interesting possible example of a biotic adaptation is the system of defense from predators shown by some cattle—for example, the musk ox. Attacked by a pack of wolves, the bulls turn to face the predators with their horns lowered, closing ranks flank to flank, while the calves flee behind them. The cows also tend to lower their horns toward the wolves, but keep near the calves, and so move off in the direction in which the calves went. But they do this some moments later, so that they are in a position between the defending bulls and the running calves, forming a sort of second line of defense (Figure 12-1). In this way, the entire group

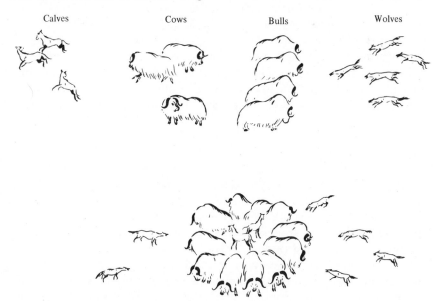

Calves   Cows   Bulls   Wolves

Figure 12-1   Two supraindividual defense patterns assumed by musk oxen.

assumes a superstructure related to defense, functioning in a way that would be beyond the power of an individual. The bulls appear to defend the cows and calves, and the cows to defend the calves. First, the sizes of musk ox herds are small and that suggests that the situation is one in which kin selection could operate, so that if a bull is sacrificed defending the herd it could die without losing any selective advantage in the process. However, the assumption that the defense position of the bulls is dangerous for them is probably not true. A running animal pursued by a pack of faster predators is in a much weaker position; continued slashing at its flanks can weaken it, and it can be hamstrung relatively easily. It is almost certainly safer for a large male musk ox to face the wolves with his horns, keeping his exposed flanks away from them. Looked at from this point of view, the action of the bulls can be interpreted not as defense of females and young, but as self-defense. The females also would benefit from similar behavior, but undoubtedly do have strong emotional ties to their calves. Their behavior can be interpreted as a compromise between impulses for self-defense and for protection of their young. They first turn to face the attackers, then follow their calves, then turn, then follow, alternately submitting to one impulse and then the other. For the calves, of course, the best thing to do in the current situation is to run. Thus, we can see how an adaptive supraindividual structure can result from a situation where individuals behave in ways best for themselves with no reference to the other individuals also contributing to that structure. Because the supraindividual structure turns out to be in fact adaptive is no reason for assuming that selection operated directly to produce it. Nonaltruistic behavior can be naturally selected because it is of direct use to the individual. If in the process an adaptive superstructure is produced, so much the better. Behavior favoring that superstructure but detrimental to the individuals making it up could not, however, continue to maintain the superstructure for reasons discussed above, except in a kin situation or unless some force such as group selection were at work.

SUPRAINDIVIDUAL STRUCTURE

This example leads into other possible examples of biotic adaptations which do not have to do with altruistic behavior as such, but are involved with the production of supraindividual adaptive structures. If selection is working directly to produce a multi-individual structure regardless of its implications for individuals, then it cannot be any sort of natural selection, but would have to be group selection. In this formulation, the group that develops traits allowing certain kinds of cooperative behavior would tend to survive longer than a group in which these traits are not favored. We have already dealt with one example, that of the musk ox. Fish schooling has been cited in this context as well. It is clear that a school of fish has properties that do not accrue to individual fishes. Not much is known concerning the possible adaptive functions of schooling. Is swimming more efficient in these surroundings? Can a school mimic a larger fish from a distance, thereby avoiding certain kinds of predators? Schooling certainly tends to occur when a group of some sorts of fishes is attacked. Perhaps it has no supraindividual meaning at all. If each individual simply tries to place another individual between himself and the predator(s), a school would automatically form. After all, there is little place to hide for pelagic fishes—the kind that usually show schooling behavior. Also, for animals that simply shed ova and milt together in a group, schooling is an obvious result of mating.

Cooperative group hunting has also been cited as something qualitatively different from individual hunting, and some reports concerning lions, Cape hunting dogs, and wolves at least imply that there is some form of planning involved in the hunt. Careful observations have failed to substantiate these tales. It seems that the only cooperation that can be discovered in group hunting by these sorts of animals is an agreement to simultaneously hunt the same prey animal. Beyond this, each individual appears to hunt as if it were alone, trying to drag down the prey itself. In this example, then, there really is no superstructure at all, simply a group all doing the same thing at the same time in the same place.

Another example involving a supraindividual structure can be seen in the spacing of individuals during colonial nesting in a heron rookery. Each pair has a small territory in which their nest is located. Any individual other than the mate who passes into this territory is attacked and driven off. Young fledglings are also not welcome in any pair's territory, including that of their parents. They occupy narrow spaces in between the territories. This places most of them in relatively safe areas in respect to attack by most sorts of predators. Their parents are no longer concerned with protecting them, but will attempt to drive off predators venturing close to their nests. If successful, they will also "unintentionally" have protected any fledgling occupying the spaces around their territory. The fledglings have no adaptations to keep them still in between the territories, but wander about. However, each time they set foot onto a territory, the owners drive them off. Thus, the fledged young are statistically most likely to be met with in a safe place between several territories, resulting in a genuine superstructure situation like that cited above in connection with the musk ox.

It seems clear that supraindividual phenomena, or superstructures, can grow out of the behavioral interactions of organisms in a group, particularly in social animals. It does not seem clear, however, that we must invoke something other than natural selection in order to explain how the interactions could come about. The primary phenomena here are these interactions between individuals, and these are what must be explained in terms of natural selection. In a very real sense the superstructures are simply byproducts or secondary effects of interactions between individuals. That the secondary effects observed tend to be adaptive simply reflects the fact that if they had been inadaptive, natural selection would have operated against the *primary* interaction, even if that in itself had had positive adaptive value (as in the case of the light-eyed mutant in *Tribolium* discussed in Chapter 11). What this amounts to is simply a more sophisticated view of the external environment. If the primary behavior of some musk oxen in response to attack by wolves was adaptive but resulted in a superstructure that was not, selection would simply remove from the population individuals that exhibited this particular class of adaptive primary behavior, leaving the field for other adaptive behaviors not associated with secondary ill effects.

To sum up this chapter so far, examples of altruistic behavior and other sorts of biotic adaptations, such as supraindividual structures, frequently turn out to be spurious on closer examination. If they cannot be explained away in this fashion, then it is usually possible to associate their evolution with either a kin group situation, or with a more fundamental primary adaptive trait, reevaluating the phenomenon in question as a secondary byproduct. In this way it is possible to avoid invoking a new and separate sort of selection from natural selection, thereby needlessly complicating our conceptual framework. (It is worth noting that at the level of prebiotic systems one may visualize a kind of group selection acting, for example, during the time when systems showing mortality of individual units became more successful than "immortal" systems that did not; this would have been intergroup selection. From this, more superficially, it is obvious that lineages of organisms may go extinct, and one would be justified in stating, for example, that ichthyosaurs as a group were relatively less successful evolutionary than crocodiles. The point here is that in this latter case the mechanisms involved in adaptation do not operate on the group itself, but on individuals within it.)

The question still remains whether we can find any genetic altruism or other kinds of biotic adaptations in man, and we will return to that question briefly after a discussion of the evolution of sociality.

SOCIALITY AND STATUS

It is clear that certain kinds of animals derive various sorts of advantages from group living, and that, therefore, sociality will be able to evolve among them. For instance, certain sorts of predators, such as dogs, that hunt by running down their prey could derive benefits as individuals from hunting in packs. This would make possible, among other things, the taking of large game animals in periods of scarcity. On the other hand, it would be less likely that cats, for example, would evolve social systems because their mode of hunting is to creep up on an unwary prey animal and leap on it from ambush. No advantage could accrue from group hunting in this situation. Only one of the living cats, the lion, shows any tendency toward sociality—and even lions do not usually hunt as a group. Group protection of the young is another obvious gain acquired by social living.

While there is no problem in understanding the adaptive nature of sociality for some kinds of animals, certain adaptations are needed in order to allow group living to occur. Without concerning ourselves as to which came first, we can recognize that group living depends in vertebrates on a status system, sometimes called a dominance heirarchy or a pecking order. Status is one of a group of values or goods (conventional rewards) which are not biologically real in the sense that food is, or mates, or an escape from a predator. Behaviors involved with obtaining, maintaining, and recognizing status are not directly involved with acquiring the three major biological necessities. However, the status system allows group cohesion and therefore indirectly facilitates the acquisition of biologically real values. Without a status system recognized by all the individuals of the group, there would be continual internecine strife regarding food, space or mates. Aggression is a characteristic of most vertebrates, and it must be subverted or channeled into constructive paths before individuals can live peaceably with one another; the dominance hierarchy is one way of accomplishing this. In such a system one individual dominates another, who in turn dominates another, and so on, so that every individual has a specific status of either dominance or submission in respect to every other individual in the population. Frequently males and females have separate dominance systems; sometimes the more dominant males dominate all individuals of both sexes. The immediate biological meaning to individuals of such a system is minimal, resulting in the dominant individuals obtaining the choicest fruits or bits of meat, the best resting places, and so on. Every individual obtains sufficient goods in such a system, but some obtain more choice tidbits. The real biological importance of this system lies in the fact that once squabbling over biologically petty matters such as the choicest apple can be peaceably and quickly obviated by one individual giving it up to another willingly, the animals are now in a position to engage in group hunting or in group nesting, or in group protection of the young. The individuals in the group obtain more selective advantage by being members of the group than they lose by occasional submissive behavior, and so submission on the part of an individual need not be seen as altruistic behavior. Furthermore, the dominance hierarchies are open-ended and fluid. An individual's status is in continual flux, both upward and downward; any male has some chance of becoming the alpha-male, provided only that he can defeat the current one in battle. If not, there is likely to be some young male he can dominate, and so he may satisfy certain psychological needs one way or the other. While mobility in the system importantly involves combat, this occurs only intermittantly so that at any given moment there will be cohesion in the group. Furthermore, the fighting is not usually "to the death," there being submission signals that can be resorted to when one is losing a fight that will immediately prevent further damage.

The status system depends upon individual recognizability. In order to know to whom one is dominant or submissive, it is necessary to accurately recognize individuals. Interestingly, while most animals are genetically individual, it is only in connection with conventional goods (goals or values only indirectly associated with survival and reproduction) that individuality becomes a necessity. This is obvious in connection with status, and is sometimes true in connection with another important conventional good, territoriality. A territory belongs to, and is patrolled by, an individual or pair of them. In this case, however, it is frequently not so much another specific individual toward whom one acts aggressively, as toward any other individual at all. The ultimate biological meaning of territoriality appears commonly to be an assured food supply and nesting place, but the direct object of territorial behavior is simply a place or a border that has no biological meaning of its own. The fullest flowering of individuality, however, is in the social context. Nonsocial animals, like the lynx, probably have not the same degree of individuality as do social animals because there is no function for such a phenomenon with them. For sexual organisms who are already genetically individuals, the evolution of salient individual traits such as those found in the human face is a small matter; perhaps more important would be the evolution of the ability to perceive the individuality already present.

ANIMAL COMMUNICATION; MULLERIAN MIMICRY

Another precondition for the evolution of status systems is a system of communication. Animal communication is widespread, and is well developed in social animals. It is based on signs or signals, which can be considered to be part of an individual's biological reaction to situations, occurrences, or other individuals (see Chapter 15). Some of the signs are inherited *in toto;* others are learned. The stereotyped responses they evoke are mainly inherited. As a rule the communication takes place between individuals of one species only, with each species having its own language. There are exceptions to this; sympatric forest-living monkeys of different species may all have the same warning calls, and the many species of cichlids in Lake Nyassa respond to each other's territorial signals. Inter-species communication is a problem related to Mullerian mimicry.

In Mullerian mimicry, several noxious species of potential prey organisms have more or less the same warning pattern, usually a bold design of spots, bars or stripes (aposematic coloration, Figure 12-2). An example would be the barred abdomen characteristic of many species of bees. Ex-

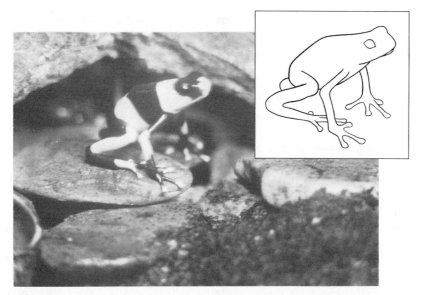

Figure 12-2   Columbian arrow poison frog, *Dendrobates sp.*, showing a bold color pattern (aposematic coloration) common to noxious animals. The frog is a vivid red and black. (Photo by the author.)

periments have shown that many kinds of birds, and also bullfrogs, are capable of learning to avoid any insect (or even models) with barred abdomens after being stung a number of times in attempts to eat bees. This means that after attempting to eat, say, two individuals of bee species A, and one of bee species C (some of which it may have maimed or killed), a bird will now avoid attacking any barred bee at all, including members of bee species B, D and E. The last three species have become protected from this individual bird even though none of their individuals were attacked. Is this altruism on the part of species A and C? No, because the next bird-instruction will be done by randomly chosen individuals from all five species, and so over a period of time each species will contribute its share of "teachers."

How can it come about that so many noxious organisms in a single region show similar aposematic patterns? Given that certain predators hunt visually, it is clear that individuals of an unpleasant-tasting species that tend to have bold color patterns will have a slight advantage after being attacked and not killed (as is frequently the case with noxious prey). The bird that attacked it will not do so again—at least not immediately, having associated the color or pattern with an unpleasant experience. In this way bold color patterns will probably acquire slight selective advantages. Now, as the number of individuals showing the *same* bold pattern increases, the selective advantage of that pattern will increase in this species by a positive feedback, because now one individual's injury or even demise can increase the selective advantage of others with the same pattern—in addition to the original effect cited above. The same process will be going on in neighboring noxious species. In order to explain why the same kind of pattern is selected in more than one species, we need only extend the above intraspecific argument by noting that in one kind of organism, say in bees, there are genetic and even phenotypic limitations on the kinds of bold patterns that can be produced. The barred abdomen is a highly probable pattern in this group, and so simply by chance some individuals in each bee species will acquire that pattern by recombination or mutation. Which particular general pattern comes to predominate in a particular group is a function of the probability of its being produced in that group. Patterns only rarely produced in a species will not likely become the species pattern. It is important to realize that Mullerian mimics are not perfect mimics (as Batesian mimics frequently are); all that appears to be required is that the pattern remind the predator somewhat of a pattern it previously had a distasteful experience with. Beyond producing sympatric species with *similar* patterns, natural selection does not proceed in this case. Ultimately each species arrives at a condition where all of its individuals have some one bold pattern, there being neither polymorphism nor limitation on the numbers of mimics in Mullerian mimicry.

Like Mullerian mimicry, interspecies communication involves a kind of coadaptation between different species. This does not imply any genetic continuity between the species; it is a coadaptation similar in a way to character displacement, discussed in Chapter 6. In that process the phenotypic properties of one species form an important part of the external environment of another species, so that the second species adapts directly to characteristics of the first. Symbiotic or mutualistic relationships (as discussed in Chapter 3) obviously involve similar concepts.

In the case of interspecies warning calls in monkey troops, individuals of one species that tended to flee or become alert when members of another species gave a warning signal would have been selectively favored over individuals that ignored such warnings. This would be equally true of members of the other species. In this way, individuals capable of responding to a wide range of warning calls would be favored in each species. However, it would not do to bolt at every sort of call one heard; there must be some degree of specificity to the warning call, as there is for all animal communication signals. As in the case of the bees, we can invoke the hereditary connections between the monkeys and note that they descended from a species probably possessing a warning call that, even with subsequent independent modification, would quite likely remain fairly similar to the original, and distinct from other classes of calls in all the species.

The main function of animal communication in vertebrates, however, is to inform members of a single-species population about the state of the environment, and, more importantly, about the emotional states of other individuals in the population. In connection with the latter, communication is an important tool for compelling the behavior of other individuals in directions required by the sender of the communication signal; thus, an anthropoid infant or a chick cries, inducing its mother to see to its needs; the female stickleback stimulates the male to continue courting her by turning her head upward; a wolf, losing a dominance struggle with another, exposes its throat to the winner's fangs, compelling him to desist from further fighting. A stereotyped sign or releaser commands or releases a stereotyped response from another individual. The result of this sort of communication is a reasonably peaceful and efficient group where individuals respond to each other's needs most of the time. The importance of species-specificity in both signal and response during courtship behavior is obvious and will be discussed in Chapter 15.

EVOLUTION OF SOCIALITY

At this point we have enough background information and terminology to suggest the sorts of steps that probably were involved in the evolution of vertebrate sociality. Both the status system and the signals involved with communicating status information probably have their foundation in the relationship between parents and offspring in animals that tended their young, as do all social vertebrates. Many submission signals have been traced to those used by the young, as, for instance, the signal cited above used by wolves to signify that they no longer wish for the moment to contend for dominance. That exposing of the throat to an aggressor is found in wolf pups, and is used by them to signal submission toward adults. Warning signals and other altruistic behaviors no doubt evolved in connection with care of the young. Thus, the primary family level of organization involved altruistic behavior toward the young associated with a necessary submission of the young to the will of the adults, and evolved various communication signals that facilitated this relationship.

The positive adaptive values of group hunting, group protection and defense, and other possibilities for group cooperation in some sorts of vertebrates, probably led to the establishment of social groups by inhibiting aggressive forces that tended to drive individuals apart after weaning or fledging, resulting in an extended family situation. The communication signals used and further elaborated were those established earlier in parent-offspring relationships, while the statuses were extensions of the parent-offspring statuses (dominant-submissive). This much seems fairly clear, certain traits developed in family situations were preadaptive to life in extended families. Kin selection could continue to operate at the extended family level to produce any further altruistic traits needed to weld together a cohesive group—for instance, the grooming behavior found in anthropoid societies, which undoubtedly forms an important part of the "social cement" of these groups. On this level, then, we would find the evolution of the conventional good, status, and the associated evolution of functional individuality, and thereby, the evolution of sociality.

The next step in this conceptual program would be the evolution of groups of relatively unrelated conspecific individuals *as* social groups. There are many cases where fairly distantly related conspecific individuals flock or herd or school together. Some of these are transitory, involving migrations, and so are not social systems of the kind we are here concerned with. In the majority of cases there is no evidence to suggest any special behavior toward other individuals—that is, as was done above in connection with fish schooling and the defense reactions of male musk oxen, behaviors in these groups can easily be explained in terms of individuals behaving independently of other individuals near them. These groups appear to be crowds rather than societies. This is true even if the crowds have functional properties not accruing to any single individual. For these groups to qualify as societies they must give rise to superstructures based specifically on social behavior—that is, behavior directed toward other individuals *as individuals*—not only as members of the same species. On these grounds it is possible to eliminate as social groups all groups of relatively distantly related conspecific individuals except for man.

Let us examine a case in more detail. While the behavior of individual fishes during schooling can be interpreted as attempts to be in the center of a crowd, there is no doubt that a number of special adaptations are involved, and that not just any sort of fish will school. There are, for instance, certain behavioral responses involving visual perception of other fishes. Experiments have shown that in a test cylinder a fish will swim head to head with models and not tail to head. If the direction of the models is reversed, the test subject also turns round. There are also chemical communication signals involved in the schooling behavior of some sorts of fishes, notably in one species an alarm substance which is secreted by a fish perceiving danger which diffuses in water to the next individual, who reacts by swimming closer to others and also by in turn secreting the substance, and so on. Many species of teleost fishes produce croaks, squeaks, and squeals which may function as communication to increase group cohesion. In order for some sorts of schooling phenomena to exist, such as the precision turning of a whole school of squids at breakneck speeds, some special adaptations must be involved. The point is that none of these adaptations need be considered to be directed toward other single, specific *individuals* as such. Each fish has adaptations that allow it to survive in a crowd; conspecific (and even in some cases supraspecific) crowds are part of the external environment to which every individual must be adapted. The alarm substance, for instance, could be considered a warning signal, but in the open ocean, where there is nowhere to hide from a predator, a substance that makes a number of other individuals swim closer to oneself can equally be interpreted as a device for placing other individuals between oneself and the source of danger. (In one case the alarm substance is apparently a by-product of a wound, and when detected by others signifies danger to them. It may be better not to treat this as communication at all.)

There is no doubt that this treatment of fish schooling is inadequate, and that the phenomenon is more interesting than here suggested, but it is not an example of social behavior.

PRIMATE SOCIALITY

Probably no primate other than man lives in societies composed of relatively distantly related conspecific individuals, and even in man this has been a relatively recent occurrence. During most of the evolution of the lineage leading to man, the social groups were probably kin groups and extended families such as those still found in "primitive peoples" today. Outbreeding is characteristic of these groups, but it does not disrupt the relatively stable gene pools because it usually involves a significant proportion of what we would call first cousin marriages. Also, the geographic distances between groups related by marriage are characteristically not so far as to preclude fairly close genetic relationships on the whole. Other primate societies are probably at a similar level of complexity, for instance, those of baboons.

Baboon societies are fairly complex, involving more than one adult male. In one sort, males and females have separate dominance hierarchies, but the most dominant males will dominate members of both sexes. One male, the alpha-male, is the dominant individual of a small coterie of dominant adult males, but the coterie acts as a unit of dominance vis-à-vis non-member males. Dominance is established by fighting, and as individuals get older they sooner or later become subordinate to younger dominant individuals. The dominant males decide when to move and where to sleep, they prevent internecine squabbling between subordinate individuals, and they do the larger part of the defensive fighting-off of predators, although all adult males will join in defense of the females and young. The prerogatives of dominant males include the choicest tidbits and stations, more frequent grooming, and more importantly from the genetic point of view, they are sought out by estrus females at about the time the latter are ovulating, and so tend to sire most of the offspring.

Besides dominance hierarchies there are other personal relationships between individuals that resemble friendships, the individuals involved forming grooming clusters and travelling together. Such friendships may occur between any individuals at the same position in the dominance hierarchy. Grooming clusters are formed as well about particularly attractive classes of individuals, such as a dominant male or a newborn infant and its mother. Even a dominant male is attracted to and will groom newborns.

Grooming is a very important social activity, and much of the baboon's leisure time is spent at this mutual activity. It is a primary source of pleasure, as well as being adaptive in the obvious sense of removing ectoparasites such as ticks.

Group functions importantly include protection of the young from predators. Upon attack, the females and young speed their progress away from the predators, while the adult males fall back and place themselves between these two groups. The dominant males will predominantly do the threatening and actual fighting, but all may join in this. Other group behaviors include occasional group hunting by the adult males, who may kill young artiodactyls or other small game. Baboons, however, are predominantly herbivorous.

In any one area there will be more than one baboon troop. They remain quite separate, and have minimal contact with each other. Probably there is occasional exchange of individuals between these troops, or perhaps merging of troops if they get too small, but over short periods of time they are endogamous. Since intense inbreeding would probably result in inbreeding depression, there must be significant gene flow between troops over longer periods of time. However, it is clear that genetic relationships in a troop must be fairly close, so that kin selection could easily be operative.

This thumbnail sketch of baboon society is presented as an example of a fairly complex vertebrate social system, not for any relationship it may have to man's society, which is probably minimal—the differences between the two being greater than the similarities. This sort of social system is based on interactions between recognizable individuals *as individuals* in contrast to insect "social systems", or probably, schools of fish.

Human society

Early human societies may or may not have been like baboon troops in any particular respect, but they must have had the same basic sort of structure, based on recognizable individuals, involving a dominance hierarchy, and being composed of relatively few, closely related, individuals (a kin group). In a number of different cultural lineages, continued social evolution eventually produced a multi-kin group social system that is characteristic of civilization. In this situation many individuals spend much of their lives among others not especially closely related to them genetically. Is genetic altruism characteristic of civilized man? Has group selection been operative between civilizations?

Clearly, personal self-sacrifice is part of the moral directives of many civilizations, including our own. Just as clearly, there are known instances of personal self-sacrifice that led to low Darwinian fitness. One can cite here the suicides of fanatics, but possibly also the complete devotion to their work that characterizes some people to the extent that it lowers their

Darwinian fitnesses considerably. These individuals tend to produce relatively few offspring, leaving the breeding to the less notable. Yet there is no evidence that there are fewer such individuals today than there were in the past. Significantly, there is also no evidence that the numbers of such individuals is increasing. It would seem that there always are just a few such individuals around in a large population. In view of this, it is probable that they represent part of those individuals selected against each generation by normalizing selection and that are generated by recombination of generally favored alleles. Other such individuals may be schizophrenics, or asthmatics, or other incapacitated individuals. The process of recombination every generation would create combinations of alleles that are inadaptive for any given polygenic traits; selection would tend to favor the current norm or mean or optimum for each trait (Chapter 9). Culturally superior individuals need not necessarily be biologically superior. Human cultures frequently have characteristics that are not biologically "sound." If the production of a certain class of biologically unfit individuals was a necessary outcome of genetic recombination each generation, there is no reason why a culture could not benefit from these individuals, providing only that it did not itself become inadaptive in the process.

We have imagined that altruistic behavior patterns evolved in the lineage leading to man in connection first with the immediate family and later in the context of the extended family. Thus, we can assume that there are in the human gene pool alleles for many traits that tend to favor such behavior. In terms of altruistic behavior *qua* behavior it makes no difference whether the individual aided is closely or distantly related. The same impulses apply, the same emotional gratifications are derived. It should not be surprising, then, that any individual occasionally sacrifices his own interest in favor of nonrelated individuals. But even in the human situation, complicated immensely by cultural overlays, an individual that does this could as a result produce less offspring than those he aided, especially at carrying capacity, and so the tendency to help others would be selected against. Most civilized men no doubt occasionally direct altruistic behavior toward distantly related individuals. This is probably made possible by the cultural systems within which man lives and can possibly be seen as a selectively neutral (or positively selected?) extension of behavior patterns whose main function is (or was) to insure reproductive success. It is no doubt the cultural context that renders these actions selectively neutral (or selected for?), but the topic is large and complex and deserves a book of its own. If there is some way in which altruistic behavior toward unrelated individuals has become selectively advantageous, it would have to be in connection with some form of group selection.

Darwin made a relevant argument in connection with the problem of the human reproductive capacity. In a population becoming overcrowded, one could appeal to the consciences of individuals to limit their fertility willingly for the good of the population as a whole. As with all traits, the presence of "conscience" varies in the population. Individuals with conscience limit their reproduction, while those without conscience do not. The result is that conscience tends to disappear from the population and the population problem remains unsolved. In our terms, we could say that alleles that tend to promote conscientious behavior (or reasonableness, or concern) will gradually be weeded out of the gene pool by this process, or, if conscientiousness is only an occasional possibility in certain rare recombinant individuals, these alleles will not acquire selection coefficients of lower magnitudes. Perhaps for these reasons even in human populations genetic altruism cannot exist. There may always be individuals who show genetic altruism, but there will never be many of them, for they will remain rare recombinants, continually selected against. Such individuals can be, and are, useful to society. It is one characteristic of human cultures that they provide a means by which these individuals can contribute to their populations even if they are biologically unfit; they may be great musicians, or legislators, or scientists. Indeed, in the cultural context such individuals can contribute importantly to the future of their populations. It is interesting to speculate on whether cultural evolution could proceed without these sorts of (frequently biologically unfit) individuals. In this connection, it should be noted that not all culturally important individuals are childless (for every Beethoven there may be a Bach), but such a category of individuals does exist, and it is these that may show genetic altruism.

It should be pointed out, too, that the fact that most human populations have been continually increasing in size for at least 200 generations may be important here. Selection automatically eliminates individuals that aid unrelated individuals to reproduce only at carrying capacity population levels—that is, when the environment is saturated with individuals. In a growing population it is not automatically true that any energy given to the reproduction of another individual must be deducted from that available to the altruist himself. The arguments developed in this chapter apply to animal populations fairly generally because these are considered on the average to be mostly at or very near carrying capacity population levels. However, altruistic behavior would not be especially more fit even in growing populations; it would at best be selectively neutral, possibly linked to some definitely advantageous trait such as child care, and so spread in the population (which must not be too large) by genetic drift.

SUMMARY

To sum up, there is, except perhaps in man, no compelling evidence for genetic altruism or for biotic traits. There is, then, no need to invoke group selection in the evolution of any known system or trait (again with the possible exception of civilized man). Altruism toward other individuals has evolved in connection with parental behavior and, probably later, in connection with extended family kin groups such as those that form the basis of typical vertebrate social systems. This altruism does not involve genetic altruism.

REFERENCES

Fisher, R. D., *The Genetical Theory of Natural Selection,* 1929. Reprint Dover Pubs., New York, 1958, 291 pp.

Ford, E. B., *Ecological Genetics.* Wiley, New York, 1964, 333 pp.

Hardin, G., "The Tragedy of the Commons," *Science,* **162:**1243–1248, 1968.

Sheppard, P. M., *Natural Selection and Heredity,* 1958. Reprint Harper Torchbooks, New York, 1960, 209 pp.

Southwick, C. H., ed., *Primate Social Behavior.* D. Van Nostrand, New York, 1963, 191 pp.

Wickler, W., *Mimicry in Plants and Animals.* McGraw-Hill, New York, 1968, 255 pp.

Wiens, J. A., "Group Selection and Wynne-Edwards' Hypothesis," *American Scientist,* **54:**273–287, 1966.

Williams, G. C., *Adaptation and Natural Selection.* Princeton University Press, Princeton, N.J., 1966, 307 pp.

———, *Group Selection.* Aldine-Atherton, Chicago, Ill., 1971, 210 pp.

Wynne-Edwards, V. C., *Animal Dispersion in Relation to Social Behaviour.* Oliver and Boyd, Edinburgh, 1963, 653 pp.

CHAPTER 13

Chance in evolution

We have already explored several important ways in which chance phenomena influence evolutionary processes, even though the sole mechanism operating in those processes is natural selection. In Chapter 4 the concept of preadaptation was explored and found to be crucial to the origination of new adaptive zones and concomitantly, of new higher taxa. Selection alone would result only in increased specialization and speciation within a given adaptive zone. Chance alone determines whether a lineage will find itself possessing the requisite equipment in an environment that allows its members at some point to begin extending their niches in such a way as to begin forming a new niche. *Archaeopteryx* was a very specialized archosaur that happened to possess the means for exploring a totally new way of life. Such cases lead to the origination of new higher taxa because, in the absence of competing forms, entry into the new adaptive zone is followed by an explosive adaptive radiation. The importance of this kind of chance effect can hardly be overstressed.

In Chapter 7, chance phenomena were found to be the source of all genetic variability, in the various types of mutation and in recombination. And, of course, Darwin first pointed out that no evolutionary change can occur in the absence of variability. Perhaps it need hardly be pointed out that from the point of view of evolving organic systems, long-term environmental change is in principle capricious since such a system can come to predict only environmental changes that repeat themselves with a periodicity shorter than a single generation. That is, individuals will have mechanisms that have evolved in the lineage allowing them to adapt as individuals (for example, by changing their physiological responses between summer and winter) only to predictable, short-term changes. All long-term changes (or all unpredictable changes) will be met by evolution or extinction of the deme or clone.

In this chapter we will focus on chance events that may affect gene frequencies in the gene pools of populations. Natural selection, gene flow and mutation are the main phenomena that are felt to initiate gene frequency changes. Under certain conditions, however, chance may enter importantly into the transmission of genes from one generation to the next in the form of a sampling error. This may assume importance when population size is small and/or in reference to selectively neutral traits or alleles. Both of these conditions will be examined in connection with genetic drift, the process of randomized (nonsystematic) changes in gene frequencies from one generation to the next.

SMALL POPULATIONS

In discussing population size in this context it must be pointed out that what is referred to is effective population size (N_e). First, it is clear that not all members of a population breed in any given season except in single-celled asexual organisms forming a clone. In other cases, some members of the population will be too young to breed (they may even be larvae). Thus, we focus on the breeding population only. Here it is found that females, in particular, in many sorts of organisms, do not breed every season because it takes much time and energy to form ova laden with food materials. Also, it may happen that there are more sexual pairs than available nesting sites, as happens in birds nesting in tree holes, in which case some sexually mature individuals do not breed because there are not enough nesting sites available for all sexually mature individuals. Thus, the effective population is that portion of the breeding age population that actually comes to form the zygotes from which the next generation will develop.

For the Hardy-Weinberg prediction (Chapter 6) of the next generation's gene frequencies to hold at equilibrium (no selection, no mutation, no gene flow), the population size must be infinitely large. Then the gene frequencies will not change from generation to generation. However, in any population of finite size, it can be shown that the gene frequency of a given allele at a given locus, $p(A)$, will not be the same in the gametes that will form the zygotes from which the next generation will be selected as it is in the total population. As just pointed out, not every individual in most kinds of populations breeds every season, and those that do will be a random sample from the point of view of the locus being examined. It can happen by chance, for example, that more *aa* individuals will breed than *AA* individuals in any given season. As an example of another sort of effect, during the maturation divisions of oocytes there is a presumably completely random choice of which allele at a heterozygous locus will come to be present in the ovum—the other one winding up discarded in a polar body. The fewer the number of eggs produced by the females (as in birds), the larger this effect will be, and this kind of sampling error will increase as the effective population size decreases.

Genetic drift is usually discussed in the following way: Given an effective population size (N_e) it can be shown that the gene frequencies of the offspring population will vary, by chance, according to the binomial expansion, $[p(A) + q(a)]^{2N_e}$, giving $2N_e + 1$ possible values of $p(A)$ in any offspring population. That is, each offspring population that is generated could have had that many different possible values for $p(A)$. The whole set of these possible values has a variance (which measures the variability or heterogeneity of the distribution of values) described by

$$\frac{p(A) \cdot q(a)}{2N_e}$$

(for asexual organisms the divisor would be N), showing that as the population size decreases it becomes less predictable what the frequencies in a given offspring population will be. Thus, while there are absolutely more possible different values that can be generated by a large population, most of them will be within a relatively small range about the mean. In a small population absolutely fewer different gene frequency values can be generated, but each one will be different from the mean to a less predictable degree.

As a rule of thumb, Sewall Wright has proposed that populations can be estimated to be "small" in the sense that genetic drift could be important in gene frequency changes at a given locus in them when $4N_e\ s$, $4N_e\ \mu$ (mutation rate), and $4N_e\ M$ (rate of gene flow) each equal less than one for that locus. It has also been suggested that a population may be considered "small" if it is less than the reciprocal of the selection coefficient of the allele whose frequency is being observed. It is, of course, very difficult to tell whether these criteria are met at any locus in natural populations. Some natural populations seem indeed to be absolutely small. House mouse populations, for example, may consist of four or five extended families of some 15 individuals each in a barn. Elephant populations can be made up of forty or so individuals. In fact, large animals need absolutely more space in terms of biomass to support them, so that all physically large species will tend to have relatively smallish population sizes. Certain isolated populations of humans are (and have been) composed of small numbers of individuals. In most cases, however, it is difficult indeed to discern the limits of demes in nature, and in only very few cases (such as cave populations) is it possible to say that they are definitely limited to some small size. The deme must be considered as defined by a high level of gene flow. If gene flow is restricted between two localities over a period of time and intermittantly reestablished, panmixis may still be present. Thus, isolated populations of pupfish in the deserts of western North America are periodically mixed when exceptionally hard rains (perhaps every ten years or so) reestablish water flow in the dry creek beds connecting the waters in which the isolates live. Such isolated populations may still be part of a larger deme composed of all of them together.

It may be recalled that, as pointed out in Chapter 6, some genetic death may be randomized even in very large populations. When a baleen whale swims through a vast school of krill, swallowing thousands in a few minutes, it is possible that the frequencies of some traits or alleles in that population will have a random component added to whatever other forces are affecting their gene frequencies, and that these frequencies, therefore, might drift to some extent. This is because certain traits—say, length of antennae, ability to taste certain substances, and so on—are not germane to, for example, escaping from whales. It may be noted that in this example, the large population sizes of krill could militate against random changes in gene frequency, inasmuch as a large sample from a large population could be representative. Another way in which large populations could be affected by genetic drift, at least at some loci, is via what has come to be called the "bottleneck effect." In this case a population of whatever size is drastically reduced in numbers by some environmental catastrophe so that, when the population begins to grow again, it will quite possibly not begin with the same gene frequencies, at least at some loci, that characterized the population prior to decimation; also in the next couple of generations effective population sizes will be likely to be "small." One kind of example is insect populations in seasonal climates. Characteristically, there is a very large decrease in numbers in the dry or winter season, so that the founders of the next season's breeding populations might not be a representative sample of last season's breeding population for certain traits *not having to do with overwintering*. This stipulation is important since it is clear that selection will be at work here distinguishing between the winter viability of various genotypes. Only traits not important in connection with winter viability will be expected to drift in this case. A further hypothetical example will make this point clearer.

Suppose we have a population of mice in a barn, and the farmer makes periodic attempts to eradicate them. At such times he employs mousetraps, cats, a shotgun on surprise visits, and so on. It is clear that during this environmental catastrophe some traits will become very important for viability—general nervousness, cautiousness, tail length, hearing acuity, fleetness, and so on. Just as clearly some other traits will be relatively less important—fur thickness, for example—which might, however, be scanned by selection more intensely at other times, as on cold winter nights. In other words, any environmental catastrophe temporarily emphasizes the importance of some traits above that of others. If the population is finite in size, and certainly if it is small, traits not germane to the current catastrophe will probably drift, in that survivors of the catastrophe, who will be the next generation's potential parents, will not have been selected in terms of them. Of course, selection may be reinstated on such traits before breeding occurs. In our example, if the farmer's attack is followed immediately by an exceptionally cold winter, fur thickness could be scanned by selection again prior to breeding, and the effects of prior drift on that trait would be minimized or suppressed. If, however, the population was very small, and most of them were killed by the farmer leaving, say, only 10 out of 100 mice, it is clear that there is a real chance that an allele could have its frequency reduced to zero. Thus, if we had 25 AA, 50 Aa, and 25 aa individuals, looking at a given locus having nothing directly to do with survival during the attack, it is clear that all individuals carrying A could have been eliminated. In this case, subsequent selection at a trait in which this locus was importantly involved would detect no differences between the survivors, and fixation of one allele would have occurred at random.

Notice that in this example, drift occurs during the life of the organisms between fertilization and breeding, not, as in the usual formulation of drift, only in reference to stochastic effects working on the gametes. If a trait or allele has a very small selection coefficient, or is virtually selectively neutral, either always or during short periods, its frequency can drift either in a gamete population or in a population of postzygotic individuals if those populations are finite in size, and especially if they are small. This aspect of genetic drift has been neglected, largely because it has not until recently seemed reasonable to consider that there might be selectively neutral or nearly neutral traits or alleles. Work with primary gene products has, however, suggested that possibility again.

SELECTIVE NEUTRALITY OF TRAITS OR GENES

As Wright has pointed out, genetic drift could occur (that is, a population would be "small") if $4Ns$ is less than one. This means that if a trait of gene is selectively neutral ($s = 0$ for all alleles impinging upon it or representing it), or if several alleles at a given gene are selectively neutral with respect to each other, gene frequencies could drift in effective populations of millions. It is, therefore, of some importance to inquire whether selective neutrality is a real phenomenon, and if so, what its importance may be in respect to evolution.

There is one class of genes and traits which undergo periodic absolute selective neutrality—those that are expressed only during a single stage in a life cycle. For example, frog tadpole hemoglobin chains are produced only during larval life. At metamorphosis loci coding for these genes are switched off and genes coding for adult frog hemoglobin chains are turned on. During egg stages, and in postmetamorphic stages, tadpole hemoglobin genes are not expressed and so selection cannot operate on them—they are selectively neutral. To demonstrate this we can do the following calculations: Beginning with 3000 eggs, mortality at egg stages would leave typically about 600 after hatching. Whatever the agents of this huge zygote mortality, none of them are concerned with tadpole hemoglobin, and so the frequencies of alleles at those loci could drift at this point away from what it was in the parents. Now a period of selection on these loci occurs during larval life, at a time when mortality is comparatively low. Selection will adjust the frequencies of alleles at these tadpole hemoglobin loci during this time. By metamorphosis we have, let us say, some 540 individuals left. Metamorphosis is a second period of large mortality and perhaps 180 individuals will pass through it. Again a period of genetic drift takes place concerning the tadpole hemoglobin loci, which are now switched off. Furthermore, all through adult life mortality will occur so that perhaps 15 of the original 180 individuals that metamorphosed will actually breed. This mortality is again accomplished without reference to tadpole hemoglobin genes, which may as a consequence experience drift in gene frequencies again. Thus, from the time of last expression of these genes, just prior to metamorphosis, to the next expression of them, just after hatching, most of the viability selection takes place on the frog population. It seems virtually certain that even in a fairly large population, this situation will result in genetic drift. Each generation of breeding adults will show slightly different frequencies of alleles at these loci, and these frequencies will be temporarily adjusted by selection during the short period of larval life. The latter effect will set limits on how far the drift can proceed in this case. Thus, two alleles, A and a, might be able to vary from 70:30 through 50:50 in the breeding adult population, drift to other frequencies being ruled out by the period of selection during larval life. This sort of drift would not be expected to lead to fixation of either allele. (Of course, if the populations are small, the usual sort of drift may occur at these loci during breeding in these populations as well.)

Another class of traits may be considered to be absolutely neutral at all times. One example that can be described concerns the spots on the backs of leopard frogs. Clones of isogenic frogs can be produced by enucleating newly formed zygotes prior to first cleavage. If such enucleated eggs are each given a blastula nucleus from another zygote, it will function as the zygote nucleus, allowing normal development. If many nuclei are removed from a single blastula and each injected into an enucleated ovum from one clutch of eggs, one would have a series of genetically identical individuals all developing at the same time. When such an experiment was performed, it was found that, after metamorphosis, each froglet had its own individual spotting pattern. That is, the particular pattern of spots on the backs of frogs does not appear to be a genetically controlled trait. [The density of spots was found to differ from group to group of these clones when each group was derived from a different blastula (different individual) and so *that* trait *is* genetically controlled.] If a trait is not genetically controlled, selection cannot affect it and it is, therefore, a selectively neutral trait. The exact pattern of fingerprints in man may be another example, since no two are exactly identical. Such a trait presumably varies at random. From the evolutionary point of view these traits may be considered to be unimportant as such. If they should become important to the lives of the individuals, they could presumably come under the control of genes and of selection.

Another group of traits that are probably selectively absolutely neutral are those that can be described as artifacts of description. Ready examples are found in terms of primary gene products. These proteins may be purified and examined by a number of means. For example, they may be placed in an electric field and their electrophoretic mobility measured and compared with that of gene products produced by other alleles at that same locus in the same population. It is clear, however, that electrophoresis at a given pH on a given substratum does not occur in the living cell and that electrophoretic classification, while capable of measuring differences between some structurally different proteins, has nothing to do with adaptive value as such. Even more clearly artifactual is immunological distance from some arbitrarily chosen reference antigen using antibodies produced in some arbitrarily chosen organism. These are useful measurements indeed, but they are not traits selection can see as such, and so evolutionary changes in terms of them must be considered to be governed primarily by drift, even though in this case the traits (or, better, measurements) *are* coded for genetically—indeed by genes that have other phenotypic effects that are scanned by natural selection. In short, not everything we can measure is "measured" by natural selection.

EXAMPLES FROM NATURE

There are a number of possible examples of nearly selectively neutral traits that have been described from nature, around which a certain amount of controversy revolves. The A, B, O blood group system in man is a good example. For a long time the presence or absence of one or another carbohydrate antigen on the surface of the red cell membrane was felt to be selectively neutral. There is no indication in the geographic distributions of these antigens of anything that might obviously suggest the action of selection. Blood group B is absent in Amerindians and Australian aborigines. Since the former live in extremely diverse environments, there is no indication of environmental selection. Also, two North American tribes have very high amounts of B in the same environment (Wyoming, Dakotas) where other Amerindians get along well without it. The feeling is that the ancestral groups that founded the Amerindian (and Australian) populations were small and subject to drift, and that the B allele drifted out (or that this was an example of the founder effect—see below). The reappearance of B in two tribes could then be ascribed to mutation, or these tribes could have had different founders from the rest.

However, the situation is more subtle and complex. First, the carbohydrate antigens involved are produced in many tissues of the body, and are secreted in various body fluids as well. It is known that bacteria have specific enzymes for attacking or hydrolysing these antigens, and so there is some suggestion that they are involved in "chemical warfare" vis-à-vis bacteria. Again, statistical studies indicate that the different genotypes at this locus are associated with certain sorts of diseases—O with gastric and duodenal ulcers, for example. A study done in Scotland indicated that there is a statistical association there of group O with male infertility. Group A is statistically associated with cancer of the stomach, colon, rectum and female genitalia, with diabetes mellitus and with pernicious anemia, and so on. These disorders tend to be associated with regions where these antigens appear in secretions—that is, they are not just any random diseases. Also, there is an overwhelming preponderance of group O in most populations, indicating that selection might be favoring it. Indeed, the frequencies of the alleles are not randomly distributed, with populations having equal chances of showing any given set of frequencies. One possibility, as pointed out by Haldane, is that selection may have been operating on this locus at some time in the past—say, in the paleolithic period—even though it may not be acting strongly at present. However, these antigens (or very similar ones) are also found in other primates, and it seems unlikely that the observed polymorphisms would survive for so many generations unless maintained by selection (the lineage leading to man and that leading to the pongids diverged from each other some million human generations ago). Thus in this case it is not clear to what extent genetic drift influences the gene frequencies in human populations.

Another case originally explained largely in terms of sampling errors concerns color and banding patterns in the European land snail, *Cepea nemoralis*. There are three different colors (yellow, pink and brown) and five different banding patterns, plus unbanded, coded for by two closely linked genes, one for color and one for banding. It was found that: (1) there can be very large differences in the frequencies of morphs from population to population, even when the latter are not separated by much distance; (2) the variance of the gene frequencies for several populations increases with the distance separating the populations compared—that is, the more distant two populations, the more likely that their morph frequencies will be different; and (3) large populations are much more like each other in frequencies of morphs than are smaller populations—or, the chance that two populations will show different frequencies of morphs is increased when the populations being compared are small.

Originally these data were interpreted as being caused by genetic drift. Nearby populations would be different because drift would work differently on their morph frequencies simply by chance. Small populations are more different from each other than are large populations because drift works more effectively in small populations. Populations become more different with distance because gene flow is increasingly restricted with distance. It has been found, however, that while drift may operate to some extent in these populations, various effects of selection more adequately explain the morph distributions. First, it is difficult to explain with drift alone why several morphs continue to occur year after year in small populations; fixation should occur. (These morphs have been found to have been maintained since the Pleistocene in mainland Europe.) In order to explain this by drift one must invoke tremendously large mutation rates ($\frac{1}{1000}$ or 10^{-3}, as opposed to the more usually-accepted 10^{-5} or 10^{-6} for measurable traits) for which there is no evidence.

Closer examination of the situation in the field disclosed that various effects of selection could easily be demonstrated. A general climatic effect was found; yellow, unbanded shells seem to be favored in open, sunlit regions, possibly because this color better reflects the sun's heat. On the contrary, pink and brown shells are favored in shaded situations. Morph frequencies have been found to change from season to season, and differential effects of temperature have been found on the rates of breeding of the different morphs. Also, it has been demonstrated that the color patterns function as camouflage in respect to bird predation. In open grassy situations yellow is favored (being more cryptic) while in woodlands the other colors produce better camouflage. Banding is favored in rough habitats and unbanded in less broken ones. These relationships have been demonstrated by actually observing which snails are more frequently captured by birds in each habitat and correlating this with the frequencies of morphs in the populations.

As with genetic drift, there is a problem with a complete selection explanation in that it does not specify why several different morphs remain in most populations. The one that is selectively favored in a given habitat should replace all the others. First, it has been demonstrated that the effects of selection differ from season to season, so that one morph may be favored on the springtime background, another on the summer background (seasonal selection, see Chapter 9). It has been suggested also that since the habitat of a given population is rarely even in one season totally homogeneous, selection in a patchy environment could favor more than one morph, with different frequencies being maintained in different micro-habitats (frequency-dependent selection; see Chapter 9). Heterosis has apparently not yet been demonstrated at these loci.

Another problem is why large populations tend to be more like each other than do small ones. The argument from the point of view of the predominance of selection over drift has been that larger populations generally cover more actual square feet of land. The probability is that as the size of an area increases, so will the number of different kinds of micro-habitats included, until in large populations there are so many micro-habitats that all or most of the morphs would come to be favored at some frequency. Small populations, on the other hand, tend to be restricted to fewer microhabitats, and so will display fewer morphs; and as the number of microhabitats decreases, the probability of the same ones being present from one population to the next diminishes.

Problematical data from real observations in nature, as in these two examples, have generally in the last decade or so been interpreted as showing a predominance of selective effects, with only minor effects of drift appearing in the smallest populations. Hence, on the whole, the importance or even reality of selectively neutral traits or genes has come to be questioned by many even though the mathematical and logical formulations of the drift arguments are sound.

SELECTIVE NEUTRALITY OF ALLELES

At another level, there is the separate question of the possible selective neutrality of different alleles at a genetic locus coding for any trait, regardless of the intensity with which selection operates on it. An extremely "important" trait, scanned by selection all through the life cycle, might be coded for in part by a gene represented by several alternate alleles some of which are selectively neutral with respect to each other, even if some of the alleles were distinguished by selection. Thus, A, A', and A'', each differing by one or two codons, may have identical effects upon the phenotype, while a, a', and a'' have different phenotypic effects from any of the first three and also between themselves, so that selection could in fact detect only four alleles at this locus. These are the kinds of possibilities we will examine now.

First, it has been felt that alleles containing synonymous codons at one or more positions, while chemically different, will produce identical gene products so that selection could not distinguish between them. Such alleles would be selectively neutral with respect to each other. It should be pointed out that the evolutionary potential of such alleles is not identical. Two different codons for a given amino acid do not have identical possibilities in respect to what other codons they could mutate to in a single step. Thus, CGU and CGA both code for arginine, but the latter can mutate in one step to a chain-terminating codon (UGA) while the former cannot. It might even be that on this basis CGA might come to be scarcer in a population than CGU as a code for arginine. Also, recent suggestions have been made concerning the nonequivalence of synonymous codons. For example, different codons might be translated into portions of protein chains at different rates. Thus, there might be a relatively low frequency of one type of synonymous tRNA so that one that recognizes CGC is rare while one that recognizes CGG is common, either because of prior mutation inactivating one allele in a diploid, or because the tRNA is differentially stable, or because different tRNAs favor different sorts of peptide bonds. There are several different possibilities of this sort, and so the selective neutrality of synonymous codons may be said to be questionable but not disproven.

Another possible type of selectively neutral alleles would be represented by two differing by a single conservative substitution not coding for a position in the active site of the enzyme or in its interior. The selection coefficients here might be on the order of 0.0003. Most other things being equal, it may be that such different alleles would not be detected by selection under any given conditions except the most stressful. A similar set of different alleles might have single codon deletions in similar noncrucial sites and even large ones at the end of the gene coding for the carboxyl end of the protein. Such differences between alleles will most certainly not be detected by selection, for example, in a growing population characterized by r-selection (Chapter 9). Indeed, during a period of population growth, alleles coding for viability traits, providing they do not disrupt homeostatic balance, will become virtually neutral regardless of what the selection coefficients would be at carrying capacity numbers.

Alleles recently arrived into a population by gene mutation or gene flow will almost certainly be recessive. All recessive alleles, or all the pleiotropic functions in which an allele is recessive, are subject to drift because recessivity is equivalent to selective neutrality when the allele in question is present at a low enough gene frequency so that no homozygotic individuals can be formed. This would be so even if the allele can function as a partner in heterosis, if heterosis accrues to *any* sort of heterozygote at that locus. Indeed, the consensus seems to be that many new mutations, whether potentially favorable or not, drift out of populations at rather fast rates. If by chance they drift in the other direction and become common enough to form some homozygotes, they might be strongly favored or strongly disfavored by selection.

Another situation where one may invoke selective neutrality of alleles is at equilibrium in frequency-dependent selection (Chapter 9). Here the selection coefficients rise along with the number of excess homozygous genotypes. At equilibrium there is no selection distinguishing the different genotypes because each has been previously adjusted by selection in conformity with the environment. If by chance the equilibrium frequency in a given environment becomes disturbed, selection will move to adjust it once again. After adjustment (at equilibrium), the adaptive value of each genotype is unity. This concept makes use of the general proposition that selection coefficients associated with any given genotype will vary from one environment to another, but it is a special case where the changes in magnitude are predictable.

If any given gene has a multitude of phenotypic effects (is pleiotropic), it is probably true that not every one of those effects is as crucial at any given stage or season as some others will be. It may be that two different alleles in a polymorphic situation, with marked differential effects on one aspect of the phenotype, may be selectively neutral with respect to each other in connection with one or more other phenotypic effects, just as the effects of a given allotype may be dominant in one system and recessive in another (Chapter 7). Thus, consider the evolution of bristle number in *Drosophila*. The mean number found on the fourth and fifth abdominal segments in *D. melanogaster* laboratory populations is 36. Directional selection experiments have demonstrated that as bristle number is modified, various aspects of adaptedness come to be upset (Chapter 9). Viability, fertility and anatomical relationships are disturbed increasingly as mean bristle number is carried further and further from 36. It may be that bristle number itself has never been the object of natural selection, and that 36 is simply the number that is most frequent given a good coadapted genome in a given environment. Indeed, in this case bristle number may even be a selectively neutral trait in itself (even though artificial selection can work on it).

Whether drift occurs because populations are very small or because the traits involved are relatively unimportant or are coded for by a series of alleles whose phenotypic expression is identical, the ultimate fate of alleles whose gene frequencies are drifting is for one of them to become fixed and the rest lost. In sexual populations the probability of fixation of an allele is $1/4N_e$ per generation, while the probability of loss of an allele is $1/4N_e$ per generation, and so the rate of total loss of variability is $1/2N_e$ per generation. This drift to lower variability is opposed by mutation and gene flow, which continually generate new alleles, and by balancing selection. In the latter case no particular allele may be favored or disfavored by selection while any sort of heterozygote may be strongly favored over any sort of homozygote. In large populations the rate of loss of neutral alleles is probably balanced by the mutation rate producing more of them. In small populations it has been the experience of most workers that variability is less than it is in large populations. Thus, whether or not genetic drift occurs in populations of all sizes, it can only have a major effect on gene frequencies in small-sized populations.

THE FOUNDER PRINCIPLE

Ernst Mayr has pointed out a probably more important sort of sampling error from the point of view of evolution, which he called the founder effect (or founder principle). If a very few individuals are chosen from a large population to found a new population, the chance of their being a representative sample of the gene pool is negligible. This is so regardless of why they happened to become founders. Thus, in the case where some small organisms are carried by winds or currents from a large mainland population to an island, it could probably be shown that natural selection had something to do with why the agent of dispersal claimed *them* and not some other individuals in the population. Perhaps they tended to frequent situations more exposed to the dispersal agent. If these individuals do start a new gene pool on the island, the very fact that selection may have determined which individuals were most likely to be carried off insures that they will not be representative of the parental population. At the same time, traits not involved with rafting, or with any phenotypic expression relevant to being dispersed, will drift simply by chance.

Now, clearly, if the founder population is successful (it could even have come from a single gravid female of some sorts of organisms, or from a seed of a self-fertilizing plant), it is clear that for a certain period of time the population will be small by any standard, and so typical genetic drift would be able to occur in the early phases of colonization, when, indeed, r-selection would predominate because a relative lack of competitors and predators in the new environment would allow the population to grow rapidly, so that even for this reason viability traits could experience drift. This is assuming that the overall structure and climate of the new habitat is within the adaptability of the species in question. If by chance it is not, selection would be very intense and survival of the small population unlikely.

When the new population has established itself, the chances are that, because its original gene pool was so unlike that of the parental population in the frequencies of alleles, and because it had so few alleles, it will not be exactly like the parental population in any of these respects or in many phenotypic expressions. It is likely to be different from the parental population in at least small ways. Also, different founders from the same parent population (say, on different islands) will for the same reasons be unlikely to resemble each other in every way. It must be kept in mind, however, that selection could favor different phenotypes on different islands, and that some of the observable differences in this case might be due to selection. Notice also that in principle it is difficult if not impossible in practice to distinguish between the sampling error produced by the founder effect and that produced by genetic drift in the initially small populations. An example will highlight the interplay of these three factors.

Bentley Glass, now of the State University of New York at Stony Brook, made a detailed study of the genotypes of a small religious isolate called the Dunkers. Fifty families of this religious sect left Germany in 1719 and settled in Germantown, Pennsylvania. They maintained a strict separation from the American population which is only in the last three or four generations beginning to break down, with about 15 percent of their children marrying outsiders and 24 percent marrying converts. The community when studied by Glass consisted of 228 individuals. A number of traits were chosen for investigation on the basis of their possible relative selective neutrality, and we can examine the A, B, O blood group data, as well as the MN blood group data (Figure 13-1).

	Numbers	O	A	B	AB	M	N	MN
Germans	3036	40.7	44.6	10.0	4.7	30.0	20.0	50.0
Dunkers	228	35.5	59.3	3.1	2.2	44.5	13.5	42.0
North Americans	30,000	45.2	39.5	11.2	4.2	30.0	20.0	50.0

Figure 13-1 Frequencies of various blood group phenotypes in two large human populations, German and North American, and in a small isolate, the Dunkers, derived from the first population and living near the second. (Data from Glass, 1954.)

As can be seen from the table, the Dunkers differ from both the current American and current German populations in these two loci. Since this has always been a small population, and since their environment has not differed from that of the American population in any major way and since the American and German populations are very similar, these data are adequately explained by genetic drift. However, they are also adequately explained by the founder effect, since it is unlikely that 50 families could carry a gene pool representative of the larger population from which they came. Indeed, there is no reason to believe that both effects have not occurred in this population. The Dunkers were found to differ from the surrounding American population in several small anatomical features as well, such as the frequency of hitchhiker's thumb, and these differences could also be due to sampling error on neutral or nearly neutral traits.

SUMMARY

Random sampling errors can occur in the gene frequencies of alleles that are phenotypically indistinguishable, and also of alleles coding for traits not being scanned by natural selection. Such effects are especially pronounced in small populations, and indeed, in them alleles at genes coding for traits being scanned by selection can drift as well. All these effects are treated as genetic drift. A special case of gene pool sampling error is the founder effect, where a small group, or even an individual carrying eggs, begins a new population with a gene pool that is not representative of that of the parent population. The latter effect is probably evolutionarily much more important than the general phenomenon of genetic drift, in that we can find examples (especially on islands) where probably this effect has led to measurable evolutionary differences between the parental and new populations.

REFERENCES

Crow, J. F., and M. Kimura, *An Introduction to Population Genetic Theory.* Harper & Row, New York, 1970, 591 pp.

Ford, E. B., *Ecological Genetics.* Wiley, New York, 1964, 335 pp.

Kimura, M. and T. Ohta, *Theoretical Aspects of Population Genetics.* Princeton University Press, Princeton, N.J., 1971, 219 pp.

Kojima, K., ed., *Mathematical Topics in Population Genetics.* Springer Verlag, New York, 1971, 400 pp.

Li, C. C., *Population Genetics.* The University of Chicago Press, Chicago, 1955, 366 pp.

Mayr, E., *Animal Species and Evolution.* Harvard, Cambridge, Mass., 1963, 797 pp.

Mettler, L. E., and T. G. Gregg, *Population Genetics and Evolution.* Prentice-Hall, Englewood Cliffs, N.J., 1969, 212 pp.

Sheppard, P. M., *Natural Selection and Heredity.* Harper & Row, New York, 1959, 209 pp.

Woese, C. R., *The Genetic Code.* Harper & Row, New York, 1967, 200 pp.

Wright, S., *Evolution and the Genetics of Populations,* vol. II. The University of Chicago Press, Chicago, Ill., 1969, 511 pp.

CHAPTER 14

Speciation

The process of species formation is perhaps the most clearly worked out area of evolutionary biology, largely due to the efforts of Ernst Mayr of Harvard University. It needs to be pointed out, however, that the clarity or precision of thought that has been achieved in this area has been at the expense of general applicability. The problem arises in connection with the definition of what a species is, which is needed before a theory of species formation can be formulated. The currently accepted species definition has arisen from the realization that populations are variable, and that, indeed, they must be variable if they are to evolve. In other words, the current species definition represents an attempt to incorporate the ideas of Charles Darwin (and data from population biology) into the concept of the species. The older view of a species was that it was a homogeneous collection of individuals with fixed attributes. They were formed just once and did not change with time. Individuals that differed from the species "concept" (a term still used in botany) were explained as accidental "sports" whose construction was flawed. If it was realized that *no* two individuals were really alike (as when comparing human faces or fingerprints), this was explained by some as a failure to realize the ideal form (in the Platonic sense) of the species in question. In short, this was a static view of the species, and their conceptual permanence was perhaps one reason why people like Linnaeus were at great pains to name every species in nature.

In contrast, many evolutionary biologists today are totally unconcerned with such a preoccupation; after all, if species are going to change (or become extinct), why bother to name them?

THE CURRENT DEFINITION OF SPECIES

The current definition of the species, then, is that it is one or more demes made up of individuals which (regularly or occasionally or potentially could) interbreed *in nature* with members of the other demes involved to *produce viable and fertile offspring*. In other words, if gene flow can (or could potentially) occur between demes, these are composed of individuals in the same species. Notice that the definition hinges entirely on the possibility of genetic exchange in nature. There is no stipulation that the individuals involved need to resemble each other phenotypically, but in practice this is more or less true. It is for this reason that the definition is not entirely impractical in dealing with the vast collections of organisms found (and labelled in Linnaean fashion) in the world's great museums, but many problems do occur in this connection.

LIMITATIONS OF THE CURRENT SPECIES DEFINITION

For example, if two demes are found living in cloud forests on nearby mountain peaks without any chance of an individual crossing from one peak to the other because all the intervening terrain is inhospitable, even if the normal curves of phenotypic measurements for both are virtually identical, there is no way to decide whether or not these two demes belong to the same species (even though they may be dealt with as such in a museum). This is because they cannot interbreed in nature. If individuals were carried from one cloud forest to the other and released, mated successfully, and their offspring competed successfully with the offspring of matings of indigenous individuals, then the two could be considered to be the same species. In actual practice it is virtually never possible to carry out such experiments—or indeed desirable in view of the possibility of disrupting a natural ecosystem by so doing. Therefore, species recognition is restricted to sympatric situations; one can consider discreet species only if they live in the same place. Any allopatric populations may or may not be the same or different species. More depth will be given to this idea below in the discussion of the process of speciation.

The species definition is also restricted to sexual organisms (organisms having gene pools). Asexual and obligately parthenogenic organisms can never be tested by the essential criterion of ability to interbreed with another group. There is no species definition for asexual organisms, which is the same as saying there are no asexual species. Every individual asexual organism may be considered to be its own species. In practice species are defined for these sorts of organisms by purely phenotypic analysis. Members of a clone will resemble each other more closely than members of another clone simply because they are genetically more closely related. A current rule of the thumb is to have before one (by means of long experience with a group) knowledge of the average amount of phenotypic difference typical of good species in a given group (or closely related group), and then to use that amount of phenotypic difference to serve as a standard for defining a species in one's asexual group. (The same procedure is used when dealing with allopatric populations of sexual organisms). In groups like the bacteria such a method is not available, and current practice is simply to divide various groups on the basis of arbitrary convenience. Interestingly, comparisons of amino acid differences in various proteins suggest that there is about the same amount of genomic difference (about 10 amino acid differences per 50 amino acid residues) between *genera* of mammals or of ascomycetes (eucaryotes), *species* of bacteria (procaryotes), and *strains* of viruses, suggesting that in future we may have to completely overhaul our current nomenclature in order to put all organisms into one system.

Another limitation of the current species definition is that it is restricted to organisms alive today. There are no fossil species, although paleontologists do name species. Indeed, a sample from a population made twenty years ago cannot, strictly, be called members of the same species as that same population today. Indeed, human beings living four generations ago cannot strictly be classified as the same species as human beings alive today. All these statements are corollaries or implications of a species definition based on capacity for interbreeding only. A gene pool is continually changing; when it has changed sufficiently so that members sharing it could no longer interbreed with some ancestral population (were this test to become possible), then it would have become a new species. This test is never possible, but on the other hand we do know that gene pools do continually change, and so, again, we make estimates about what to call species based on the average amount of phenotypic difference between established (living) species in a related group. Actually, it sometimes almost becomes possible to refer to fossil species. Some deposits have been found to be exceptionally rich in specimens of all ages. It is sometimes possible with such material to distinguish two or more clearly different but closely related populations that were alive simultaneously and sympatrically in the past. While there can be no proof of species standing (the different-seeming organisms might, for example, simply represent growth stages, different sexes, or different ecotypes, of one species), it can often be extremely probable that one is observing different species from the past.

ISOLATING MECHANISMS

From the above definition it can be seen that the gene pools of different species are isolated from each other. Little or no gene flow takes place from one species to another (an occasional hybridization might occur, but if it is infrequent the gene pools will remain separate). Although there is no way to tell whether allopatric populations are of the same species or not, it is certain that one means of establishing isolation of gene pools is geographic separation. Indeed, as will become clear below, this is perhaps the most frequent initial step in the speciation process. If organisms cannot meet, they cannot interbreed. This sort of isolation has been called geographic isolation. All other sorts of isolating mechanisms are generally referred to as reproductive isolation.

If two species are sympatric, some of these other isolating mechanisms may come into play to keep their gene pools separate. If any of the following mechanisms can be shown to exist in nature between two populations, then they are different species. Some sympatric organisms are not syntopic. If two populations occupy the same geographic region, but one is found in the treetops and one on the ground, as in some lizards in a single genus, they may never meet, and are considered to be ecologically isolated. The effect of ecological isolation is the same as geographic isolation. Another possibility, found often among frogs, is that two populations may be syntopic during most of the year, but not during the breeding season, which is when it counts. A special case of the latter sort, again found in frogs, and in plants, is where the populations are syntopic all year but have different breeding seasons. Again, they could not meet to interbreed; this has come to be called seasonal isolation.

Supposing that none of these sorts of isolation are present. Two populations might breed in the same place at the same time and still be isolated by a number of means. For example, there may be no sexual attraction between the members of two animal populations. Most frequently it is the females that exercise the final choice of mates, but courtship might be broken off at any point by either the male or female. This has been called sexual or psychological isolation, and the slightest difference in odor, color, or courtship procedure might be sufficient to cause it. If, however, there are no psychological barriers to mating with members of a different population, and attempted copulation does occur (or, in plants, pollen does fall onto the stigma of a member of another population) there is the possibility of mechanical blockage of fertilization. In many insects copulation is attempted freely with members of different species, but the male intromittant organ is of a size or shape that does not fit into a female of a "wrong" species. In this case the male falls off and wanders off in search of a new female. In plants the surface of the stigma may be so made that certain types of pollen cannot stick to it, and this would accomplish the same end in conceptually the same way. This type of isolation is known as mechanical isolation.

All of the above kinds of isolating mechanisms can be gathered under the rubric of premating isolating mechanisms, while the ones to be yet discussed are referred to as postmating isolating mechanisms, in reference to their time of action. The first postmating isolating mechanism that may come into play if there are no premating mechanisms is gametic isolation. Here the gametes cannot interact so that the sperm or pollen DNA cannot enter the ovum or ovule. For example, the spermatozoans may not be able to penetrate the egg membranes or capsules, as happens in some frogs and many marine invertebrates. Or, as in many plants, the pollen tube may be unable to grow into the style of the flower, even though the pollen was able to stick to the stigma. This isolating mechanism, (and any of the others so far discussed) is sometimes referred to as a kind of prezygotic isolating mechanism, as distinguished from postzygotic isolating mechanisms, comprising all the ones still to be considered.

If the gametes from two different populations do interact so as to activate the ovum or ovule and begin development of a new organism, there are still some events that may occur to establish effective isolation of the two populations. The hybrids may simply be inviable, dying at various stages of development, or they may be unable to compete with either sort of parental organism in an ecological sense. In the first case it is felt that the gene products of the two species have such dissimilar requirements and kinetic properties that a system composed of both sorts simply is not coadapted. The degree of actual genomic difference between the parental species would be crucial here, the effect (hybrid inviability) being worse the more different they are. On the other hand, if the parental forms are very close genomically, the hybrid offspring, physiologically healthy, may not be able to compete with either of them. Most hybrids are intermediate phenotypically between the parental forms. As such they are not as specialized for a way of life as are the parental forms, and could compete well with neither. It does occur, however (especially in plants) that a third niche may be possible in the given environment, for which the hybrid is by chance better adapted than is either parental form. In this case gene pool isolation might not exist between the parental forms.

Should the latter situation prevail, with a hybrid population existing in a neighboring niche, there are still further postzygotic isolating mechanisms that may come into play. The hybrids may be sterile, a frequent occurrence (hybrid sterility). This effect is common since meiosis importantly involves synapsis of homologous chromosomes. Slight differences (such as inversions or translocations) between the chromosomes of the two species might make it difficult for even a single pair of chromosomes to synapse successfully, even though the alleles found at virtually every gene in both species are capable of forming a good coadapted (and even heterotic) phenotypic system.

Another, and very peculiar, postzygotic isolating mechanism, as yet unexplained, is hybrid breakdown. Here, the F_2 or back-cross progeny are unfit in terms of viability and/or fertility. In this case one might continue to find F_1 hybrids, as one might also in hybrid sterility, but the two populations would remain effectively isolated genetically.

THE PROCESS OF SPECIATION

In order for gene pools to become different enough so that when the populations carrying them become sympatric they remain isolated, there has to have been a prior period of time during which the genomic differences between them accumulated—that is, a period of time when the gene pools were physically separated. That, in short, is the basis of the current theory of speciation, also called the allopatric theory of speciation. This theory insists that there is only one reasonable way for gene pools to become physically separated during this crucial period, and that is geographic separation resulting in geographic isolation. We will return to this question below, but will proceed here to discuss the processes involved in speciation, invoking only some unspecified means of establishing the initial physical separation of a gene pool into separate components.

If the separation into two populations lasts only for a "short" period of time, then, when the populations again become sympatric, there will not have been enough time for enough genomic differences to accumulate so that some form of isolating mechanism will be present, and the two populations will meld together forming what has been called a hybrid swarm. There are many examples in plants where such a process seems to be occurring. Among frogs, there is the example of the American toad and Fowler's toad in the eastern United States. It is rapidly becoming difficult to find places where these two remain distinct. The isolating mechanism previously separating them appears to have been ecological only, in that they bred in different types of water in the same season. Fowler's toad favored upland, sandy ponds, while the American toad tended to breed in slow streams and stream-fed ponds. Evidently the massive alteration of the environment carried out by man has obliterated much of one or the other habitat, or destroyed the differences between them, so that both species were driven to breed in the little remaining available water. We feel that these "species" must have been separated from each other only for a relatively short period of time inasmuch as the genomic differences between them proved so slight that hybrids between them could easily be formed and could compete well with members of both parental populations.

The period of isolation of these two groups of populations is considered to have been "short" *because* they are being submerged by what is called introgressive hybridization, or introgression, into a hybrid swarm. What such a period of time would actually be is not clear. On the one hand we know that it would take some 300 generations to replace one allele by another with moderate selection pressures. On the other hand, the species of catalpa trees in the eastern United States and in China, for example, are completely interfertile and produce apparently perfectly viable hybrids, even though perhaps some million generations have passed since their ancestors were last members of the same gene pool. There is probably no general rule here. A single change, if it affected some vital part of a courtship procedure, for example, could be enough to set up an isolating barrier. On the other hand, large amounts of genomic change might accumulate without effect on isolating barriers, even though simply by chance, the more such differences accumulated, the greater would be the probability of reproductive isolation resulting from them.

If, instead, the two populations remain isolated for a very long period of time before again becoming sympatric, it is possible, simply by chance, that so much genomic difference will have accumulated between them that some form of reproductive isolation will prevent the production of hybrids between them should they again become sympatric. Such populations would have "speciated" during the period when they were allopatric. The speciation would be completely insured if the isolating mechanisms that had arisen were of the premating type (excluding ecological, which can be destroyed by drastic environmental change). In this case, the complete speciation process would have taken place during the time when the gene pools were physically separate. If, however, only postmating isolating mechanisms had arisen by chance, hybrid formation will take place with some frequency, and this will lead to selection for premating isolating mechanisms leading to character displacement in secondary sexual features.

Thus, suppose two populations come into contact after a long period of isolation, and the hybrids are sterile or inviable. (The inviability can range from actual death of hybrids at an early stage to inability to compete with parental forms ecologically—the result is the same). All individuals of both populations that mate with members of the other population waste their gametes, and the alleles responsible for their lack of discrimination will not be passed on to future generations of either population. On the other hand, individuals with the capacity to discriminate between individuals from the two populations will pass on those traits that allow them to do so. Also, individuals that are discriminated against by members of the other population will tend to pass on to future generations the alleles responsible for their distinctiveness.

For example, suppose two populations of fishes found themselves sympatric after having diverged to this point. Suppose there is a courtship procedure involving a red thoracic region in males. The redness ranges from purplish to orangish, with a different mean for each population. Suppose there is female discrimination adjusted to the mean for each population. Thus, very orange males in one population will not attract females from the other population, while very purplish males of the latter population will not attract females of the former population. There are, therefore, in each population some individuals with traits that prevent them from hybridizing with members of the other population, and these will have traits that will be successfully transmitted to the future populations of both lineages. The male sexual coloration in this case will diverge in these two populations, leading to character displacement, and to the establishment of premating isolating mechanisms between the two populations (Figure 14-1). In fact, natural selection will have produced premating isolation where before there was only postmating isolation.

One may ask at what point speciation had occurred in an example

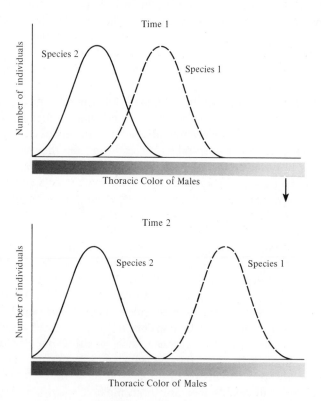

Figure 14-1 Character displacement of courtship color in two species after they find themselves in the same community. The shaded bar indicates the breeding coloration of the males—darker to the left, lighter to the right.

like this, and the answer is that it is a matter of taste. Clearly, the presence of postmating isolation leads directly to premating isolation as well, and so one might as well see the former condition as having established species distinctness, in which case the crucial part of the speciation process is not the final character displacement of secondary sexual features, but the chance acquisition of postmating isolation during the period of gene pool isolation.

SYMPATRIC AND ALLOPATRIC SPECIATION

The above sequence of events plausibly follows a period of gene pool isolation between two populations, and forms a brief sketch of all modern theories of speciation. Controversy, however, arises in connection with the concepts of how the gene pools become isolated in the first place. Most evolutionary biologists follow Ernst Mayr, who insists that in virtually every case this initial isolation must be brought about by geographic isolation. This is the burden of the allopatric theory of speciation. It is clear that splitting a gene pool into two separate ones by some physical barrier (water, mountains, desert, distance) would be sufficient to allow genomic differences to accumulate that could ultimately lead to some form of reproductive isolation. The controversy centers around the relative degree of importance one allows to some form of sympatric speciation.

One clear example of sympatric speciation (accepted as such by Mayr as well) arises occasionally in plants in association with the condition of allopolyploidy. The following sequence of events is an example: Suppose two fairly closely related plants live sympatrically and have different chromosome numbers. Suppose that hybrids can be formed between them and that there is a niche available for them. The difference in chromosome numbers prevents the hybrids from being fertile, however, because synapsis during meiosis is disturbed. Now suppose that an individual hybrid zygote undergoes endomitosis (mitosis without cytokinesis) so that it becomes a tetraploid. There are then two copies of each chromosome, so that chromosome synapsis can occur during meiosis. The plant is then functionally diploid, and operational pollen and ovules will be formed. Suppose, further, that these sorts of plants are capable of self-fertilization (selfing), and that this occurs to the tetraploid in question. It will drop many seeds, all of which are tetraploids. These will develop in the niche available to them, and when blossoming occurs, they will be able to fertilize each other, but not members of either of the parental species. They will be reproductively isolated from these parental forms because the back-cross hybrids have different chromosome numbers, and, indeed, a kind of "instant speciation" will have occurred in sympatry with the original gene pools (Figure 14-2). Evidently just such a process has occurred with some of our food plants, such as the bread wheats.

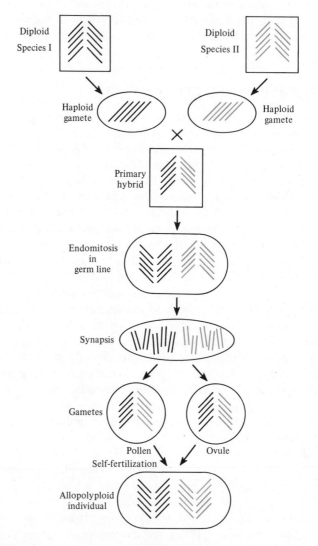

Figure 14-2 Instant speciation in plants via allopolyploidy.

Other possible examples of sympatric speciation have been suggested from time to time, all of them involving several highly improbable events. For example, suppose we have a species of insect that feeds exclusively on hawthorne trees, and deposits its eggs thereon. Suppose that by accident some eggs fall onto a nearby apple tree. Suppose that the eggs hatch and that the larvae can feed upon apple leaves. Several of them will mature on the apple tree. Suppose, now, that these insects normally become imprinted (or programmed) to return to the food tree to mate and deposit eggs, and that those that developed on the wrong tree could become imprinted to it in the same way. Suppose, in addition, that two of the siblings that developed from the misplaced eggs actually find themselves on the same apple tree (after having dispersed during a feeding stage), so that a mating can occur between two individuals wrongly imprinted. If the female now deposits many eggs on the apple tree, and they all develop, a small population of apple-imprinted insects would be present in this region (that is, if the degree of homozygosity imposed by sibling mating allows any of the offspring to be free of genetic disease). Perhaps there could even be enough of them to establish a population, which would be ecologically isolated from the parental population. It may well be that such events have occurred from time to time, as, evidently in the case just cited in North America sometime between 1600 and 1860. There is no theoretical reason why they should not. But such events are probably rare for most kinds of organisms relative to the physical division of gene pools by climatic and geophysical means, or by chance establishment of waif or founder populations in isolated habitats (Chapter 13). It is this probability estimate that is at the heart of the allopatric theory of speciation—it simply seems improbable (and, therefore, rare) that the initial stages of speciation should occur in the absence of geographical separation.

Other possibilities for sympatric speciation arise from the action of disruptive selection or of modification of development by environmental influences. For example, suppose a population of small mammals lives in an environment that has two major kinds of food particles available, each requiring different tooth shapes or sizes for most efficient utilization. Beginning with a unimodal curve of tooth variability in young individuals, selection might produce in adults a bimodal curve (Figure 14-3). Or, there

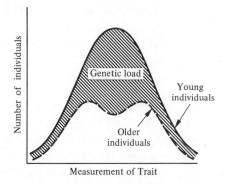

Figure 14-3 Disruptive selection, producing a bimodal curve of variability in the survivors of a cohort from an initial unimodal one in the young.

might be two different sorts of nest site, each tending to elicit (via some variable) quite different tooth sizes or shapes, preadapting the offspring from each differently for the two kinds of food particles. Suppose, now, that the two sorts of food particles are located in different microhabitats—one under bushes, the other among grasses, for example—and that mating tends to occur near the food source. It seems at least possible that in such a situation the gene pool of the population could become cleaved in time.

SEVERAL PROBLEMATICAL SITUATIONS

THE RING SPECIES

There are many widely distributed species whose ranges cover entire continents. Clearly, a leopard frog in Canada has no chance to mate with a leopard frog in Costa Rica, and yet they might be considered members of the same species. This decision would be based upon an estimate as to whether or not significant gene flow could occur now, or has been occurring in the recent past, between these populations. If, for example, it could be shown that every adjacent population, starting in Canada and ending up in Costa Rica, was interfertile, then the entire string of populations from north to south would be considered to form a single species. Even then, however, it seems doubtful that alleles found in northern populations would be selected in southern ones, even if they could arrive there by gene flow. Natural selection would preserve a certain degree of differentiation between such populations. Indeed, in the case of North American leopard frogs there is known to be potential reproductive isolation (discovered by laboratory matings) between populations found in different geographic regions, so that if the intervening populations were to be eliminated, there would be two species where there had previously been one (if they should subsequently become sympatric).

This sort of example is sometimes presented graphically when populations at the extremes of a species range actually meet in nature, as when, by range expansion, the ranges of these populations converge, say, from east to west, as in some cases known from Eurasia. Thus, a certain species of seagull has a range extending from Eastern Siberia around Europe, through the Middle East and India, around China and back into Siberia (forming a ring—hence, ring species). Where the two end populations meet, in eastern Siberia and northern China, they behave as if they were "good species," the individuals being unable to produce fertile offspring. Presumably, however, there is gene flow all through all the other populations, although this has not been conclusively demonstrated.

Clearly, widely dispersed populations of a single species are being differently molded by natural selection. They may be considered to be potential species or protospecies. Many of them certainly have genomic differences at a level characteristic of good species. It only remains for the interconnecting populations to disappear, and allopatric speciation would result.

SIBLING SPECIES

Recently diverged species often resemble each other closely, or, at least that is what we would expect. There are now many examples known of sympatric (and allopatric) populations composed of phenotypically very similar organisms that are known not to interbreed in nature (or to be reproductively isolated in laboratory experiments). These have been given the name sibling species. As pointed out by Paul Ehrlich of Stanford University, however, natural selection could easily maintain similar phenotypes in different populations in similar environments. Thus, there is no necessity to suppose that all sibling species are only recently diverged from each other. There is no one-to-one correspondence between genotype and phenotype, as pointed out in Chapter 8 and elsewhere.

HYBRIDIZATION BETWEEN SPECIES

It has long been known that in plants hybrids are frequently formed between good species. It is now clear that hybridization is common among animals as well. If the hybrids are infertile or inviable, they simply represent part of the genetic load of the parental populations and have no evolutionary significance. Indeed, even with the production of many hybrids, there may be no gene flow between the populations in question. Such a situation frequently leads to character displacement in the two species; in the zone of interbreeding the two species may be phenotypically more dissimilar (especially in secondary sexual features) than they are in geographic regions where they are not in contact. This pattern has now been discovered in many sorts of organisms, and is usually taken to mean that hybridization is occurring or has occurred in the recent past. There is some evidence that zones of interbreeding contact between species may be maintained for many generations, indicating that character displacement of secondary sexual features is not always efficient in reducing hybridization. It may be that this mechanism is sometimes even incapable of completely eliminating hybridization.

In some cases in plants hybrid populations are known to exist side by side with the parental populations in nature, but in different niches. Even a population of sterile hybrids could exist in this way as long as hybridization continued at some minimal rate. On the other hand, some cases of hybridization, in both plants and animals, may be considered transient situations occurring during a period of introgression, the outcome of which will be a hybrid swarm of intermediate phenotypes.

SUMMARY

Species are defined as groups of interbreeding demes. As such, the species definition is restricted to sexually reproducing, contemporary organisms. In addition, it is generally not possible to determine whether isolated allopatric populations are conspecific or not. The usual process of speciation probably involves an initial period of geographic isolation of the populations during which some sorts of isolating mechanisms develop between them by chance, depending only on the amount of genomic difference (more or less proportional to time since the populations were last united) between them. In rare cases gene pool isolation may arise sympatrically. Premating isolation mechanisms may arise via natural selection in cases where postmating isolating mechanisms alone arose by chance during the period of gene pool isolation.

REFERENCES

Cain, A. J., *Animal Species and Their Evolution.* Harper & Row, New York, 1954, 190 pp.

Dobzhansky, T., *Genetics and the Origin of Species.* Columbia, New York, 1951, 364 pp.

Mayr, E., ed., "The Species Problem," *American Association for the Advancement of Science Publication No. 50,* Washington, D.C., 1957, 395 pp.

Mayr, E., *Animal Species and Evolution.* Harvard, Cambridge, Mass., 1963, 797 pp.

"Systematic Biology," *National Academy of Sciences, Publication No. 1692,* Washington, D.C., 1969, 632 pp.

Simpson, G. G., *Principles of Animal Taxonomy.* Columbia, New York, 1961, 247 pp.

Stebbins, G. L., *Variation and Evolution in Plants.* Columbia, New York, 1950, 643 pp.

CHAPTER 15

The ascendency of man

The discussions in this chapter will be organized around the fossil record of primate and human evolution, which is depicted in Figures 15-1 and 15-2. Each stage in the evolution of the lineage that eventually produced man is characterized by traits that are recognized by men as being of importance to them, and, therefore, of importance in their evolution. Again, the method utilized here is quite the opposite of the process of natural selection that has operated to produce the fossil record. We are familiar with a current "end product," that is, ourselves, and we look back into the fossil record in search of the remains of organisms that may have been part of our lineage. What we find in the fossil record is determined in large part by what we are seeking. Natural selection was seeking nothing, and simply operated mechanically to produce populations of organisms adequately adapted to produce more organisms. Thus, we will find trends leading from early primates to ourselves, and we will describe them. These trends have a reality only now and only for us (see Chapter 5).

This chapter presents a coherent, internally consistent story of the evolution of man, such as one day we hope to have before us on more concrete grounds. Today there are many other possible interpretations that could be advanced that would be concordant with the fossil record. Every interpretation given here can be (and no doubt has been) challenged. The author has chosen from among many possibili-

Millions of years ago

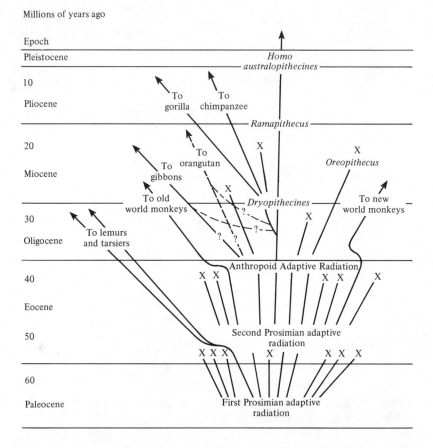

Figure 15-1 A simplified primate phylogeny, showing one current view of relationships in this group, together with the proposed dates of divergence of the various sublineages. Lines ending in X are extinct.

ties those ideas that work well together to make a flowing story. In Chapter 1 it was pointed out that the process of reconstructing a given evolutionary history is not itself science, even though scientific data may be used—this should be borne in mind. The facts are only a few fragmentary fossils and their radioisotope dates.

THE VERY EARLY RECORD

No members of the suborder Anthropoidea (the anthropoids) appear in the fossil record until the Oligocene, some 30 million years ago, at which time we find the three superfamilies of anthropoids already separate. From our point of view all three can be discussed jointly in terms of the kinds of adaptations they probably had, and which were of importance for the later evolution of man. The Oligocene anthropoids seem to have been forest-dwelling unspecialized herbivores, feeding on fruits, nuts, buds, shoots and so on, probably supplemented with occasional invertebrates, a diet not unlike that of monkeys today. It is thought that most were climbing forms, and some may have been semi-brachiators with occasional

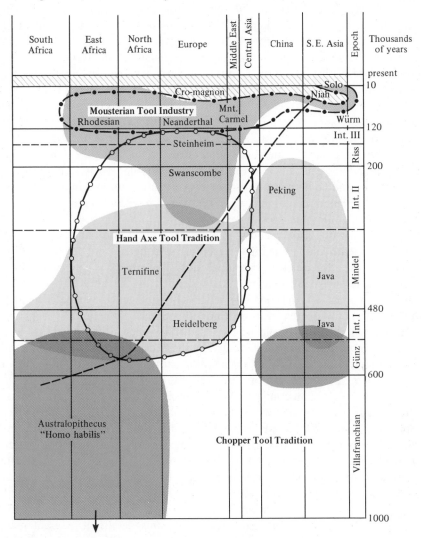

Figure 15-2 Summary of human evolution. The different shades of gray indicate the different taxa: Australopithecines ■ ; Pithecanthropines ☐ ; Neanderthalers ▦ ; modern *Homo sapiens* ▨ . Associated tool traditions are delineated by distinguishing black lines. The chopper tool tradition, indicated as occurring in the western Pithecantropines, refers to the Clactonian and Levalloisian Flake tool traditions.

forays to the ground. As in all known living anthropoids, the trunk was probably held upright as often as not; monkeys of today sit, feed, groom, and even sleep while sitting upright. The arms were fairly long, the hands mobile and capable of grasping objects and of performing such tasks as peeling fruits; the thumb was in some degree opposable. It is generally agreed that life in trees necessitated an emphasis on the visual sense, and that the development of binocular vision in primates is related to life in this habitat. In jumping from branch to branch it is of the utmost importance to be able to gauge accurately the distances involved, and that is possible with binocular vision (although birds and squirrels manage these feats without binocular vision). On the other hand, the sense of smell is of less importance to this way of life, and it is usually considered that a change in the importance of olfaction associated with living in trees had something to do with the loss of abilities in this direction in anthropoids. However, supposed lack of need for some structure or function is usually felt to be in principle insufficient to explain an evolutionary loss. Nevertheless it is clear that in man smell is not of very great importance, and it has also been demonstrated that it is not very important in other primates either except for some prosimians.

Along with visual acuity, movement in trees probably favored the ability to learn and remember direction, safe limbs, and landmarks that could be used for quick navigational aids. This ability is related to intelligence as it is usually defined. When defined as the ability to learn (or as the presence of conditionable reflexes), intelligence can be seen to be present in some degree in most kinds of animals. Sometimes it is present only at specific ontogenetic stages, as, for instance, when hatchling birds become imprinted on their parents. In these cases what is learned is not subsequently modified—that is, learning stops. Learning of this kind is often not really learning at all, in the sense that what can be learned has strict limitations. Thus, males of some species of birds must learn the species-specific song or they will not perform it "correctly." Experiments have shown in some of these cases that these birds cannot, however, learn anything *but* the correct species-specific song. In this case the song pattern is really inherited, but needs a specific external stimulus in order to develop normally. It is probably best to restrict the term learning to a process whereby the organism possessing the ability to learn can acquire and build into its nervous system information, patterns, or abilities that are not closely specified by its genome. Even with this definition, a great many different kinds of animals can be seen to show intelligence; that is, they can learn to perform tasks such as running a maze.

In view of our lack of knowledge concerning the genetics of the ability to learn, it may be fairly objected that we should not postulate that learning is a partially unspecified or open-ended trait (Chapter 8). Perhaps even the human ability to learn is more strictly determined than we are aware.

Are there, for instance, predetermined neural patterns in our brains which correspond to mathematical operations such as addition? We are certainly restricted in learning to only four dimensions. Perhaps this is why we say there *are* only four dimensions outside of mathematical abstractions. Octopi cannot tell the difference between a line slanted to the left and one slanted to the right, presumably because of the pattern of neuronal connections in their nervous systems; for this reason they could never learn to differentiate the two—indeed such differences do not exist for the octopus. What is there that does not exist for *us* outside of the framework of genetically determined sense organs and neuronal patterns? Still, we can differentiate *degrees* of unspecifiedness in what can be learned, and anthropoids appear to have comparatively wide latitudes in this respect. Man, of course, is the most versatile in this way, as an example will demonstrate. None of our senses can detect ultraviolet radiation. We cannot see it, but bees can, and many floral adornments are of this color. But man has learned to detect and photograph patterns reflected in these wavelengths, so that a dimension of the world completely outside of the range of his ancestors is now accessible to him indirectly. This means that man's sensory and mental equipment was such that prior to achieving the ability to detect ultraviolet patterns he had the potential to do so. Is there any portion or aspect of the universe that man does not have the potential to learn about? If there is none, then his ability to learn is truly an open-ended trait. In any case, it *appears* to be relatively open-ended.

It is reasonable to hypothesize that the Oligocene anthropoids had some sorts of social systems, in view of the fact that all living primates with one or two exceptions in the prosimians have them. Primate societies, and those of mammals generally, tend to be based on recognition between individuals of the deme, and to involve dominance hierarchies (see Chapter 12). Each individual has at any given moment a specific status vis-à-vis each other individual. This status is not fixed in time, there being both upward and downward mobility. We have already examined, in Chapter 12, the selective significance of this sort of social system, and will not pursue that further here. It should be emphasized that mammalian social systems are intimately tied up with individual recognition, involving as they do individual status. While complex civilized human societies clearly are not adequately described by invoking the dominance hierarchy alone, it is probable that early human societies were grounded in this ordering principle, just as are those of other primates. Many insect societies, on the contrary, (such as those of ants or bees) are organized around classes of morphs which are not functionally individuals in the sense of having individual roles.

Primate societies involve communication by both visual and auditory means. Vocal-auditory means, in particular, are considered to have been important for life in trees or forests where members of the troop would be out of each other's sight frequently. Many living anthropoids keep up a more or less constant chattering as they move about and feed, and this no doubt helps to keep the troop together and in touch. This chatter can go on even while the body is occupied in many different pursuits, and, of course, the signals can penetrate leaves and other barriers. Neither of these advantages is associated with visual signals, which, however, have other virtues, and are also utilized in primate societies. Signalling visually absorbs the entire consciousness of both sender and receiver, and is used during important interactions such as dominance conflicts. Visual signals can usually be more precise than auditory ones, and so can probably convey a greater range of meaning at the level of signs and signals, although the point is perhaps debatable. In any case, both sorts of signals are used in anthropoid societies and probably were used in the Oligocene.

It is possible that at least some of the signals and signs used in communication are also learned. Living anthropoid societies demonstrate cultural traditions (in the sense of having learned conventions; it may be noted here that some anthropologists restrict the word culture to what will here be termed symbolic culture, characteristic of man alone). For example, different troops of mountain gorillas living in the same general area feed on statistically different food items, each troop having its own particular mix of available items. Each individual learns from other members of its troop what to eat. In macaques it has been observed that youngsters exploring new food items are punished for doing so. Here no food item can become part of the tradition until it is passed on by a dominant (alpha) male.

Thus, our (heavily extrapolated) picture of the Oligocene hominoid that gave rise to the lineage leading to man is of a smallish, more or less monkey-like herbivorous-omnivorous anthropoid possessing binocular vision, living in forests in some sort of society of small troops involving both visual and vocal communication as well as a learned tradition. By Miocene times the hominoid lineage (that of apes and man combined— recognized by their having 5-cusped molars instead of 4-cusped ones as do monkeys) is represented in Eurasia and Africa by the dryopithecines, for whom the general description just given would not be inappropriate. Nothing new of substance appears of interest here except to note that some definitely ape-like features appear in some of them, suggesting that the pongid-hominid divergence was already under way.

RAMAPITHECUS

By upper Miocene times (some 13 million years ago) we find the first distinctively hominid fossil, a kind of dryopithecine known as *Ramapithecus*. It is represented in both India and Africa by parts of upper jaws and teeth. It is considered to be hominid because the canine teeth are relatively small (as in man) compared to those of other anthropoids. A contemporary of *Ramapithecus* was the "swamp ape" *Oreopithecus,* for which an almost complete skeleton is known. It too has relatively small canines and other hominid features, but also shows many details unlike any other known hominid or hominoid, and has been classified in a group by itself, one that is considered to have become extinct. It has, for example, a peculiar combination of ape-like limbs and limb girdles with quite monkey-like vertebral column and teeth. One wonders whether more detailed knowledge of *Ramapithecus* would force us to exclude it, too, from the lineage leading to man. We should note, however, that even if only the same amount of material was present for *Oreopithecus* as we have for *Ramapithecus,* it would still have been relegated to a group of its own on the basis of the way in which the canines occlude, which differs from the condition in all other hominoids.

The evolutionary decrease in the size of the canines in the lineage leading to man (or in any primate lineage) needs explanation. All primates use their canines for two very important functions. First, they are used as tools in peeling fruits and nuts, and secondly (and perhaps more importantly) they are used as weapons for intraspecific strife, including dominance quarrels, and for defense. Characteristically, the canines of males are relatively larger than those of females in living primates and this is related to the male dominance system, as well as to defense against predators, which often is a male specialty. The canines of females, however, are of ample size too. In view of the widespread importance of the canines in feeding, defense, and in the social systems, how can one visualize selection favoring a decrease in their size? The only suggestion that seems to have been made concerning this is a change in food habits leading to the necessity for increased mastication of food. Typically, other primates cannot chew as well as we can; they cannot chew (or grind) as well from side to side because the canines occlude in such a way as to prevent that. This is perhaps as plausible an explanation as any.

The implications of decreased canine size are of major importance. Perhaps most importantly, it suggests that *Ramapithecus* was capable of using sticks and/or missiles for defense, and perhaps for intraspecific fighting as well. It is hard to see how canine reduction could have occurred in the absence of some sort of defensive and aggressive weapons. The implication here is that the use of tools of various sorts antedates *Ramapithecus,* arising among some dryopithecines sometime in the mid-Miocene. Further, defense with tools like sticks is probably more effective in a group, and so one might speculate the *Ramapithecus* lived in troops involving more than one adult male, as for instance baboons do today. Still further, the use of weapon-tools grasped in the hand would certainly set up selection pressures in favor of bipedalism. Perhaps *Ramapithecus* was more capable of remaining on his hind limbs than any living ape. It has been suggested that the decreased canine size may have been associated as well with a decrease in the ape-type sex dimorphism in body size (huge males), which is not so extreme in man. The use of artificial weapons ("equalizers") would tend to weaken selection for extra-large body size in males. The decreased size of the canine appears to be associated with a delay in its eruption rather than with any other factors since man's canine does have a fully developed, robust root, but is among the last teeth to erupt. Thus, during the Pliocene, of some 13 million years duration in which no hominid fossils have been found, we can imagine the lineage that was going to give rise to man as having begun to move toward bipedalism, and to already have achieved the use of artificial tools and weapons. They may in addition have lived in some kind of multimale extended family society, and eaten foods that required much chewing, suggestive of a more granivorous diet.

Tool using is found sporadically throughout the more advanced vertebrates. One of the finches on the Galapagos Islands uses cactus spines held in the beak to probe cracks in tree bark for insects; shrikes impale insects on thorns in order to hold them so that they can be dismembered; the California sea otter uses stones to crack open mollusc shells, as do some species of thrushes feeding on land snails. There is even a kind of wasp that uses pebbles to tamp down the soil around its nest. Of course, there are examples of tool use by living anthropoids as well, some of which may pick up and throw objects at predators. Tool use as such is not remarkable among animals.

THE AUSTRALOPITHECINES

The fossil record of hominids resumes some 10 million years after *Ramapithecus,* some 2 to 3 million years ago, primarily in Africa but also slightly later in southeast Asia, with the appearance of the australopithecines. Relatively large portions of the skeletons of a few of these have been found and studied, and they are rather well-known by the standards of the human fossil record. The really important advance toward the human condition shown by them is the acquisition of some degree of bipedalism; they may have stood and walked completely erect, more or less as we do—or perhaps they were fully erect only while running. The upright stance involves many changes from the ape grade of organization, primarily in the pelvis and foot, but also in the entire hind limb and its musculature. One of the muscles, the gluteus maximus, for example, has changed from a flexor to an extensor, forming the buttocks so characteristic of man. The foot has acquired an arch. The hands are now completely freed from involvement in progression.

Bipedalism must have been achieved sometime in the Pliocene, perhaps in connection with life on open plains. Widespread drying tendencies in the Pliocene correlate well with this formulation. As forests were diminishing in size a semierect primate could have been driven to adapt to savannah life, and was indeed preadapted for it. One can imagine that only a change in behavior might have been needed to make the initial transition, selection then beginning to work on running ability in the context of an uprightness that allowed continual scanning of the horizons.

Australopithecus remains have been found in cave sites in South Africa and in open sites in eastern Africa. Thus, we know that at least by 2 million years ago our ancestors were using caves for shelter. Stone tools are found in sites of the same age in the same regions, some associated with the fossils. These are of a style known as pebble tools and are roundish stones with one edge sharpened by chipping away some flakes of the stone. We are clearly in the presence of a tool-maker as well as a tool-user. Tool making is very much less common among animals than tool using. Chimpanzees in nature do fashion some crude tools. An example is when they take straws and break them off to certain sizes and then use them to probe termite nests, licking off the termites that have attacked the intruding straw and are clinging to it with their mandibles. In the laboratory it was found that a chimpanzee could learn to fit together two small sticks provided with ferrules to make a single, longer, stick in order to knock down fruits placed out of its reach. Tool making outside of the primates is very rare. Male bower birds construct elaborate mating stations out of colored fruits, feathers, flowers, and so on to attract females; birds in general construct nests, as do many other kinds of animals. Whether nests qualify as tools is perhaps a matter of taste; construction of tools *other* than nests is quite rare outside of the hominoids.

What the pebble tools were used for is not clear; they look like choppers, scrapers, crude hacking instruments. It is clear that the ability to make tools has entered into the learned (cultural) tradition by this time. Almost certainly the hand-brain complex would not have developed to the point where enough skill was available to make these tools until the hands had been largely freed of a function in walking—another reason for postulating bipedalism in this form. The thumb was presumably now fully opposable as it is in us.

Australopithecus appears to have lived at an economic level described as hunting-gathering (or perhaps scavenging), as we know from finding the bones of slow game in association with the tools. These bones include those of lizards, snakes, birds, small mammals, young (including suckling) giant mammals, and young crocodilians. These are the kinds of animals caught by today's hunter-gatherer peoples, including the women; it is for this reason only that we assume that *Australopithecus* also gathered fruits, nuts, and shoots. Whatever food specialization in the past led to a decrease in the size of the canines in the lineage leading to *Australopithecus,* this prehuman form was probably as omnivorous as people living today in most cultures are.

Australopithecus is associated with another hominid in both Africa and Asia, a much larger form given the name *Paranthropus*. *Australopithecus* was relatively small, an adult weighing some 60 to 75 pounds; *Paranthropus* weighed more on the order of 130 pounds. The latter has very peculiarly specialized teeth. The molars are huge and form an efficient grinding battery, while the incisors and canines are very small. It is thought that the differences in their dentitions indicate that these two hominids did not compete with each other in terms of food supply; *Paranthropus* presumably fed only on very coarse vegetal matter, possibly of a kind not utilized by *Australopithecus*. The lineage leading away from *Parathropus* is considered to have become extinct, while that involving *Australopithecus* continued to survive, ultimately giving rise to man.

Australopithecus is not considered to be a member of the genus *Homo* primarily because of its small cranial capacity, ranging from 435 to 700 milliliters in adults, which is about the same range as found today in gorillas (420–752 milliliters). However, it is now considered by some that the brain size:body size ratio is more significant than absolute brain size. In these terms the gorilla is 1:420, modern man 1:47, and *Australopithecus* (with its small body size) about 1:42. The two latter have, then, more associative neurons per gram of body mass than do other anthropoids. On the other hand, it is still considered plausible that the mental faculties of modern man depend on some lower limit or critical mass of associative neurons—for example, a small monkey with a brain:body ratio of 1:40 almost certainly would not show a mental capacity comparable to that of modern man. *Australopithecus* would probably be considered relatively stupid by modern men (with cranial capacities ranging from 850 to 2100 milliliters in adults). Of course, these calculations do not take into account specializations in the structure of the brain that may account for some of modern man's mental ability. Thus, a three-year-old child has an absolutely smaller brain than that of *Australopithecus,* but one could not call the average three year old stupid—in fact, he is learning faster than will be the case at any later age. Specializations in the brain will be taken up again below.

Probably at the level of the australopithecines wisdom began to assume a degree of importance, and with it, perhaps a kind of group selection (see Chapter 12), and also the phenomena of senescence. We can perhaps picture *Australopithecus* as living in extended families or even in multifamily groups. In the world of the hunter-gatherer experience counts for a great deal, if not in the day-to-day training of the young to perform tasks related to survival, then in respect to infrequent natural occurrences—volcanic eruptions, exceptionally bad winters, surprise contact with a distantly related foreign human group, locust outbreaks, eclipses, and other physically or psychically dangerous occurrences of which some of the oldest members of the group only will have had previous experience. The presence of such an experienced (wise) individual could mean the difference between survival or nonsurvival of the group. A group including such an individual stands a better chance, as a group, than one that does not when some infrequent crisis occurs. If the group is an extended family, then kin selection can be invoked to favor longevity of postreproductive individuals; if the group is multifamily, group selection is a possibility, providing that population sizes were below carrying capacity. In times of expanding population size (which have occurred quite obviously in many human groups from time to time) it is not automatic that any energy expended to aid unrelated individuals is necessarily deducted from the altruist's own reproductive effort. This effect *must* occur only at carrying capacity numbers. Selection could tend to favor genotypes that produce, in whatever fashion, individuals that live longer—*providing* that some of the individuals benefited by an individual's longevity are close relatives (see Chapter 12).

Selection for wisdom, then, may also have been selection for long life, which undoubtedly would have interacted with the phenomenon of senescence. In most organisms the usual path to individual destruction is by way of being harvested as a food resource by some other living system. This almost invariably occurs long before any signs of old age. Probably man is one of the few organisms for which senescence is an important biological factor, although it can occur in other organisms under laboratory conditions. Indeed, it seems that most living systems are capable of senescing if artificially protected from predators. The most widely accepted definition of senescence is an increased force of mortality, or a decreased capability to survive environmental insults *of all kinds* with increasing age— that is, the probability of death increases with age (Figure 6-4). This is an actuarial or population-based definition, and does not consider the familiar gray hairs, loss of memory, and arthritis found to accompany the increased probability of dying in aging human individuals. Perhaps these familiar manifestations can be seen as the scars resulting from an increasingly less efficient repair mechanism, but it must be admitted that the causes of physiological aging in humans are not understood. What is known is that *all* systems break down with age, so that the acquisition of a new heart by surgery would be followed by the need for a new kidney, and so on. The average upper limit on human age appears at present to be around 100 years; by that age the probability of any individual dying becomes 100 percent. Senescence, then, appears to be a trait potential in all living systems but which has actually appeared to a significant extent in human populations (and perhaps in a few others).

Now, senescence can be viewed as a force that has been opposed to selection for longevity. The mean maximum age in *Drosophila* flies has been increased by selection in a laboratory study, so that we know that such a process is possible, but there appears to be some intrinsic limitation on it; postreproductive individuals inevitably break down later if not killed off sooner. Actually, selection for longer life would not in itself uncover the aging process; it would only delay its onset. However, in humans, the same situation that gives rise to selection for longevity (say, the increased survival value of wisdom) would also favor the development of traits involved with nonphysiological means for living longer (increased protection from predators and inclement weather, and so on) which in human beings is associated with the development of culture. The development of such means for living longer is conceptually the equivalent of placing an animal population in the laboratory and artificially protecting them from predators, bad weather, and other sources of mortality. Both result in uncovering the aging process. An aged human individual can still be of great value to his or her group in terms of the wisdom stored away, providing only that aging has not progressed too far (to senility). We can, then, imagine the beginnings of a distinctively human-style culture with (or soon after) the australopithecines.

Selection for wisdom may have had simultaneously another effect on the history of man. In mammals there is a fairly good correlation between longevity and body size; for some unknown reason larger mammals tend to live longer—perhaps for some reason involved with the fact that the metabolic rates of larger mammals tend to be slower. The next group of fossil hominids to be found in the record has a larger body size than the australopithecines, and the process under consideration may have contributed to this change, although selection working to decrease the surface area:volume ratio was perhaps also involved during the Günz glaciation (Bergmann's rule).

HOMO ERECTUS

In summary, as we approach *Homo erectus* in Eurasia during the first Pleistocene interglacial period, some 550,000 years ago, we may speculate that selection favoring accumulated wisdom may already have been producing (1) increased body size, (2) increased longevity, and (3) the beginnings of a human-style (symbolic) culture (see below) with its concomitant senescence, while selection to curtail loss of body heat may have contributed to the increased body size.

Homo erectus (*"Pithecanthropus," "Sinanthropus," "Atlanthropus"*) was about the same size as modern man, with a somewhat smaller brain (on the average); cranial capacities were 775 to 1125 milliliters as compared to modern man's 850 to 1700 milliliters. These were what are popularly conceived of as "cave men" or "ape men." They had larger brow ridges than modern man, and less chin. Evidently during the 300,000 years of their existence natural selection produced physical man as we know him today (see below).

There are some 18 individual pithecanthropines known from Java, 2 from North Africa, 4 from southeastern Africa, and 40 from China. Some of the individuals are represented only by single teeth, however, and the record for these is poor indeed. Opinion is divided as to whether the contemporaneous parent populations formed a continuous gene pool with each other or not. The degree of variation found among all of them is not greater than that found between modern human races, and so they are considered to have formed a single polytypic species.

The most important single occurrence at this level from the point of view of producing modern man was an apparent change from the economic status of hunter-gatherer to that of big-game hunter. This change was dependent upon the presence of huge herds of very large mammals during middle Pleistocene times, and probably selection for lower body surface: body volume ratios among mammals in general was partly responsible for the presence of so many large species (see Chapter 5). Bones of big game are associated with the fossil finds, and with contemporary finds of tools. Evidence has recently come to light to the effect that large-scale hunting by man may have led to the eventual extinction of most if not all the large Pleistocene mammals in Eurasia—the mammoths, mastodons, woolly rhinoceri, cave bears, giant deer, and so on.

In the type of hunting in question, a party of several males drives a stampeding herd into a cul-de-sac where they can be stoned en masse from surrounding cliffs, or the herd is driven through a narrow gorge fitted with pointed stakes onto which the lead animals are driven by the crush from behind, or a herd is driven off the edge of a cliff and slaughtered later at leisure (as the American Indians are known to have done with bisons, and which may have been done to elephants in Africa at around the same time). Initially, big-game hunting was probably not carried out on such a scale, and may have been simply a cooperative venture whereby several males selected some individual to surround and kill by the handiest method. The important point is that big-game hunting must have been a *cooperative* adventure, given the sorts of weapons that were available. Cooperative big-game hunting occurs among other mammals as well, for instance, among wolves, lions, and the Cape hunting dog of South Africa. In none of these cases can any planning be detected; each individual hunts as an individual and large animals can be brought down simply because more than one individual is simultaneously hunting it. Here the cooperation only involves everyone selecting the same target. It is probable that by the time man takes to this adaptive zone he has not yet achieved sufficient communicative skill to be able to plan a complex hunt in advance, but he was probably at a stage beyond that of the wolf. Selection working in the context of big game hunting would certainly tend to favor traits allowing the formation of multifamily groups, and socializing tendencies must have been greatly enhanced during these times (see Chapter 12); for the first time in the record we *must* postulate the presence of more than one male per functioning group, and any means favoring their peaceful coexistence was favored by selection. If they were not already present, dominance systems would have had to evolve now; probably they were already present in at least rudimentary form (see Chapter 12). In this connection it may be pointed out that male beards and large penises have been cited as probably evolving as aggressive signals that were used in dominance struggles. Not only is cooperation necessary in big-game hunting, so is sharing. A mastodon would contain more meat than could be consumed by a single family before spoilage. This fact, too, would tend to favor the development of socializing tendencies, and the federation of different families.

The arduous nature of big-game hunting is thought to have probably excluded the females from more than passing participation, involved as they would have been with bearing and nursing children. Thus, big-game hunting may well have been the original impetus for the development of very different social roles for males and females, emphasizing as it would the division of labor already implicit in female child-bearing and rearing. Women probably continued to gather vegetal matter and slow game.

Big-game hunting has been considered to have affected sexual patterns in important ways. If the males are away on hunting parties, they could return at unpredictable times, and with renewed contact with the women would probably be stimulated into sexual activity. This means that the women would need to be sexually excitable at all times, not just during estrus as in other anthropoids. Since women have a fertile period as short as in other animals, an increase in sexual activity in times other than near estrus would not increase the probability of becoming pregnant, but the increased sexual activity could function as a means to orient males toward the home site. It may be pointed out that some other anthropoids show continual female sexual receptivity—some baboons, for example—and so this feature is not unique to man. However, in these cases estrus still occurs, and at that time the females actively seek out the dominant males. This being so, the evolution of the menstrual cycle may go back to prehominid times. It has been postulated that the development of permanent secondary sexual characters in women, such as prominently displayed nipples, characteristically wide hips, characteristic voice, and so on, was associated with the loss of estrus. In other anthropoids secondary sexual characteristics do not appear except during estrus. Thus we may have had in this phase of human evolution a tendency to disengage sexual activity from completely cyclic hormonal control and to associate it partly with cortical control and reflexes based on external cues and stimuli (with sight of breasts, genital titillation, and so on). Solving one problem frequently involves raising another, and the change in sexual arousal from mostly intrinsic to mostly extrinsic control associated with a need for continual receptivity in women could have led to the need for increased social controls over sexual activity, which could now take place at any time. This need no doubt would have had its effect too on the socializing process going on during these times.

The fact that big-game hunting is a very arduous way of making a living is thought to have left its mark on man's physical structure too. Many big-game-hunting animals hunt at night or during dawn or dusk; man tends to hunt in the daytime (a possible heritage from arboreal ancestors). This would mean that the hard work of the chase and the kill had often to be accomplished in the heat of the sun. Under such conditions it has been suggested that selection could have favored individuals that lost metabolic heat faster than their companions, so that during these times man lost thick body hair and developed increased numbers of sweat glands. Other anthropoids, like dogs, can accomplish evaporative cooling mostly at the tongue; man can effectively evaporate and cool all over his body. (Other animals selected by man for hard work, like race horses and greyhounds, also have developed an increased density of sweat glands in the skin, as well as short hair.) Some hair, on the other hand, was elaborated as part of a spectrum of permanent secondary sexual traits (axial and pubic hair).

The pithecanthropines show significantly increased skill as tool makers compared with the australopithecines, a fact possibly associated with their larger brains. It should be mentioned that larger brain sizes had to have had their effect on the female pelvis along with the sexual selection mentioned above. A wider birth canal is needed, and its development widens the dimensions of the female pelvis. It is, of course, not unusual for more than a single selection pressure to simultaneously favor a single evolutionary development. A secondary sexual characteristic can be almost anything. In the case of pelvic width, selection for a larger birth canal would have involved many cases of death in childbirth and/or infant mortality due to too narrow birth canals, while at the same time fathers sexually excited especially by wide hips would leave more children (as would the owners of the hips). Thus, genotypes would be favored that were capable of producing wider birth canals in the presence of sex hormones in concentrations typical of females; the alleles making up such genotypes would be the ones preferentially projected into the next generation. At the same time alleles of other genes that are capable of encouraging sexual stimulation at the sight of wide hips in the presence of male-level sex hormones, and other alleles at still other genes more capable of producing genotypes that would preferentially deposit fat on the hips and buttocks in association with female internal environments, would be simultaneously favored by selection. This hypothetical example demonstrates how, in a sense, the entire gene pool of a species would evolve as a unit, the alleles present at any given locus tending to be coadapted with those of all other loci (Chapters 7 and 8).

Returning to pithecanthropine tools, there can be distinguished at the time of *Homo erectus* a curious division of tool industries east and west. The earliest tools (associated with the australopithecines) are what are called pebble tools. Here a stone has one edge sharpened by chipping, either on one side or alternately on both sides. Possibly the chips themselves were used for some sorts of tools as well. This basic style of tool making is found in more advanced form (more variety of tools better made) associated with the eastern pithecanthropines (Java and Peking man), and was the basic style of the recent immediate ancestors of the Australian aborigines as well, suggesting no basic change in tool making throughout the transition through *Homo erectus* to *Homo sapiens* in that region, excepting refinements. In western Eurasia and Africa, however, this basic style of tool making—referred to as the chopper tradition—evolved in two directions. On the one hand, we get a flake tool tradition, in which stone flakes struck off a core are prepared by further chipping, often on one side, thus emphasising the chips. On the other hand, we have the development of the biface core tool, or hand axe, where the central core from which the flakes are chipped away is more or less carefully shaped on both sides to form a chopping or cutting tool. Tools of both kinds are often found at a single site, but frequently one or the other predominates, possibly reflecting the sort of stone available in a given area, since each method is better expressed with different minerals. There is also some belief that the hand axe tends to be present in relatively tropical sites, while flake tools predominated in more boreal regions. For example, during the last glaciation hand axes disappeared from Europe, possibly reflecting migration south by their makers. The predominantly flake industries tend throughout the old stone age to be concentrated in northern or glaciated regions. This may reflect ecological differences as much as anything else, and it has been suggested that flake tools are better for skinning animals and for preparing skins than are hand axes, and that they were thus favored by northern peoples who wore furs. (It may be noted in this context, however, that a modern western scientist was able to skin and butcher a small antelope in 20 minutes with a relatively unspecialized pebble-tool chopper; the skin was not scraped and cleaned, however.)

The interesting aspect of the geographical distribution of hand axes is that they never have been found east of India—this over a period of minimally 300,000 years during which they were common, especially in tropical regions to the west. Inasmuch as physical contact between human groups usually involves cultural exchange, this raises an interesting question as to whether this distribution reflects a period of isolation between eastern and western populations of *Homo erectus*. This does not automatically follow, of course, since cultural contact need not extend to every artefact. Again, cultural isolation might be maintained even though gene flow occurs at a level high enough to prevent major genomic differentiation. The maximum period of potential isolation here would amount to some 14,000 human generations. Recall that it took only some 300 generations of peppered moths to replace one allele with another under the action of quite severe selective forces (Chapter 10). This means that during this period there would have been ample time for race formation in *Homo erectus*.

Such race formation has been claimed by Carleton S. Coon, of the University of Pennsylvania, who has advanced the thesis that the racial differentiation observed in modern people originated at the level of *Homo erectus*. The major evidence he has cited for this is that there are certain similarities between Java man, Peking man, and modern mongoloid peoples in the shapes of certain bones, and that these traits are not shared by other pithecanthropines or other modern racial groups. The major merit of this idea is that it proposes a time and place for racial differentiation that also unites pithecanthropines and modern man, forming an overall picture of the gross evolution of the genus *Homo*. The objection has been raised that it is difficult to see how such long periods of isolation could fail to lead to speciation. But one must consider that both *Homo erectus* and *Homo sapiens* were and are widespread polytypic species with some gene flow between the various populations at all times, differing in rate at different periods of time, but always with enough to have prevented speciation.

A further objection that has been raised to this idea is that it implies that whatever evolutionary events led from *Homo erectus* to *Homo sapiens* would have to be conceived of as happening in different populations independently and in parallel. This, however, is really overstating the case inasmuch as we may visualize a series of populations connected by a degree of gene flow that prevented speciation. Some traits might become differentiated in adapting to environmental conditions, but not all; in some respects the species would evolve as a unit—notably, in this case, in various facial and cranial traits, resulting in the simultaneous evolution in the same directtion of various features of the skull in all populations. Further information will be needed before this idea can be properly evaluated.

Another tool used by *Homo erectus* was fire. Remains of fires in association with human artefacts are found in both Europe and Asia dating to the old stone age. How and when man first tamed this wild, fearsome, and unpredictable force is not known. Fire has many uses. Its importance as a source of warmth at a time when body fur was probably being eliminated is obvious. Its ability to scare off large carnivores was probably of use. Its ability to soften tough meats and vegetables is important to an organism with such a generalized tooth row as man has. Its role in the development of social behavior may have been considerable in that it would mark the home, the hearth, the warm place where the women and children stayed together as a unit while the hunters were away. Fire could be used to harden pointed wooden stakes, converting them into usable spears. It is difficult not to imagine that it might have been the religious center of the early societies as well, in association with lightning or the sun, with the stars and comets and the moon.

HOMO SAPIENS

Homo sapiens first appears in the fossil record about 200,000 years ago, just before the Riss glaciation (Illinoian of North America) with Swanscomb man in England and, slightly later, Steinheim man in Europe. During the next interglacial period very fine flake tools (Mousterian) are found in Europe that undoubtedly were made by *Homo sapiens*. Thereafter remains are found in Europe, Asia, and Africa to which continents man was confined until the discovery of America perhaps some 40,000 years ago by the ancestors of one group of American indians, and the discovery of Australia by the ancestors of the Australian aborigines at least by 16,000 years ago.

The physical differences between *Homo erectus* and *Homo sapiens* are relatively slight, and are bridged by the earliest known group of *Homo sapiens*—Neanderthal man. Some authorities have even placed some of the Neanderthalers in *Homo erectus* from time to time. *Homo sapiens* has a somewhat larger and rounder brain case, smaller brow ridges, a more pronounced chin, and so on, in comparison with the pithecanthropines, and the Neanderthalers are more or less intermediate. They evidently had on the average an even larger cranial capacity than modern man. The Neanderthalers were present on all of Eurasia and in Africa, and it is not clear whether they represent a stage or grade in the evolution of modern human beings from pithecanthropines, or whether they were only another (and now extinct) race of modern man. Only during the later part of the Würm glaciations are both modern man and the Neanderthalers found contemporaneously, but since this is so at one point, it is evidence for the position that the latter represent only another modern race.

Neanderthalers are found in association with a rather fine tool industry—the Mousterian. Some of this is found even as early as the latter part of the Riss glaciation, and gradually comes to replace the earlier hand axe cultures in all the subglacial regions during Würm times. Just before this, during the third interglacial period, there was in Europe a dichotomy of tool traditions, with the hand axe cultures occupying North Africa, Spain, France, Italy, and extending into Britain. Overlapping them in Britain and France, and extending eastward to the Caspian Sea, were various flake tool cultures. It is felt by some that the Neanderthalers developed their

Mousterian industry from the latter, northern cultures, and became increasingly successful with the gradual cooling of the climate, slowly replacing the hand axe people further and further south. Relatively suddenly, at the end of the Würm glaciations, typical modern man appears on the scene and replaces the Neanderthalers—some have even suggested violently.

It is usually considered that mental abilities comparable to that of modern man do not appear until after the *Homo erectus* level (or grade?). As suggested above, it is improbable that man's mental abilities are associated simply with a certain mass (or number) of associative neurons, but involves also some specializations in the organization of the associative centers. For instance, the speech center in most people is located only on the left side of the cerebral cortex. Damage to this area can result in a situation in which the individual recognizes objects but cannot name them—he no longer associates their names with them. This asymmetry of brain function has not been discovered in animals other than man. Possibly associated with the asymmetry of the speech center is the right-handedness of most people; the right hand is controlled by the left side of the brain due to a crossing over of nerve tracts from one side to the other. There is no doubt that the continued evolution of skill as a tool maker involved both the brain, the hands, and the nerve connections between them. The right hand becomes the tool for precision manipulation controlled by the left side of the brain at the same time that the speech center evolves on that side in connection with naming the objects manipulated. Again, the impression is that of whole blocks of genes becoming coadapted to a single major adaptation.

SYMBOLIZING

Man's major specialization (or adaptation) is symbolic thinking as expressed in symbolic languages. It is on this foundation that human culture is built, and it is this that differentiates that culture (symbolic culture) from those found among animals. *Symbols* need to be contrasted with signs and signals. Symbols are arbitrary (conventional) sounds or ciphers that *stand for something else*. Thus, the word "tree" or "arbre" represents a tree but *is* not a tree. Spoken and written languages are all symbolic—as is the Amerindian "sign language," or that used by the deaf. Mathematics is a logical system of manipulation of quantitative symbols; musical notation is a symbolic representation of music. A sign or signal, in contrast, does not represent something else, although it does convey information. For instance, a blush is a sign of a certain emotional state; a baring of the teeth in a grin universally among anthropoids signifies that the grinner means no harm, is ready to be reasonable or sociable; a direct stare accompanied by a furrowing of the brow (or frowning) signifies anger or disapproval, and so on. A traffic "sign" with the word *stop* is not a sign in this sense, but could become one. Thus, a dog conditioned to become hungry and salivate at the sound of a bell is responding to the bell sound as if it were a sign—it has meaning for the autonomic nervous system. If one were to become so conditioned as to respond to the sight of the traffic sign by immediately and unthinkingly coming to a halt, then that traffic sign would be functioning as a sign rather than as a symbol. Many signs are inborn—that is, their performance and the response they elicit are both inherited as reflexes. The study of animal behavior is largely a study of systems of inherited signs and their use in communication. Animal communication, then, is a matter of specific situations releasing appropriate signaling by the organism, which functions to communicate its condition to others of its own kind who then respond reflexively with another appropriate inherited pattern of behavior. In all of this there is no thinking because there are no words. An analogous human situation might be a fencing match, where many of the movements are stereotyped and learned so well as to become "second nature." In this situation a "thrust" is a certain clearly defined thing and *means* the same thing, just as a distress call from a nestling not only means distress, it *is* distress in the sense of its being one of a set of obligatory inherited responses to being disturbed. There can be no displacement of meaning with signs; they always and only occur in the presence of the specific stimulus that releases them. Symbols, being abstractions, can be handled and played with at any time. Another important distinction between languages based on signs and those based on symbols is that the latter show productivity, that is, they can be used creatively. Thus, we have a concept-word "tree" and a concept-word "gold," and these can be combined into "gold tree" or "tree gold"; neither represents things in nature, and both are in fact entirely new concepts. In this way a symbolic language can be used to produce or create something unrestricted by the natural world. Needless to say, the elements of animal sign languages cannot be so manipulated, perhaps mainly because there can be no displacement from the appropriate stimuli because the signs are part of the biological response to a given situation.

It can be seen that symbols having displacement, and rules for generating productivity (grammar), forming together a symbolic system or language, have the quality of being open-ended in the genetic sense. An organism that has the ability to symbolize programmed into its genome can use that ability to generate a virtually unlimited and genetically unspecified series or number of qualitatively different concepts. We can now refine our notion of genetic open-endedness in learning. There is no doubt that what any organism can learn is strictly limited by its sensory apparatus. In addition, the ability to name things and states and processes may be limited by the number of things and states and processes an organism can perceive, but the words so derived can be combined grammatically to produce an almost unlimited number of new entities (concepts). The quality of unlimitedness resides in the new creations, not in the sensory world. However, some of the new concepts can (and do) suggest things about the real (sensory) world that could not be apprehended in any other way. For example, "stones fall; little stones fall—big stones must fall too; little stones fall on bug and kill it—big stones fall on rhinoceros and kill; little stones fall from my knee—big stones need high place to fall from—cliff is high place;" and so on. It was no doubt this quality that led to the selective advantage of symbolic thought; this quality allows planning. Notice that the generation of new concepts and planning both involve a playing with the symbols, a juggling of them in a way that cliffs and rhinoceri cannot be juggled.

It seems clear now that only man has the capacity to symbolize, whatever that means neurologically. Other organisms can learn new things. For example, in England an enterprising individual of a species of titmouse learned how to remove the tops from milk bottles and then drank the cream. This behavior spread rapidly from one population to the next until in only a few years it occurred all over the country. It is clear that birds can learn to do new things with their beaks by observing another bird. Other organisms can "think on their feet." For example, a chimpanzee is capable of putting together two sticks to make a long one in order to obtain a piece of fruit placed out of its reach. It can do this if all the elements of the situation are together in one place, and there is no reason to believe that it could not come up with the idea of stoning leopards from a cliff *if* it were on the cliff and the leopard was down below and the stone was obviously ready, and so on. But in the absence of abstract words, it could not conceive of the situation, say, at night in the dark just before it fell asleep. (Recent work with the chimpanzee appears to show that this, our closest relative, also may have the ability to symbolize in some degree, although it may not use this ability much in nature.)

It is not too much to say that with the ability to symbolize man entered a new adaptive zone. Instead of dealing with direct stimuli from the external environment, symbolizing results in man dealing with his own parallel, analogous world of words triggered by these stimuli. In this situation the concept of snow, for example, is more important (and, of course, more varied from one language to the next) than the white, wet, cold, soft stuff one might flounder in. For man, concepts constructed of words according to the rules of grammar are at least equally as important as (and perhaps more important than) raw sensory data unfiltered through verbal consciousness. Man lives within his world of symbols every bit as much as a mollusc lives in the shell it secretes. People need to make extraordinary conscious efforts to detach themselves from the conceptual world in which their personal development took place, using methods such as those devised by mystics in various religions (Zen, yoga). The state of consciousness achieved by such methods (satori, samadhi, nirvana) is frequently described as being beyond verbal consciousness, in a state that in the present context could perhaps be referred to as prehuman or animal.

Symbolic language is the foundation on which human cultures are built. The institutions of these cultures (political structures, religions, arts, economic organization) could not possibly exist without symbols. Interesting examples are the kinship and marriage relationships characteristic of various peoples today, which might possibly reflect the kinds of systems devised very early by *Homo sapiens*. All these systems promote outbreeding in that they involve incest taboos, and stipulate precisely the degree of relationship of marriageable individuals. Most allow only one type of cross-cousin marriage, either to mother's brother's offspring or to father's sister's offspring. Prohibited are marriages to father's brother's offspring (one might call this descent via the Y chromosome—patrilineal descent) or, in some cases, mother's sister's offspring (matrilineal descent). Note that the greatest degree of homozygosity this system allows is for one-fourth of the chromosomes to be identical. This is the greatest degree of homozygosity that could be achieved under *any* cross-cousin marriage, and, therefore, we see that the distinctions made by peoples as to who is marriageable and who is not has no biological sanction whatever. These distinctions are completely artificial in respect to biology, as are such distinctions as father's sister and father's brother in respect to who can be one's parent-in-law. Now, in fact, without symbolic concepts such as "mother's brother" it would not be possible to reckon who is who. Without symbolic language the best one could do, perhaps, is to distinguish one's mother and siblings. The importance of these kinship systems lies not in biology but in economic relationships. Outbreeding promotes new ties with new families, allowing broader trade and better defense, and so on. Thus, symbolizing allows man to construct a kinship system which then becomes the agent that defines reality in connection with mates; so we see how the biologically important function of breeding is dealt with not for itself on its own terms, but in terms of economic and social needs. Whatever symbolic system is erected, of course, must not be grossly at odds with the needs of biological nature. For instance, if a kinship system were erected involving marrying one's mother, the group in question would probably become extinct fairly rapidly. Thus, the needs of biological man define limits for the cultural institutions, but do not otherwise determine them.

Symbolic languages frequently come to be written. The invention of writing has fairly important implications. No longer is the content of the symbolic culture limited to what can be carried about in the minds of living persons. Things can be written down, forgotten, and retrieved again. This increases the content of any culture, and promotes accuracy. With the promotion of accuracy there sets in a kind of conservatism in cultural evolution that can be exemplified by the codification of the French language by the Académie Française during the 17th century. Until this time that language was changing fairly considerably with time (as do most languages), but it has almost not changed since; not surprisingly, present-day French is known for its precision. Another important result of writing is that it tends to eliminate the function of the elderly wise ones, and, indeed, converts them into simply the elderly. Can anyone seriously maintain that there is any function for wisdom in twentieth century America? This result is achieved because no individual can possibly hold in his mind the knowledge of a single culture once it begins to accumulate as writing instead of being transmuted, condensed, and generalized in the minds of men. No one can know enough about enough problem areas to qualify as being wise; authority passes from living men to books; indeed, the generation of new knowledge itself often becomes a team task. Of course, today's very large populations also contribute to the erosion of the importance of wisdom in that only a very small proportion of the population would have access to a wise individual even if one could exist; books are accessible to all (television does not qualify as personal contact).

TROPHIC LEVEL AND POPULATION SIZE

Increase in total population size has been an important aspect of the history of *Homo sapiens,* and is importantly involved with his ecological status. *Homo erectus* was a big-game hunter. Since he could capture only about 10–20 percent of the energy present in the animals he killed, his biomass was limited to roughly a tenth of the biomass of the herbivores he ate. Now, in a situation where there is an unusually rich source of these herbivores a reasonably large population can be maintained, as seems to have been the case with the neolithic peoples of France and Spain who produced the famous cave art, and, in later times, the Amerindians of the Pacific coast of North America. In the first case, the basis of the large populations seems to have been large herbivorous mammals, in the latter a combination of fish and cetaceans. Such people are secondary consumers (predators, carnivores). Only in the rare cases mentioned above is the source of game so rich that some surplus can come into existence, allowing for further specialization of social roles (allowing for priests, perhaps artists and/or artisans) because not every able-bodied male will have to take active part in the hunt. Now, as the hunting becomes increasingly efficient as a result of symbolic culture, the source of food becomes threatened with extinction. This way of life for man, who is culturally capable of rapidly increasing his efficiency, is ultimately self-destructive.

Some peoples met the crisis of game extinction by becoming herdsmen, by domesticating certain kinds of animals, on which they continued to prey, but which they now also bred. This gambit has never resulted in truly large populations or significant cultural innovations. (It is of interest that those other great "societies" on earth today, those of the ants, have also produced species that herd other sorts of insects, and aid in their breeding.)

The really significant way out of the game-extinction crisis was for man to return once again to the status of herbivore—to become himself a primary consumer. This now allows him to increase his biomass (and also population sizes) to the levels characteristic of the animals he used to kill. He can now store about 10 percent of the energy contained in the food plants in his own biomass instead of, as with hunters, only about 1 percent. The ability to return again to mainly herbivorous status was possible because all along man continued to be omnivorous, eating vegetable matter gathered by the women along with the meat of the game animals. In order to maintain really large populations as an herbivore an efficiency at least equalling that of the later hunters must be present in the gathering of food plants. This efficiency is possible if one determines where the food plants will grow, and if one can place all of one kind together so that they are more easily tended, guarded, and harvested. That kind of activity would be farming or agriculture.

Agriculture appears in the archaeological record about 10,000 years ago in the Middle East, and slightly later in Mesoamerica, in the Andes region, in China, and in southeastern Asia. From these nuclei it later spreads throughout the world. The idea of planting seeds where one wants them appears to have a high probability of occurring to man in no matter what cultural context. The ability to produce surpluses is vastly increased with agriculture, and all the great civilizations (large populations, luxury items, urban centers, specialized roles) were and are based on it. (Incidentally, here too there are parallels with ant societies, there being some that cultivate fungi—plant its spores, feed it with decaying matter, gather it, store it, and feed on it.)

ADAPTATION IN THE CONTEXT OF SYMBOLIC CULTURE

The sizes of human populations have continued to grow, partly because of the development of industry and science. Improved methods of farming and distribution depended on industry; death control depended on medicine, both preventive and curative. Together, these forces allowed more and more individuals to live through successful breeding and child-rearing, and by improving vigor, to begin breeding at earlier and earlier ages. All this can be seen within the framework of man's unique ability to control, change, and modify his environment. Instead of adapting to it in the biological way that *Homo erectus* was largely still doing, *Homo sapiens* has come to adapt to it culturally. If it gets colder, clothes are invented instead of growing fur.

The capture of fire by *Homo erectus* was perhaps the first important step in the direction of adaptation by culture rather than by biology. It allowed a broader choice of foods and it allowed man to live in places otherwise too cold. By thus modifying his immediate environment man did not have to evolve in certain directions. The process of dealing with environmental challenges by means of cultural adaptations itself undergoes evolutionary change analogous to that of biological evolution. The invention of the wheel leads to, and determines, the invention of the axle, and so on, while at the same time the nature of the wheel itself is gradually changed to conform to the needs of many moments. By a process akin to preadaptation, the wheel can become a pulley or a gear. The fact that conscious planning is involved does not invalidate the analogy with organic evolution because all through history the choices made were choices of the moment; there has never been any foresight or control over the future.

It has been noted that much of any culture is irrational. Probably irrationality in man is in any case usually adaptive; rationality (reason, logical thought) is present in man as only one special adaptation involving the cerebral cortex, and is, as far as we know, the only reasonable process in the universe. Furthermore, it has undoubtedly had only a very small effect on cultural evolution, more or less restricted to the economic sphere. For instance, one could make a plausible argument to the effect that clothes were invented, not to preserve body heat, but as a decorative or magical adornment; only later did it turn out that clothes were a preadaptive cultural trait for the more prosaic purpose sometimes found among us. (It should, however, be noted that magic in most cases must be considered as rational behavior.) In this sense, a successful cultural trait is one that does not lead to the destruction of the culture it is part of; it need not, however, be a trait tending to produce an effect that would be judged laudable by a western scientist. We simply do not know enough about man's psychic needs to be able to pass judgment on any cultural trait that does not seem to us to represent a utility; perhaps in the long run it will turn out that an overemphasis on rationality (should this ever occur) would also reveal itself to be inadaptive.

CULTURAL EVOLUTION

Attempts have been made to define progress (or discover it) in cultural evolution. Is there some sense in which any particular culture can, above and beyond simply remaining more or less adapted, actually progress from a lower to a higher state? This all depends on one's criteria concerning what is higher—on value judgments. One such value chosen by some western scientists is efficiency. It has been proposed that western cultural history shows an increase in the energy gained in respect to energy expended in the economic sense. Thus, big-game hunting allowed a certain small surplus; the invention of agriculture provided even more; the development of modern industrial technology has allowed still larger surpluses; the expectation is that the use of new fuels, or direct use of solar energy, will allow even greater returns on energy expenditure. One problem with a notion of this kind, which may be descriptively valid, is that the trend so described occurs in only a small portion of the total culture. It may not be the aspect of a culture that is most important in terms of survival at any moment, and may itself lead ultimately to inadaptiveness. For example, after the human population of the earth has produced the (hypothetical) maximum density tolerable given the nature of the endocrine and nervous systems of man, further expenditure of even small amounts of energy to increase even further the economic efficiency of a culture would be wasteful and destructive. Any sort of presently definable progress would ultimately become a liability. In any case, an Australian aborigine or a Hindu mystic or a Detroit housewife would not necessarily agree that some particular discoverable trend represents progress.

Another aspect of cultural evolution that is potentially self-destructive is that cultural adaptations are frequently acquired which are adaptive to the culture itself rather than to the external environment. Thus, for example, in our own culture industrial technology has progressed to a point where industrial wastes, and other by-products such as noise, are creating major ecological and emotional problems. The processes which have led to this situation involved appropriate adaptations to existing cultural situations (the need for, and the simultaneous presence of, large and growing markets), but they have created a situation where rapid further cultural evolution must occur. Whatever cultural adaptations may be acquired in respect to this challenge may bring further critical problems with them, and so on. None of this need refer to the biological needs of man and his associated flora and fauna, but may be ultimately limited by those needs (say by the ability of lungs to withstand pollution).

It should be pointed out that one of the advantages of partially replacing biological evolution with cultural evolution is the relatively faster rate of change that can be achieved. Biological evolution is a relatively slow and cumbersome process in comparison with cultural evolution, which can on occasion occur at revolutionary speeds. This difference in rates allows man to adjust temporarily to temporary changes in climate or other environmental parameters. This difference in rates is exactly the reason why in some respects cultural evolution has supplanted the biological; if a cultural adjustment is made before a biological one, the latter has no way of occurring since the challenge has been removed. In this respect culture has removed the geophysical environment from contact with man's genome, or uncoupled the relationship usually found between the two.

NATURAL SELECTION AND MAN

While it is true that man's major adaptation to the external world is symbolic culture, and that many biological challenges have been met with cultural rather than biological adaptations, it is not true that man has stopped evolving biologically. Natural selection is as active as ever, but in different ways. To take an exaggerated example, the Plains Indians of North America had cultures that necessitated the presence of leaders with the psychological traits of visionary warrior mystics. Men without such traits were probably relatively unsuccessful in siring and rearing children; in our culture men who have these kinds of psychological traits probably end up in institutions for the mentally ill. Among the Hottentots in Africa feminine beauty involves having very large buttocks; in our culture sexual selection might leave a woman with this trait childless. In our culture selection may be increasingly favoring psychological and endocrinological polygene combinations that allow individuals to remain relatively calm in the face of daily onslaughts of bad news, psychedelic advertising, and industrial ugliness; among the Vikings such polygene combinations would probably have relegated an individual to a low status.

There is, nevertheless, an important way in which the intensity of overall natural selection relents somewhat in technological civilizations such as ours. For instance, the ability to produce insulin cheaply allows some sorts of the genotypes that are susceptible to diabetes to breed where they would not have done so in the past. The invention of eyeglasses has no doubt allowed alleles to accumulate in civilized populations that are inimical to good vision. However, while some genetic loci have had the selection coefficients of their alleles considerably reduced during civilization, other loci probably have had them increased. For instance, the alleles at whatever loci are involved in the predisposition to schizophrenia probably acquired large differences in selection coefficients as civilized life became increasingly frenetic. In order to eliminate natural selection altogether three things are needed. First, all women must bear and raise (or have raised for them) two and only two offspring, and all men must sire two and only two offspring; if this were biologically impossible in individual cases, medical answers would have to overcome the problems; any means including force would have to be used to see that every individual had two offspring and that no individual had more than two. Second, reproductive age would have to be regulated such that no family would begin to reproduce faster than another by shortening its age of reproduction. Third, a total medical and psychiatric control over illness would have to be present such that every prereproductive death was due only to chance (accident). It is clear that we have a long way to go before reaching that stage.

At present the human species is evolving by adapting biologically more to its own cultures than to the external world. The symbolic cultures themselves are evolving, of course, and *that* change with time has largely replaced the geophysical one as a stimulus for continued biological evolution. But, while the cultures must remain adapted to the geophysical world, individual persons in them have less and less need to do so as civilization advances. How far this process can go depends on how closely the above requirements for the elimination of natural selection can be met without producing a culture that has become itself inadaptive to conditions on the earth.

From time to time a quite opposite point of view is raised concerning man's gene pool—that is, the question of eugenics. Should not a society decide what kinds of individuals it wishes to have reproduce, and then enforce that decision? Immediately the problem of inbreeding depression looms, as well as the problem of maintaining polygenic traits when by and large mostly only individual genes can be inherited with certainty. Beyond these lies the problem of deciding what kinds of individuals are desired, and what to do with possible wishes to reproduce on the part of undesirables. Beyond this lies our total ignorance of the genetics of important human traits; selection for mathematical ability might inadvertently be selection for schizophrenia.

SUMMARY

In this chapter an attempt was made to reconstruct some aspects of human evolution by conjecturing on the known scanty fossil record. Other interpretations are of course possible, and the author has no wish to champion the particular course here charted, which simply was the one that seemed most plausible to him. The first step was when a social anthropoid with mobile hands and binocular vision acquired reduced canines. The next step was the development of bipedal locomotion. The next step was the development of big-game hunting which stimulated the last major phase of purely biological evolution (gene pool adjustments made in response to changes in the biological and climato-physical worlds), including the evolution of a relatively larger brain. Next, and finally, came the development of symbolizing ability and the invention of symbolic culture which projected man into a new adaptive zone, involving adaptation via cultural evolution.

REFERENCES

Bresler, J. B., ed., *Human Ecology: Collected Readings.* Addison-Wesley, Reading, Mass., 1966, 472 pp.
Buettner-Janusch, J., *Origins of Man: Physical Anthropology.* Wiley, New York, 1966, 674 pp.
Campbell, B. G., *Human Evolution: An Introduction to Man's Adaptations.* Aldine, Chicago, 1966, 425 pp.
Coon, C. S., *The Origin of Races.* Knopf, New York, 1962, 724 pp.
Dobzhansky, T., *Mankind Evolving: The Evolution of the Human Species.* Yale University Press, New Haven, Conn., 1962, 381 pp.

Jay, P. C., ed., *Primate Studies in Adaptation and Variability.* Holt, Rinehart, and Winston, New York, 1968.

Korn, N., and F. W. Thompson, eds., *Human Evolution: Readings in Physical Anthropology.* Holt, Rinehart, and Winston, Inc., New York, 1967, 466 pp.

Kraus, B. S., *The Basis of Human Evolution.* Harper & Row, New York, 1964, 384 pp.

LeGros Clark, W. E., *The Antecedents of Man.* Edinburgh University Press, Edinburgh, 1959, 374 pp.

Marler, P., and W. J. Hamilton III, *Mechanisms of Animal Behavior.* Wiley, New York, 1966, 771 pp.

National Academy of Sciences, National Research Council, *Time and Stratigraphy in the Evolution of Man.* N.R.S. Publication no. 1469, 1967.

Scientific American, "The Human Species," *Scientific American,* **203:**62–217, 1960.

Southwick, C. H., *Primate Social Behavior.* Van Nostrand, New York, 1963, 191 pp.

Glossary

ACCIDENTAL ENZYMES Abiotically formed proteins or proteinoids that act as catalysts in many chemical reactions. They have no specific sequence of amino acids and their specificity and rate of catalysis is low. They are conceived of as the ancestors of later enzymes.

ADAPTABILITY The ability of an individual organism to adjust (homeostatically, physiologically, behaviorally) to an environment which has become altered from that to which the kind of organism is optimally fit. Also, the ability of a structure to assume more than one form and function.

ADAPTATION The process of evolving structures, functions, or behaviors which allow the organisms (or lineages) to continue to exist and function in some community. As a noun, any structure, function, or behavior that is involved in defining the specific ecological niche of a species of organism.

ADAPTIVE RADIATION The relatively rapid production by divergent evolution in some lineage of many species and genera exploring a variety of specific ecological niches within some adaptive zone.

ADAPTIVE VALUE (w) The amount of reproductive success of an individual or genotype relative to the success of the most successful individual or genotype. Precisely, it is $1 - s$ (where s is the selection coefficient).

ADAPTIVE ZONE A generalized ecological niche. It may be subdivided into less general categories. Thus, one can describe a bird adaptive zone. Within that there can be a hawk-eagle adaptive zone. Within that we find specific ecological niches associated with different species of hawks, kites, and eagles. Some boundaries are less clear; thus, the penguins may be considered to have left the bird adaptive zone for a new zone, created by them.

AGGRESSION Negative or belligerent behavior directed toward a member of one's own species or toward a member of a closely related species. This behavior usually appears in connection with conventional goods such as territory or status in a dominance hierarchy in social animals.

ALLASTHETIC TRAIT A trait that has been produced by sexual selection, usually involving secondary sexual characteristics.

ALLELE A specific sequence of codons at a given genetic locus—or any other

410

alternative sequence that may be present in the population at hand. This is the actual form of a gene as it appears on a given chromosome.

ALLEN'S RULE   A statement of the frequently observed cline in the length and attenuation of projecting appendages within a single species of animals, or between members of closely related species with latitude or altitude: Individuals in colder climates have relatively shorter and stubbier appendages.

ALLOMETRIC RELATIONSHIP   A relative growth relationship between two parts of an organism; one of them increases during individual growth more rapidly or slowly in size (or number) than does the other. In a given population, a comparison of different-sized individuals will show the relationship. The same relationships are frequently found by comparing individuals from closely related species or genera of different sizes, even when some of these are fossil forms. The relationship is then considered to be a reflection of an important part of the genetically determined developmental program of the organisms, and may partially determine the limits of their evolutionary potential.

ALLOPATRIC   Populations or species are allopatric with respect to each other if they form parts of different communities, and therefore do not interbreed in nature.

ALLOPOLYPLOIDY   The condition of having a chromosome complement composed of chromosomes from different species (via hybridization) which have been reduplicated by endomitosis so that the organisms are tetraploid. Since the reduplicated chromosomes can synapse, such organisms are functionally diploid.

ALLOSTERIC TRANSITION   A change in the folding of a protein and in its association with other molecules caused by reversible interaction with another molecular species; if the protein is an enzyme, there are changes in its affinity for substrates and/or coenzymes. Probably this is an important aspect of homeostasis at the cellular level.

ALLOTYPE (or ALLOTYPIC VARIANT)   The specific polypeptide coded for by a given allele, as contrasted with other allotypes coded for by other alleles at the same locus.

ALTRUISTIC TRAIT   A trait (usually behavioral) that functions to aid other individuals in a population.

AMPHOTERIC MOLECULE   Any molecule having the property of solubility in water at one point and solubility in organic substances at another. Such molecules accumulate at the interfaces between water and other media, and are the fundamental structural units of all biological membranes.

ANABOLIC SYSTEM   A system of chemical reactions that results in the synthesis (rather than the breakdown) of organic materials. All such systems must have a supply of chemical energy (are endergonic).

ANLAGE (pl. ANLAGEN)   A primordium. The relatively undifferentiated area of an embryo that will later develop into some organ, usually referring to a stage when the primordium is visible as a nubbin or bump.

APOSEMATIC COLORATION   Warning coloration; bold patterns of bright colors characteristic of noxious species.

ASSORTIVE MATING   Any mating structure that would falsify the assumption of random mating. Preferential mating. Its occurrence in one generation, or a few, is insignificant if the patterns it forms change within a few generations.

ATMOSPHERE   The gaseous envelope around the earth.

AUTOTROPH   Any kind of organism that does not need a carbon source more complex than CO_2. Autotrophs are not independent of their environment and need to obtain from it energy, water, minerals, heat, and a host of other requisites.

BACK MUTATION   The mutation of an allele or codon back to an ancestral form. Thus, if an allele mutates at one codon and this new allele becomes common in the population, a mutation back to the original allele would be a back mutation. The rate of back mutations is very low because it is a product of the rate of any sort of mutation at that locus and the inverse of the number of different possible ways a given gene can mutate.

BALANCING SELECTION   (1) Most generally, any sort of selection that preserves a balanced polymorphism over many generations, such as heterosis, frequency-dependent selection, and seasonal selection. (2) Sometimes used as synonymous with heterosis or the combination of heterosis and frequency-dependent selection.

BALDWIN EFFECT   The change in status of a form of a polygenic trait from being a rare possibility in some relatively rare genotypes under unusual environmental conditions to being a common form of that trait under normal conditions, *when* this transition is effected by the accumulation of new alleles in the polygenes responsible for the trait. (See also Genetic assimilation.)

BATESIAN MIMICRY   The situation wherein a palatable species mimics a noxious one. The mimicry is often almost perfect. One might generalize the definition further to include floral mimicry by restating it as an impotent species mimicking a potent one.

BERGMANN'S RULE   A statement of the frequently observed cline in body size in animals within a single species or between members of closely related species with latitude or altitude: Individuals in colder climates tend to be larger than those in warmer ones.

BIOGENESIS   (See Biopoesis.)

BIOPOESIS   The origin of life. This is not usually conceived as occurring instantly; it is preferable to consider biopoesis to be an extended process consisting of a number of different stages, beginning at nonliving levels of organization, traversing systems neither alive nor purely inorganic, and gradually arriving at fully living systems.

BIOMASS   The physical mass (or weight) of the biosphere or any specified portion of it. The biomass of an elephant would be equivalent to that of many thousand mice.

BIOSPHERE   All the material and energy involved in living systems on the earth at any given time.

BIOTA   The sum total of all species of living organisms in a given geographical subdivision. The flora are the sum total of all plants in a given region, while the fauna are the sum total of all animals.

BIOTAL REPLACEMENT   The phenomenon, perceived in the fossil record, wherein one group of organisms replaces another in time, either geographically or in terms of adaptive zone.

BIOTIC ADAPTATION   An adaptation that is good for the species or population but not advantageous for individual members of the group.

BOTTLENECK EFFECT   The decrease in genetic variability caused by intermittent or periodic drastic reductions in the number of individuals in a population. Genetic drift might occur at some loci in these survivor populations. On the other hand, natural selection clearly determines which individuals will survive the catastrophes involved by scanning traits important in surviving them.

BRACHIATION   Swinging from branch to branch by using the arms only, as in gibbons.

BRADYTELIC EVOLUTION   Significantly slow taxonomic evolutionary rates.

CANALIZATION   The ability of a developing polygenic trait to regulate so as to produce a normal form of that trait in the presence of a few alleles of the polygenes involved whose gene products tend to disrupt that development or in the presence of a somewhat abnormal external environment. It is a property that arises because the trait is coded for by more than on genetic locus; phenotypic stability is achieved by means of genetic complexity. Also, the ability of a trait to become actualized in one form even when genotypes differ.

CARRYING CAPACITY   The amount of biomass of any given kind of organism that can be supported by its community. It is a property of the community in terms of a species (or sometimes of genotypes, as in frequency-dependent selection).

CATABOLIC SYSTEM   A system of chemical reactions, resulting in the breakdown of organic material, and liberating energy (exergonic).

CATALYST   Any agent that is capable of speeding up some chemical reaction. The reaction must be possible on thermodynamic grounds in the given environment.

CHARACTER DISPLACEMENT   The reciprocal evolution of a trait in two (usually closely related) species in opposite directions, due to interspecific competition. It may also result from the inviability or infertility of hybrids between two populations significantly different genetically, but with no premating isolating mechanisms operating. In this case it is often associated with secondary sexual characters or courtship behavior.

CHROMOSOME MUTATION   Any alteration of a chromosome other than a gene mutation which changes its genetic information. Chromosome mutations involve breakage of the chromosome and repair in a form different from the original.

CLADE   A lineage or group of them.

CLINE   A gradient in the magnitude of some measurement of a trait or in the frequency of a present-absent trait or allele across some geographic region.

CLONE   The descendants, by mitosis and cell division, of one individual. Procaryotic populations are of this kind, and also, by secondary loss of sexuality, a few eucaryotic populations.

COACERVATE   Any solution that has, at certain concentrations and temperatures, separated into two or more distinct phases, usually involving a gel or liquid crystal phase in which most of the solute is concentrated and a sol phase composed mostly of solvent.

COADAPTATION   The process whereby various organisms in a community, or organs in an organism, or genes in a genome, gradually evolve adjustments to each other, resulting in a fairly smoothly operating system. The word is used instead of adaptation in order to emphasize the aspect of evolution whereby several units adjust to each other rather than each one adjusting separately to the inanimate environment.

CODOMINANCE   The condition which obtains when two alleles present in a heterozygous diploid are both expressed equally. Primary gene products are usually codominant (with few exceptions), while multigenetic traits frequently show dominance and recessiveness.

COMMENSALISM   A relationship between two kinds of organisms from which one derives benefit and the other is unaffected.

COMMUNITY   The system of relationships (food chains, food webs, and so on) which functionally unites all the various kinds of organisms in a given habitat. The members of the community can be said to have become coadapted.

COMPETITIVE EXCLUSION   A principle that describes one aspect of community structure—that no two species will share the same ecological niche or even large portions of each other's niches. It is also a prediction that if two species with very similar niches from different communities should find themselves together in a single one, they will not long continue to compete before either both of them have adjusted to each other by coadaptations or one of them outcompetes the other, which becomes extinct in that community.

COMPLEMENTATION   The production of a normal phenotype by combining two genotypes, each of which is deficient in a different way.

CONFORMER   (1) A kind of organism whose individuals have little or no powers of physiological regulation if placed in an environment significantly different from the one it is adapted to. A species of this kind has become highly specialized during the process of becoming adapted in its community. (2) Slightly different forms of a specific protein or polypeptide caused by chemical reaction with substances in the environment. These are chemical alterations not encoded in the genome; they probably lead frequently to loss of activity of enzymes, but occasionally may cause only an alteration of catalytic activity.

CONSERVATIVE SUBSTITUTION   This is the result of a codon substitution such that an amino acid is substituted for one of similar physico-chemical properties. In some positions on a polypeptide, these might have very little effect on the functioning of the polypeptide, and so become associated with little or no selective advantage or disadvantage.

CONSPECIFIC   Belonging to the same species.

CONSUMER   Consumers are organisms that feed on other organisms. In a way the term is equivalent to heterotroph. It is used in order to distinguish between different positions in a food chain. Thus, a primary consumer is one which feeds directly on the producers, the plants (or autotrophs). These are the herbivores. The secondary consumers feed on the herbivores and are carnivores or predators. The tertiary consumers feed on the secondary consumers, and so on.

CONVENTIONAL REWARDS   Any goals that are the direct or proximate aim of the behavior of animals that are not directly involved with the obtaining of food or mates, or with escaping from predators. Two very important such rewards are territory and social status.

CONVERGENCE (CONVERGENT EVOLUTION)   The process whereby two genetically very different living systems evolve very similar adaptations in response to similar demands from the environment. Usually convergence affects only a minor portion of the organisms involved, so that it is always clear that they are not closely related.

COPE'S RULE The assertion that most lineages tend to produce organisms of larger sizes during an adaptive radiation. This is a rule with so many exceptions that it is probably not worthwhile retaining as a rule.

COUNTERSELECTION A kind of homeostatic mechanism regulating and interrelating different selection pressures; it is the relationship between different selection pressures resulting in reciprocal limitations on the possibilities for change. In nature (as opposed to the laboratory or the farmyard), selection acting upon some trait cannot be unresponsive to the total needs of the organism. Increased fecundity, for example, will finally be stopped, however desirable, when it comes to interfere with other processes.

CRITICAL PERIOD (of development) The developmental stage during which the form of a fixed trait can be more or less easily modified by altering the external environment. Presumably this is the period of activity of the products of genes important in coding for the trait in question.

CRITICAL POPULATION SIZE The size of population below which a population could not survive several bad years or seasons in a row. It is partly an inverse function of r.

CROSSOVER An event in diploid sexual organisms during which a part of one chromosome becomes associated with part of another (usually the homologous one), resulting in two new chromosomes. In the germ line these events occur importantly during meiosis.

CULTURAL EVOLUTION The adaptive change with time of human symbolic cultures. Nonadaptive changes will result in extinction. There are many parallels between organic and cultural evolution, but they are analogous processes, not homologous. The term "evolution" is preferred instead of just "change" because the changes that take place during the history of a culture are adaptive and restricted by the nature of the antecedent cultures. Cultural evolution can be more rapid than organic evolution in terms of absolute time, allowing man to make very quick adjustments to his environments.

CULTURAL TRADITIONS Learned traditions (not predetermined by the genome). They are present in animals other than man and do not necessarily involve symbols.

CULTURE The totality of institutions and traditions found in any given human society, including language, religion, architecture, and so on. The term may be used also for learned traditions as found among nonsymbolizing animals.

CYCLICAL SELECTION Any selection that periodically reverses its direction, such as seasonal selection, or sometimes, density-dependent selection.

DARWINIAN FITNESS (See Adaptive value.)

DELETION The removal from a chromosome of some portion of the genetic information. One or a few codons may become deleted from a genetic locus; if this change does not result in a frameshift, it could become preserved by selection, and would represent a new gene mutation. Successful deletions of larger portions of chromosomes—perhaps one or several loci—would be a form of chromosome mutation.

DEME   The local breeding population of a sexual species. It is the carrier of the gene pool.

DENSITY-DEPENDENT SELECTION   (1) The situation where one form of a trait (or one allele) is selectively favored under crowded conditions and another under conditions of sparse population. (2) Sometimes, this refers to the fact that selection for viability is not possible (other than in the sense of weeding out genetic diseases) under conditions of less than maximal population size, leading to r-selection.

DEVELOPMENTAL FIELDS   Nongenetic influences postulated to emanate from a group of developing cells or an anlage, resulting in a gradient with a high point at the source and falling off on all sides. Such models can help to simplify what we need to postulate as resulting directly from a coded developmental program. Thus, with models of this sort it is not necessary to have a specific code for a canine tooth, or chela, or a petal, but simply a pattern of gene action. Thus, say, enzymes capable of helping to produce dentine are switched on in the cells of a particular region, possibly in the presence of more than a threshold amount of some field substance, while the shape of the dentine will be determined by various other developmental fields in the vicinity.

DIRECTIONAL SELECTION   The mode of selection that occurs during or after environmental change when the measurement of some trait, or a gene frequency, changes more or less in one direction over a fairly long period o time. In effect, it is balancing selection with removal of more individuals from one end of the normal or binomial curve than from the other end. The result is a change in the mean value of the trait in the population.

DISPLACEMENT (in language)   The capacity of a word or concept to exist in the absence of the object, situation, or process to which it refers. This is only possible with symbols.

DISRUPTIVE SELECTION   (1) As used here, the process whereby viability selecttion leads to a polymodal frequency distribution of the measurement of some trait even though genetic recombination returns the measurement to a unimodal distribution in the next generation of young organisms. Much of the genetic load is then made up of individuals with intermediate measurements. (2) Often used to refer to any selection that maintains a distinct phenotypic polymorphism in the population regardless of the details of the process. Thus the maintenance of two sexes in dioecious organisms is considered by some to be produced by disruptive selection even though in every generation many intersexes are not born and later weeded out.

DIVERGENCE (DIVERGENT EVOLUTION)   The major mode by which evolution has occurred. An ancestral form gives rise to many different kinds of descendants, each of which retains some ancestral features but has acquired its own unique features. This process has given rise to the increasing complexity of the biosphere perceived in the fossil record.

DOLLO'S LAW   The principle of evolutionary irreversibility (which see) as observed by examining phenotypic change in the fossil record.

DOMINANCE HIERARCHY   A system whereby each individual in a social system has a definite relationship of either dominance or subordinance in respect to each other individual in the social group. The systems are open-ended, and an individual may move up or down the hierarchy in respect to any other individual. At any given time the hierarchy will be intact, allowing peaceful cooperative behavior among the members of the social group.

DOMINANT   An allele may be dominant over another (or be a dominant in respect to this other one) if in a heterozygote some aspect of the phenotype with which the locus in question is involved has the same properties as it has in an individual homozygous for the allele in question.

ECOLOGICAL CATASTROPHE   Any change in any aspect of an ecosystem that occurs at a rate too fast for at least one member species of that ecosystem to adjust to it homeostatically. This initial change will eventually begin to affect all members of the ecosystem. Probably such changes are common in nature and are an important aspect of the continually altering conditions on earth, eliciting more or less continual evolutionary change as a response from the living systems.

ECOLOGICAL DETERMINISM   The hypothesis, derived from paleontological and biogeographical data, that, given enough time and a large and diverse enough geographical region, a given basic stock is bound to produce by divergent evolution certain predictable adaptive zones. One aspect of this is parallel evolution, which see.

ECOLOGICAL ISOLATION   The condition that obtains if two populations are sympatric but not syntopic during the time of breeding.

ECOLOGICAL NICHE   Everything a species does, when it does it, and where it does it. Usually the concept is used in comparing closely related species and frequently it is restricted to trophic relationships. The process of adaptation is the process of forming or shaping the ecological niche which is, then, a continuing process. Often the simpler term "niche" is used to refer to ecological niche.

ECOSYSTEM   The abstract concept of a functioning community system in an environment. A community is an actual representative of this concept in the physical world.

ECOTONE   A narrow transitional region between two communities which frequently has its own characteristics and is often thought of as a separate community.

EFFECTIVE POPULATION SIZE (N_e)   The size of the population that will actually give rise to the next generation. It is less than the total breeding population, and can be estimated by:

$$\frac{4 \cdot \text{number of males} \cdot \text{number of females}}{\text{number of males} + \text{number of females}}$$

EFFECTOR MOLECULES   Any molecule, usually of small molecular weight in comparison with proteins, that can act to change the affinity of an enzyme for its substrates or coenzymes, or that can selectively initiate or repress the transcription or translation of a genomic message. These processes are important in cellular homeostasis; effector molecules frequently are hormones, substrates, or end products acting as negative feedback.

ENANTIOMORPH   Either the right-handed (dextro) or left-handed (levo) form of an asymmetrical molecule. Enantiomorphs are also known as racemates. A number of sorts of molecules involved in living systems on earth are represented in those systems by only a single enantiomorph, and therefore, living systems are molecularly asymmetrical.

ENDEMIC SPECIES   A species or other taxon native to an island, lake or continent and found in no other place.

ENZYME INDUCTION   The process whereby an effector molecule stimulates or initiates the transcription of a previously inactive genetic locus, or initiates the previously blocked translation of the RNA message produced by some locus. The process is known to occur with certainty only in some single-celled organisms. It frequently involves the induction of an enzyme appropriate to deal with the effector molecule as a substrate. The process is reversible and depends on the concentration of effector molecules; when this is very low, no induction occurs.

EOBIONT   The earliest system in a biopoesis that would be defined as fully living, or almost so, by the criteria used to define living systems (metabolism, biological growth and reproduction, distinct separation from the surrounding environment).

EPIGENESIS   The process of development. Use of this term is warranted when one wishes to stress the fact that any developmental phenomenon depends on those that have occurred previously. Thus, in order for the lens to develop, the optic cup must develop first; in order for nerves to innervate a structure, the structure must have developed first. One stage of development arises out of a previous stage which largely determines it. Epigenesis is that aspect of ontogeny that results from the possibility of a single gene product responding differently in different environments. The epigenetic effects are cumulative: Once a region of the organism has become slightly different from another region, further change is most likely to make the two still more different, partly because the accumulation of biochemical products will be different.

EUCARYOTES   Primarily sexual organisms having a nuclear membrane, mitochondria and a mitotic-flagellar apparatus. They form the bulk of the biosphere today.

EVOLUTIONARY POTENTIAL   The ability to become modified or adapted as a response to ecological catastrophes. Evolutionary plasticity.

EVOLUTIONARY TREND   An observation that a particular lineage, or group of them, has been evolving in a certain direction over a fairly long period of time. The discovery of such a trend has associated with it no predictive value; trends may be terminated at any point if they become selectively inexpedient.

EXTENDED FAMILY   A social group composed of members of the immediate biological families of siblings, their parents, and offspring. All the members are fairly closely related genetically and tend to have the same spectrum of alleles at any genetic locus or in connection with any polygenic trait.

EXTINCTION   The disappearance from the historical record of all members of some lineage, leaving no descendants. The case wherein some group leaves the fossil record because it evolves into something else is not usually considered to be an extinction since the lineage itself survives.

FIXATION   Fixation occurs when a single allele becomes the only one to represent its genetic locus in a population. Minimally, it must reach a gene frequency of 99 per cent.

FIXED TRAITS   Traits that develop only once during an individual ontogeny and are not subsequently modified. Examples are the number of teeth in mammals and the color patterns of butterfly wings.

FOOD CHAIN A linear series of organisms involved in energy transfer, beginning with the producers (plants), continuing with primary consumers (herbivores), secondary consumers (predators or carnivores), tertiary consumers (predators or carnivores), and so on. Each step in the chain can capture some 10 to 15 percent of the energy stored in the preceding step, the efficiency usually increasing slightly at higher levels.

FOOD WEB The intricate mesh of trophic relationships that results from the fact that a given heterotroph species usually feeds upon more than one prey (or food) organism.

FOUNDER EFFECT The difference in gene frequency between a parental population and one that was begun by founders taken from it. The difference is due to sampling errors that occur because the founders did not carry a representative gene pool.

FRAMESHIFT The result of a deletion (or insertion) of genetic material from (or into) a chromosome, involving one, two, or three bases or some multiple of these (if three bases—beginning *within* a codon); all the information beyond it in that locus will be converted to nonsense. Such an event will almost never be preserved by selection.

FREQUENCY-DEPENDENT SELECTION The process whereby an allele or trait acquires an association with increasingly larger adaptive values as its frequency nears an optimum for the population. In connection with alleles or gene arrangements in inversions, this is often found as part of a heterotic situation. It appears to occur fundamentally because habitats are broken up into microhabitats each better coped with by different genotypes.

FUGITIVE SPECIES A species that either physically follows its food supply or reproduces at a rate fast enough to produce significant numbers only during the time that its specialized food supply is abundant. At other times it may become locally extinct, only to reappear again when the food supply reappears by immigration of a few individuals from nearby areas.

GAMETIC ISOLATION The condition which obtains when, in the absence of premating isolating mechanisms, the gametes from some individuals from two different populations meet but cannot interact to effect fertilization.

GENE FLOW The obtaining of new alleles or chromosomes from other populations. These might be other demes of the same species or other species. A small amount of interbreeding in either of these cases could bring new genetic material into the gene pool of a deme when the hybrids backcross to individuals in the deme. The degree of this would be the degree of gene flow. If gene flow between populations of a single species is very common, it is better to consider that system to be a single deme.

GENE FREQUENCY The frequency of an allele at a given genetic locus in the gene pool of a deme.

GENE MUTATION The process of miscopying or misrepairing that results in changing one or more codons within a gene, or in inserting extra codons into the gene, or in deleting some codons from the gene.

GENE POOL The sum total of all alleles at all genetic loci carried in a deme (or, rarely, in a species).

GENERALIZED (as opposed to specialized) A species or other taxon is said

420   GLOSSARY

to be generalized (1) if it inhabits a comparatively broad ecological niche; (2) if it has traits held in common with all or most members of its taxon. A structure is generalized if it can assume many forms and functions.

GENERATION TIME   The length of time from birth (or hatching or germination) to reproduction. In animals this applies to females only if they differ from the males in this regard.

GENETIC ALTRUISM   The behavior of an individual who aids an unrelated individual (other than its mate) to reproduce. At carrying capacity, alleles promoting such behavior disappear from the gene pool, making it a condition impossible to maintain by natural selection. In a growing population, such behavior could be ignored by selection if it did not in some way directly inhibit the altruistic individual from reproducing as well as other individuals.

GENETIC ASSIMILATION   The change in status of some specific form of a polygenic trait from being a rare possibility in some relatively rare genotypes under unusual environmental conditions to being a common form of that trait under normal conditions, *when* this transition is accomplished by recombination of alleles existing in the population prior to the beginning of the process, as well as by the elimination of some of these alleles from the population. (See Baldwin effect.)

GENETIC BACKGROUND   The allelic structure of all the genes present in an individual other than the one (or few) being considered or manipulated.

GENETIC DEATH   The lack of successful reproduction by (or failure in reproductive competition of) an individual.

GENETIC DISEASE   An incapacity in an individual to maintain adequately functioning physiological homeostasis when this is caused by possession of an allele whose gene product is faulty in the given intracellular environment. The possession of a genetic trait that results in gross malformation of anatomy is also a genetic disease. The criterion is that an individual possessing a genetic disease has an adaptive value of near zero in any of the normal environments of the species, and cannot, therefore, engage in ecological competition with other individuals.

GENETIC DRIFT   The result of a sampling error that causes random changes in gene frequencies from one generation to the next. This can occur, if the phenotypic effect is slight, at any genetic locus if the population is small. It can also occur in a large population if the intensity of selection directed at a given locus is small, or between alleles of a single locus that do not cause phenotypic differences.

GENETIC LOAD   Originally, the percentage of individuals in a population that suffered from homozygosity of recessive lethal and semilethal alleles caused by a buildup of the frequency of these alleles in the population undetected by selection because they are recessive. Their frequency increases with the mutation rate per locus. This aspect of genetic load (called mutational load) is today of less interest than segregational load (which see) and substitutional load (which see).

GENETIC LOCUS   The gene.

GENETIC MAKE   In developmental genetics, make is the amount of genetic determination a group of polygenes provides in a given genotype. Since the environment also influences the degree of development of any polygenic trait, it modifies the effect of the genetic make either upward or downward.

GENOME   The hereditary material or information as opposed to its phenotypic expression (phenotype) in some specific environment. One can speak of the genomes of individuals, or of species.

GENOTYPE   The specific combination of alleles possessed by a single individual organism, or the particular alleles that an individual carries at any given genetic locus. One can speak of the genotypes of individuals only.

GERONTOMORPHOSIS   The process whereby a few traits of some kinds of organisms have their ontogenetic development speeded up relative to the rest of the organism during the evolution of the lineage. This is purely a descriptive term and implies nothing about mechanisms.

GLOGER'S RULE   A statement of the frequently observed cline in mammals and birds in the lightness or darkness of the skin, fur or feathers within a single species, or between members of closely related species, with latitude or altitude: Individuals in colder climates are lighter colored than those in warmer climates. It has been found that the phenomenon is better observed if only moist habitats are included in the comparison.

GRADE   A structural level, frequently associated with some major adaptive zone. Many of the higher taxa of the recent past are turning out to be grades rather than clades—that is, more than one related lineage has evolved in parallel to produce them. In other words, polyphyletic taxa are grades rather than clades.

GROWTH   Biologically, an increase in the amount of living matter. More specifically, it is an increase in the amount of those polynucleotides and proteins characteristic of the system that is growing. Thus, increase in amount of the proteins of a virus invading a cell is not growth of that cell.

HABITAT   For animals, the inanimate and plant portions of its community —that is, places where it can walk and hide and so forth. For a plant, the inanimate portion of its community and the other plants—that is, any factors affecting in a passive way its ability to take root, contact sunlight, and so on.

HABITUS   The anatomical structure of an organism, particularly the external anatomical features.

HETEROSIS   Heterozygote advantage. This is the condition that obtains when heterozygotic genotypes produce more fit phenotypes than do homozygotic ones. Under these conditions alleles will be maintained in a population indefinitely regardless of the magnitudes of the selection coefficients against their homozygotes, resulting in a kind of balancing selection.

HETEROSTYLY   The phenomenon wherein two distinctly different floral types are maintained in a population of plants. These floral types function much like the different sexes of dioecious organisms in that they can fertilize each other but not themselves. They appear to be genetically based upon supergenes.

HETEROTROPH   Any organism that needs a source of carbon more complex than CO_2. Prebiotic heterotrophs fed on the hot dilute soup, modern ones on plants and animals.

HOMOLOGOUS   Said of structures or genes, implying that two entities have descended from the same ancestral genes. Homologous structures often look alike and usually have similar developmental histories. The wings of bats and birds are homologous, while those of bats and insects are not. The latter are analogous, which means that they function similarly but derive from different genetically determined pathways.

HYBRID BREAKDOWN   A form of postmating reproductive isolation in which the offspring of F_1 hybrids between different populations (either F_2 or backcross to parental species) are inviable or infertile.

HYBRID INVIABILITY   A form of postmating reproductive isolation in which the F_1 hybrids between members of different populations are inviable relative to individuals of the parental populations.

HYBRID STERILITY   A form of postmating reproductive isolation in which the F_1 hybrids between members of different populations are sterile.

HYBRID SWARM   The aggregate that results if, upon becoming sympatric, two recently diverged populations freely interbreed to produce viable, fertile hybrids. Typically, such populations are composed of individuals phenotypically intermediate between the parental populations.

HYBRIDIZATION   The breeding of members of two species (or two formerly allopatric populations of one species) in a zone of contact when no premating isolating mechanisms are operating, and when the postmating isolating mechanisms, if any, do not involve outright lethality of the hybrid individuals.

HYDROSPHERE   The oceans, rivers, lakes, streams, ponds, brooks, swamps—in short, the free water present on the earth's surface.

IMAGINE (or IMAGO)   The final, reproductive (adult) stage of a life history involving larval stages. The transition from larval stage to imagine is usually very rapid and has been termed metamorphosis.

IMPRINTING   The process whereby a young animal becomes, at a certain stage in its ontogeny, attached to its parent. The process is functionally akin to programming a computer. It can also be considered a kind of learning *if* the imprinting can experimentally be made to occur with other than the natural parent.

INBREEDING DEPRESSION   The destruction or impairment of viability and/or fertility resulting from increased homozygosity of the genome, often a result of intense artificial selection in the laboratory, or the result of there being too few different alleles at various loci in a very small population. It can be thought of as a physiological result of the elimination of some degree of heterosis.

INFORMATIONAL MACROMOLECULES   Polymers composed of a number of different kinds of monomers, the sequences of which are not specified thermodynamically or chemically; they are random unless ordered by other forces. In biological systems on earth, nucleic acids and proteins are such molecules, the monomer sequences of which are ordered by natural selection. These kinds of molecule are also known as semantides.

INGRAM EFFECT   Restrictions placed on the evolutionary plasticity of individual gene products because of the number of other gene products with which they have to interact such that molecules having to interact with many different gene products evolve more slowly than do those interacting with fewer different gene products. It is a molecular result of the need for coadaptation of gene products.

INSERTION   The incorporation of extra genetic material into a gene or a chromosome. It results in an increase in the size of the genetic material, and could result in duplications of genetic information. It could also result in frameshifts.

INTELLIGENCE   The ability to learn.

INTERSPECIES COMPETITION   Competition between species (for food, nesting sites, and so on); this is a transitory condition that may occur if two species with overlapping ecological niches meet. The competition will be resolved by both species evolving away from each other (character displacement) or by one becoming extinct where they meet.

INTERSYSTEM SELECTION   A form of selection determined by the relative duration of different systems or their descendants. Unlike natural selection, this is not considered to be a specific force, but rather the result of any number of different factors, including chance.

INTRACISTRONAL CROSSOVER   A crossover event in the middle of the message contained in a single gene, resulting in two new recombinant alleles.

INTRASPECIFIC COMPETITION   Competition between individuals sharing the same gene pool (restricted to eucaryotes). This competition occurs at both the ecological level and at the reproductive level. Individuals or genotypes successful in this competition have a high Darwinian fitness.

INTRINSIC RATE OF NATURAL INCREASE (r)   The number of females (or appropriate plant equivalents) produced per female per unit time under optimum conditions. This can never be more than roughly approximated; optimum conditions are in practice unknowable since they vary from genotype to genotype and from population to population even within a species.

INTROGRESSION   The process of introgressive hybridization.

INTROGRESSIVE HYBRIDIZATION   The process of successful interbreeding between two populations that results in the production of a hybrid swarm.

INVERSION   The result of an unknown process whereby a segment of a gene or chromosome becomes reversed from its normal order.

IONOSPHERE   The upper level of the atmosphere, where ultraviolet radiation interacts with various molecules and is thereby prevented from penetrating closer to the surface.

ISOENZYMES   Isoenzymes are enzymatically active multimers whose subunits can be chosen from among the products or more than one gene in a given system. They probably often are the results of gene duplication, with subsequent independent evolution of the duplicates.

ISOTYPES   Protein multimers whose subunits can be made up of the products of more than one gene in a given system.

k-SELECTION   Selection working at maximum population density. Selection under these circumstances is predominantly concerned with viability as opposed to fertility. Such selection undoubtedly involves the production of increased abilities in terms of utilization of resources and in terms of intraspecific ecological competition.

KIN SELECTION   A specific case of natural selection when it is not sufficient simply to dump gametes into the environment, but when the young must be cared for in some way. In this case altruistic behaviors (parental behavior) will evolve whereby an individual will aid another individual (closely related to it) to reproduce.

LEARNING   The process of incorporating into the nervous system information not coded for by the genome.

LINEAGE   (See Phyletic lineage.)

LITHOSPHERE   The stones, earth, and sediments making up the solid substratum of the earth.

MECHANICAL ISOLATION   The condition that obtains when, in the absence of other premating isolating mechanisms, there is some mechanical bar to the union of gametes from individuals of different populations.

METAMORPHOSIS   Also known as transformation, the rapid and drastic changes that occur during the transition from larva to imagine. The changes are usually coordinated by hormones which are produced at appropriate times as responses to stimuli arising in the external environment and/or when the organism reaches a certain stage of development.

MICELLE   One form of liquid crystal formed spontaneously at certain temperatures and concentrations by amphoteric molecules.

MICROHABITAT   The specific subareas actually frequented by members of a given species: for instance, the underside of leaves, the tunnels of worms, the outer top branches of a spruce tree. Closely related species in a single community seldom frequent the same microhabitats at the same time of the year or time of the day—that is, they are not syntopic.

MONOMER   The subunits from which chemical multimers and polymers are built. The monomers of proteins are amino acids. Many active enzymes are multimers made up of from two to several polypeptide monomers that are not active by themselves; these are referred to as protomers.

MONOPHYLETIC   The status of a higher taxon that accrues to it because it has been produced by just one lineage. The birds appear to be monophyletic, for example.

MORPHOLOGICAL EVOLUTIONARY RATES   Rates of phenotypic change in some given trait, structure or functional complex of structures.

MOSAIC EVOLUTION   An evolutionary pattern whereby different traits, structures, or functional complexes in one lineage evolve at different rates during a given period of time.

MULLERIAN MIMICRY   The situation wherein several species of similar noxious organisms display similar aposematic coloration. The similarities of pattern and color do not necessarily look very close to human eyes.

MULTIGENETIC TRAIT   Any trait whose actualization involves the (presumably coordinated) activity of primary products from many different genes acting in concert and/or sequentially.

MULTIMER   A combination of more than one polypeptide protomer, but not of very many (which would be a polymer). Multimers are typically combinations of two (dimers) or four (tetramers) subunits. Most enzymes are multimers.

MUTATION   Any alteration of the genetic material.

MUTATIONAL LOAD   (See Genetic load.)

MUTUALISM   A relationship between two or more species from which all of them derive benefit. At any given moment in geological time, the community relationships can be taken to be more or less mutualistic at the species level, but not necessarily at that of the individual organism.

NATURAL SELECTION   Differential reproduction of sexual individuals or of clones.

NEOBIOGENESIS   Any biogenetic process that has occurred on earth since the origin of the dominant living system.

NEOTENY   The process whereby some few traits of a given kind of organism have their ontogenetic development slowed down relative to the rest of the organism during the evolution of the lineage. This is purely a descriptive term and implies nothing about mechanisms. Synonym for paedomorphosis.

NEUTRAL ALLELE   Said of an allele vis-à-vis another one when under the given environmental conditions their phenotypic expressions are identical, or are identical from the point of natural selection (regardless of whether or not man can measure phenotypic differences caused by them).

NEUTRAL GENE   A gene whose phenotypic expression is not accessible to natural selection. Some genetic loci may be selectively neutral intermittently or at certain stages in a life history, yet scanned by selection at others.

NEUTRAL TRAIT   A trait that is not accessible to natural selection. Some traits may be selectively neutral intermittently or at certain stages in a life history but scanned by selection at others; still others may be completely neutral at all times.

NICHE BREADTH   The degree to which a population or species is unspecialized in the ecological sense. A broad-niched species may occupy several subniches (different microhabitats) within the broad niche. A species with a narrow niche is a relatively more specialized species, being restricted to one or a few microhabitats.

NONHOMOLOGOUS CROSSOVER   The crossing over of nonhomologous chromosomes, or of nonhomologous genes within homologous chromosomes, or of nonhomologous segments of a homologous gene. Some amount of gene duplication could be achieved via the second process.

NONSENSE MUTATION   The substitution of one of the three chain-terminating codons for a codon of an amino acid or initiator, so that the polypeptide is not completely synthesized prior to being released from the ribosomes. This eliminates the activity of the gene product, and would most often be disfavored by selection.

NORMALIZING SELECTION   A term introduced by C. H. Waddington to denote the negative sort of selection that weeds out continually produced individuals with extreme measurements; many authors call it stabilizing selection (which see).

OMNIVORE   A heterotroph that feeds on both plant and animal substance of many kinds.

OPERATOR GENE   A gene or segment of a gene adjacent to the portion of a structural gene coding for the amino terminal end of a polypeptide. Its function in the regulation of gene activity is the ability to be affected by effector molecules in a such a way as to initiate or prevent transcription of the adjacent structural gene.

OPERON   In procaryotes, a set of genetic loci linked to a single operator gene so that initiation of transcription leads to transcription of all of them. They seem to lie adjacent to each other on the chromosome, and frequently, if not always, produce enzymes active in the same biochemical pathway.

ORGANIC ADAPTATION   Any sort of adaptation functioning to further the reproduction of individuals. It is a synonym for adaptation, but used to contrast with biotic adaptation.

ORGANIC EVOLUTION   The process of divergence that has occurred since the living system containing all organisms alive today became the dominant one.

OROGENY   Deformation of the earth's crust, resulting in continental drift, mountain-building and deepening of oceans.

OVERSPECIALIZED   Said of an organism or lineage that has achieved a relatively narrow niche breadth, because this condition will make it difficult to adjust to an altered environment.

PAEDOGENESIS   The reproduction of larvae. The gonads, germ cells, and primary sexual characters develop to maturity in an organism that is otherwise larval.

PAEDOMORPHOSIS   (See Neoteny.)

PANMIXIS   Random mating.

PARALLEL EVOLUTION   The evolution by similar, fairly closely related lineages of very similar adaptive types and adaptive zones. The phenomenon need not be restricted to simultaneous evolution of similar forms. It involves the production of organisms similar overall, not just in one or a few features.

PARASITE   Usually defined as a specialized predator, often small in relation to its prey (which is called the host instead of the prey). A better definition would be an organism that feeds on only a single other species of organism (or on a very few organisms) at any given stage in its life cycle.

PARENTAL BEHAVIOR   Altruistic behaviors shown by individual organisms toward their offspring when it is not sufficient to scatter gametes in the environment. These include nest-building, suckling, defense of young, and so on.

PARTHENOGENESIS   The secondary asexual reproduction of some eucaryotes (often on a temporary basis), whereby haploid or polyploid ovules or ova are not fertilized but are in some other way unblocked so as to develop.

PHENOCOPY   The production in individuals, by the action of an abnormal environmental factor during the critical period of the development of some structure, of an alteration of the form of that structure very similar to that produced by a known genetic mutation.

PHENOTYPE   The expression of the hereditary material in some particular environment. The cumulative result (via ontogeny) of the interactions of gene products with a given environment or sequence of them.

PHENOTYPIC PLASTICITY   The empirical fact that a given genotype can produce more than one phenotype, depending on environmental conditions.

PHYLETIC LINEAGE   An evolving gene pool or group of closely related gene pools.

PHYLOGENY   1) A plausible series of ancestor-descendant groups of fossil organisms. 2) A dendrogram connecting species or taxa, with not more than two branches separating at any point of divergence.

PHYSIOLOGICAL HOMEOSTASIS   The ability of individual organisms to regulate their internal environments so that shocks received from the external environment or from internal processes do not unbalance the system, leading to death.

PLEIOTROPHY   The fact that the gene product of a single locus may have widespread influences on many structures and/or processes.

POLYGENES   Genes coding for some quantitative characteristic of a trait in a more or less additive fashion. Polygenic characters are measurements of some part of a structure. All multigenetic features have a set of polygenes that can be defined as coding for some linear or numerical dimension.

POLYMER A macromolecule made up of a very large number of chemical subunits or monomers, such as proteins and nucleic acids.

POLYMORPHISM Any difference between individuals that is clear cut and without continuous variation in the population. At the level of genes and gene products all variability is polymorphic. One either has allele x or does not have it; there are no different degrees of having it. Polymorphism at the phenotypic level occurs when a single gene has an overriding effect on a trait so that one monitors the alleles present predominantly at that locus in examining the trait; this is called a single gene trait.

POLYPHYLETIC The status of a higher taxon that accrues to it because it has been produced by more than one lineage evolving in parallel. The mammals and the teleost fishes, for example, are polyphyletic. Polyphyly is a direct result of parallel evolution.

POSTMATING ISOLATING MECHANISM Any means of reproductive isolation that acts after mating has occurred, including gametic isolation, hybrid breakdown, hybrid inviability, and hybrid sterility.

POSTZYGOTIC ISOLATING MECHANISM Any means of reproductive isolation that acts after the hybrid zygote has been formed, including hybrid breakdown, hybrid inviability, and hybrid sterility.

POSTZYGOTIC SELECTION Natural selection occurring in the zygote or at any subsequent life history stage.

PREADAPTATION (or PROSPECTIVE ADAPTATION) A trait which, occurring in some ancestral group, we know through hindsight to have been crucial in allowing that group to explore some particular new adaptive zone. Also, the state of having such preadaptive traits, or of being preadapted for some particular adaptive zone.

PREBIOTIC EVOLUTION A process of intersystem competition before the time when the biological system characteristic of currently living organisms had assumed dominance by winning that competition. The changes involved would primarily be the weeding out of incompetent systems by negative selection, but the origin of new systems either *ad hoc* or by fusion or splitting of previous systems can be considered a part of this process, at least in its early phases.

PREDATOR A heterotroph that feeds on the bodies of animals that it has killed. All predators are carnivores (some are parasites).

PREMATING ISOLATING MECHANISM Any means of reproductive isolation that acts to prevent mating, including ecological, seasonal, and sexual isolation.

PREY A kind of animal that is frequently killed for food by carnivorous organisms of one or more kinds.

PREZYGOTIC ISOLATING MECHANISM Any means of reproductive isolation that acts to prevent the formation of hybrid zygotes, including ecological, seasonal, sexual, and gametic isolation.

PREZYGOTIC SELECTION Natural selection occurring among gametes.

PRINCIPLE OF EVOLUTIONARY IRREVERSIBILITY A probability estimate to the effect that evolutionary change at the genetic level can not be reversed effectively to any significant degree (beyond one or two codons).

PROCARYOTES Single-celled, asexual organisms lacking a nuclear membrane, mitochondria, and a mitotic-flagellar apparatus, represented today by the bacteria and blue-green algae.

PRODUCTIVITY (in language)   The capacity of symbols to be recombined into novel concepts. This is the basis for the creative use of symbolic language and depends on the property of displacement (which see).

PROGRESSIVE SELECTION   (See Directional selection.)

PROTEINOID   Amino acid polymers that, unlike proteins, are composed of branched chains that are not antigenic. Being abiotically formed, they have only random sequences of amino acids. They show weak catalytic activities.

PSYCHOLOGICAL ISOLATION   (See Sexual isolation.)

QUANTUM EVOLUTION   A rapid transition from one adaptive zone to another.

QUATERNARY STRUCTURE (of proteins)   In the association of polypeptides with each other to form multimers, the particular shape taken by the multimers.

r (See Intrinsic rate of increase.)

r-SELECTION   Selection working at less than maximum population density. Such selection cannot be involved with viability differences (except for genetic diseases), and so is inevitably concerned with fertility, favoring higher population growth rates and higher productivity.

RADICAL SUBSTITUTION   The result of a codon change causing an amino acid to be substituted for one of different physicochemical properties. These are usually, though not always, associated with functional changes in the primary gene product, so that selection will be able to detect the mutation. Also, any substitution at a codon coding for a portion of the active site of an enzyme, or for some crucial configurational site. More loosely, any substitution that gives rise to phenotypic difference from the wild-type.

RANGE OF RESPONSE (of the genome)   The limits of phenotypic plasticity. It is conceptually important that the range of response of a given genome is probably not infinite, even if it could be subjected to an infinite number of different environments.

RECESSIVE   Said of an allele in respect to another allele if in a heterozygote there is no indication in the phenotype of its presence, but a phenotypic difference is seen in individuals homozygous for that allele.

REGULATOR   A species whose individual members have the ability to regulate physiologically so that they can survive for longer or shorter periods in environments significantly different from those to which the species is best adapted. All species are probably regulators in respect to some traits.

REGULATOR GENE   A gene whose product, either alone or in concert with an effector molecule, acts to unblock or block an operator gene.

RELEASER   A sign or signal communicating information to another individual (or other individuals) that elicits, compels, stimulates, or releases appropriate behavior from it (or them).

REPLICATION   The synthesis of a replica of some specific informational macromolecule. Errors in this process result in mutations.

REPRODUCTION   The production of copies of itself by some system, so that, where there was previously only one system, after reproduction there are two or more.

REPRODUCTIVE ISOLATION   In contrast to geographic isolation, gene pool isolation between members of sympatric populations of different species. Some form of reproductive isolation is thought to arise by chance during any extended period of geographical isolation, and if it does not, speciation will not (or did not) take place. Good species are always reproductively isolated.

RING SPECIES   A special case when the most strongly divergent populations of a widely distributed species meet in nature and are reproductively isolated. Active gene flow may occur between these populations by means of all the interconnecting populations, but it is not sufficient to prevent species-level genomic differences from existing among them.

SEASONAL ISOLATION   The condition that obtains when sympatric populations of different species breed in different seasons. Individuals of such populations could not interbreed.

SEASONAL SELECTION   The situation in which the direction of selection alters with the seasons, favoring one allele or form of a trait in one season and another form in another season. This process can result in the maintenance of a polymorphism.

SEGREGATIONAL LOAD   The portion of the genetic load produced by balancing selection—for example, homozygotes in the case of heterosis and heterozygotes in some postulated forms of frequency-dependent selection. It is caused by the presence of less fit genotypes being generated by sexual recombination.

SELECTION COEFFICIENT   A measure of the intensity of selection against some individual or genotype. Precisely, it is $1 - w$ (where w is the adaptive value), and so is the opposite of the relative reproductive success.

SENESCENCE   The gradual breakdown of homeostatic responses in multicellular organisms, partly at least because natural selection has not been able to scan postreproductive individuals. The increasing probability of death with age, accompanied by, and possibly caused by, a breakdown in the ability of all organ systems to recover from environmental insults.

SEXUAL ISOLATION   The condition that obtains when members of different sympatric populations are not attracted to each other sexually.

SEXUAL RECOMBINATION   A process involving union of gametes in diploid sexual organisms whereby the chromosomes of a parent become associated in many different combinations with chromosomes from the other parent (or parents), thereby generating many new combinations of chromosomes, or many new genotypes.

SEXUAL SELECTION   Natural selection working on traits concerned with sexual attractiveness.

SIBLING SPECIES   Different species that resemble each other phenotypically very closely. There may be significant genomic difference between sibling species.

SIGN   A signal. A specific, stereotyped unit of communication that is elicited only in the presence of the situation about which it carries information; that is, it does not have the property of displacement (which see). In one sense it is actually a part of the biological interaction of the organism with the situation that elicits it.

SOCIAL SYSTEM   A system, evidently always in vertebrates based upon a dominance hierarchy, evolved as a means to allow individuals to live together harmoniously as a group.

SOCIALITY   The characteristic of living together in a social system shown by individuals of some kinds of animals.

SPECIALIZED   Said of a species if its adaptations have produced an ecological niche with very narrow possibilities for regulation; for instance, if it can feed on only a single other kind of organism. A structure is said to be specialized if it can assume only a single, or few, forms, and therefore can perform only very few functions. A species or taxon is specialized also if it has traits that distinguish it from other closely related, species or taxa.

STABILIZING SELECTION   The negative sort of selection that weeds out continually regenerated less fit genotypes in sexual organisms. (See Normalizing selection.) Waddington has subdivided this concept into normalizing selection and canalizing selection. (See Canalization.)

STANDING CROP   The biomass tied up in (or making up) some part of the biosphere at a given moment.

STATISTICAL PROTEINS   Proteins made up of random sequences of amino acids.

STATUS   Social standing. The temporarily agreed-upon relationship of either dominance or subordinance in respect to some particular other individual in a population of social animals. The sum total of the statuses of each individual to each other individual makes up a dominance hierarchy.

STROMATOLITES   Concentrically laminated deposits of calcium carbonate deposited outside the cells of certain tidal blue-green algae. Vast deposits of these make up some of the earliest indications of living systems in the geological record of the earth.

STRUCTURAL GENE   A gene whose transcription results in a messenger RNA. Such genes hold the genetic information coding for the primary structure of all polypeptides (or all primary gene products).

SUBSTITUTION   A mutational event in which one amino acid residue in a gene product is changed to another.

SUBSTITUTIONAL LOAD   The aspect of genetic load generated by the fact that it costs a certain number of selective deaths (or genetic deaths) to replace one allele by another.

SUPERGENE   A battery of genes coding for some complex trait (mimetic color pattern, heterostyle condition) that is inherited as if it were coded for by a single gene. It is postulated that the different inversions found in fly chromosomes represent such batteries of genes, inasmuch as each gene arrangement seems to be an adaptation to a different environmental situation and inasmuch as they would be prevented from being destroyed as units by crossing over.

SUPPRESSOR MUTATION   A mutation at one genetic locus that restores the wild-type (or nearly wild-type) phenotype to an individual whose ancestor had previously had a mutation at another locus that changed its phenotype away from wild type.

SYMBOL   A specific, stereotyped unit of communication that stands for something, else and which can be elicited or utilized in the absence of the thing, situation, or process to which it refers (it has the property of displacement). Symbolic languages consist of rules for the manipulation of symbols.

SYMBOLIC THINKING   The recombining of symbols in certain patterns according to the rules of grammar of a particular symbolic language. This process is capable of generating genuinely new entities (concepts) and is, therefore, potentially creative.

SYMPATRIC   Said of species living in the same community.

SYNTOPIC   Said of species spending significant portions of their time in the same microhabitats. Syntopic individuals could physically meet each other.

TAXON (pl. TAXA)   Some subdivision of the biosphere; for instance, a species, a genus, a kingdom. It is a specific entity, not an abstract concept. Thus, the word or concept "species" is not a taxon, but "man" is.

TAXONOMIC CATEGORY   An abstract category such as "species" or "subclass." Actual examples of these, such as *Homo sapiens* or Anthropoidea, are the taxa themselves.

TAXONOMIC EVOLUTIONARY RATES   The rates of production of genera or higher taxonomic categories per unit time by some lineage, measured in a number of ways. These give an overall view of the tempo of evolutionary change in some given lineage.

TELEOLOGICAL   Concepts are fully teleological if they involve the notion that functional considerations are necessary and sufficient to explain a structure or phenomenon. Also, goal-directed processes (if such existed) would be teleological.

TERRITORY   A particular geographic area from which an individual or mated pair excludes all other individuals of the same or sometimes a closely related species. It functions as a means to insure a food supply and nesting place and materials in the face of an intrinsic rate of increase that is capable of driving the population numbers above the carrying capacity of the environment.

TERTIARY STRUCTURE (of proteins)   The folding of a polypeptide chain into a three-dimensional shape.

THERMOCLINE   A cline in temperature. In a large body of water that is not circulating vigorously, the sun will warm the upper layers so that there will be a gradient of temperature (and, therefore, of density) in the water from the surface to the bottom.

TRANSIENT POLYMORPHISM   A state which may occur after or during an important environmental change when one allele may be replaced by another because of directional selection. At some intermediate stage of replacement both alleles can be found in the population, which is then fleetingly polymorphic for that trait.

UNEQUAL CROSSOVER   A crossover event involving different loci of homologous chromosomes or different segments of homologous genes.

WILD-TYPE ALLELE   The common allele, or one of them, found at a given locus in a given deme. Its presence in high frequency in the gene pool is the result of natural selection in recent generations of that deme. It is the allele that helps to produce the fittest phenotype given the existing genetic background and existing external environment(s).

WILLISTON'S RULE   The observation that in many lineages there has been a long-term trend toward decrease in numbers of similar serial structures (such as teeth) and a concomitant increase in structural difference between those that remain.

WISDOM   The knowledge and experience of a lifetime stored in the minds of individual organisms.

WORD   A symbol used in spoken and written languages.

Index